# Enzymes in the Valorization of Waste

Enzymes play a vital role in the enzymatic hydrolysis of waste for its conversion to useful value-added products. *Enzymatic Hydrolysis of Waste for Development of Value-added Products* focusses on the role of key enzymes such as cellulase, hemicellulases, amylases, and auxiliary enzymes (LMPOs), used in the hydrolysis step of the biorefinery setup. Further, it discusses the role of enzymes in the generation of reducing sugars and value-added compounds, with major emphasis on recent advances in the field. The mechanism, importance, type, evolution, and role of enzymes in hydrolysis constitute the crux of this volume, which is illustrated with examples and pertinent case studies.

Features:

- Explores the role of hydrolyzing enzymes in the breakdown and transformation of biomass hydrolysis.
- Discusses the potential of auxiliary enzymes (LPMOs) for enhancing hydrolysis potential.
- Covers recent developments in the field of enzymatic-assisted hydrolysis of waste for conversion of waste to value-added products.
- Deliberates all possible products that can be generated from enzymatic hydrolysis of waste and their potential utilization.
- Elucidates the limitations and advantages of enzyme-based hydrolysis and possible strategies for moving from the laboratory to large scale industries.

This book is aimed at graduate students, researchers and related industry professionals in biochemical engineering, environmental science, wastewater treatment, biotechnology, applied microbiology, biomass-based biorefinery, biochemistry, green chemistry, sustainable development, waste treatment, enzymology, microbial biotechnology, and waste valorization.

**Novel Biotechnological Applications for Waste to Value Conversion**

**Series Editors: Neha Srivastava,** *IIT BHU Varanasi, Uttar Pradesh, India and* **Manish Srivastava,** *IIT BHU Varanasi, Uttar Pradesh, India*

Solid waste and its sustainable management is considered as one of the major global issue due to industrialization and economic growth. Effective solid waste management (SWM) is a major challenge in the areas with high population density, and despite significant development in social, economic and environmental areas, SWM systems is still increasing the environmental pollution day by day. Thus, there is an urgent need to attend to this issue for green and sustainable environment. Therefore, this proposed book series is a sustainable attempt to cover waste management and their conversion into value added products.

**Utilization of Waste Biomass in Energy, Environment and Catalysis**
*Dan Bahadur Pal and Pardeep Singh*

**Nanobiotechnology for Safe Bioactive Nanobiomaterials**
*Poushpi Dwivedi, Shahid S. Narvi, Ravi Prakash Tewari and Dhanesh Tiwary*

**Sustainable Microbial Technologies for Valorization of Agro-Industrial Wastes**
*Jitendra Kumar Saini, Surender Singh and Lata Nain*

**Enzymes in the Valorization of Waste**
Enzymatic Pre-treatment of Waste for Development of Enzyme based Biorefinery (Vol I)
*Pradeep Verma*

**Enzymes in the Valorization of Waste**
Enzymatic Hydrolysis of Waste for Development of Value-added Products (Vol II)
*Pradeep Verma*

**Enzymes in the Valorization of Waste**
Next-Gen Technological Advances for Sustainable Development of Enzyme based Biorefinery (Vol III)
*Pradeep Verma*

**Biotechnological Approaches in Waste Management**
*Rangabhashiyam S, Ponnusami V and Pardeep Singh*

**Agricultural and Kitchen Waste**
Energy and Environmental Aspects
*Dan Bahadur Pal and Amit Kumar Tiwari*

For more information about this series, please visit: www.routledge.com/Novel-Biotechnological-Applications-for-Waste-to-Value-Conversion/book-series/NVAWVC

# Enzymes in the Valorization of Waste

## Enzymatic Hydrolysis of Waste for Development of Value-added Products

Edited by

Pradeep Verma

CRC Press is an imprint of the
Taylor & Francis Group, an **informa** business

First edition published 2023
by CRC Press
6000 Broken Sound Parkway NW, Suite 300, Boca Raton, FL 33487-2742

and by CRC Press
4 Park Square, Milton Park, Abingdon, Oxon, OX14 4RN

*CRC Press is an imprint of Taylor & Francis Group, LLC*



ISBN: 978-1-032-03509-3 (hbk)
ISBN: 978-1-032-03510-9 (pbk)
ISBN: 978-1-003-18768-4 (ebk)

DOI: 10.1201/9781003187684

Typeset in Times
by SPi Technologies India Pvt Ltd (Straive)

# Contents

# Preface

The modern era can be called an era of waste. Rapid population growth has increased the needs of people and accommodating these requirements of a modern fast-paced life has led to the generation of huge amounts of waste. Different anthropogenic activities result in the generation of waste from non-biodegradable plastics to biodegradable organic or biological products (for example, from agriculture, animal product processing, forest product processing, marine and municipal waste, and food, feed, textile, pulp, leather processing industries etc.). Waste disposal has been considered one of the most critical issues in current times. Nevertheless, humankind always surprises with a great ability to turn a crisis into an opportunity.

With this aim in the past few decades, several processes have been developed to convert these wastes into value-added compounds such as: biobutanol, bioethanol, and other high-value essential compounds. Several physical and chemical methods have been developed to facilitate these conversions and resource recoveries from waste. But the cost involved, environmental concerns and the non-sustainable nature of physical and chemical methods has led the focus to shift toward biological methods. Among several biological methods, the role of enzymes in these important steps of bioconversion of waste to value has been critical.

Waste-based biorefineries focus on an integrated system for complete valorization of wastes generated by generation of value-added compounds. The waste biomass consists of lignin, hemicelluloses, cellulose, starch, pectin, lipids and so on that can be used as precursor compounds for generation of several building block chemicals as well as fuels. The production of suitable liquid or gaseous fuel and value-added compounds occur through a series of steps. The role of different enzymes is important to the biorefinery concept with a critical role in the two most rate-limiting steps of biorefinery: pretreatment and hydrolysis. These enzymes by their action on different components of the waste can lead to generation of specific oligomers and monomers. These oligomeric and monomeric compounds can be subsequently converted to biofuels or can be used as the building block for commercially important chemicals. In order to understand the role of enzymes in these two critical steps, their source, structural and mechanistic properties, and technological advancements in the field of enzymology for their application in waste-to-value generation need to be understood. Thus, this book as a part of the book series Novel Biotechnological Applications for Waste-to-Value Conversion consists of nine chapters under the thematic areas associated with the hydrolysis step of biorefinery. This book provides an insight into enzyme-assisted waste biorefinery via presenting basic information on waste biomass saccharification and the development of the enzymatic hydrolysis process. This book also gives mechanistic insight into enzymatic hydrolysis during waste biomass valorization, with special emphasis on production, purification and overexpression of cellulases and pectinases, and their implementation in waste biorefinery. Furthermore, recent advances in enzyme-assisted hydrolysis of waste biomass to value-added product generation has also been highlighted in this book. Additionally, the role of extreme microbes in the valorization of recalcitrant feather waste and the bioremediation of saline waste using halophilic enzymes have been systematically presented. This book can provide insight to researchers, students, academicians, scientists, and professionals working in the area of enzyme-mediated waste biomass hydrolysis, valorization, and bioremediation.

# Acknowledgments

First, I would like to convey my gratitude to the series editors, Dr. Neha Srivastava and Dr. Manish Srivastava for inviting us to submit this book in the book series Novel Biotechnological Applications for Waste-to-Value Conversion. I am thankful to CRC Press for accepting my proposal to act as editor for the current book volume. The current volume of this book series is only possible because of the support from all the researchers and academicians who contributed to this book, therefore the editor is thankful for their contributions. I would also like to thank my PhD scholar Dr. Bikash Kumar, currently working as Post-doctoral Researcher at the Indian Institute of Technology, Guwahati for providing me with editorial assistance and technical support during all stages of book development. I am also thankful to the Central University of Rajasthan (CURAJ), Ajmer, India for providing infrastructural support and a suitable teaching and research environment. The teaching and research experience at CURAJ has provided the necessary understanding of the needs of academicians, students, and researchers that was greatly helpful during the development of this book. I am also thankful to the Department of Biotechnology for providing me with funds through sponsored projects (Grant No. BT/304/ NE/TBP/2012 and BT/PR7333/PBD/26/373/2012), for setting up my laboratory the Bioprocess and Bioenergy Laboratory. I am always thankful to God and my parents for their blessings. I also express my deep sense of gratitude to my wife and children for their support both during the development of this book and in life.

# Editor

**Prof. Pradeep Verma** completed his PhD at Sardar Patel University Gujarat, India in 2002. In the same year he was selected as UNESCO Fellow and joined Czech Academy of Sciences Prague, Czech Republic. He later moved to Charles University, Prague to work as Post-doctoral Fellow. In 2004 he joined the UFZ Centre for Environmental Research, Halle, Germany as a visiting scientist. He was awarded a DFG fellowship which provided him another opportunity to work as a Post-doctoral Fellow at Gottingen University, Germany. He moved to India in 2007 where he joined Reliance Life Sciences, Mumbai and worked extensively on biobutanol production which led to the attribution of a few patents to his name. Later he was awarded the JSPS Post-doctoral Fellowship Program and joined the Laboratory of Biomass Conversion, Research Institute of Sustainable Humanosphere (RISH), Kyoto University, Japan. He is also a recipient of various prestigious awards such as the Ron Cockcroft award by Swedish society, and UNESCO Fellow ASCR, Prague. He has been awarded Fellow of the Mycological society of India (MSI-2020); the Prof. P.C. Jain Memorial Award, Mycological Society of India 2020; and Fellow of the Biotech Research Society, India (2021).

Prof. Verma began his independent academic career in 2009 as a reader and founder head at the Department of Microbiology at Assam University. In 2011 he moved to the Department of Biotechnology at Guru Ghasidas Vishwavidyalaya (A Central University), Bilaspur, and served as an associate professor till 2013. He is currently working as professor (former head and dean, School of Life Sciences) at the Department of Microbiology, CURAJ. He is a member of various national and international societies and academies. He has completed two collaborated projects worth 150 million INR in microbial diversity and bioenergy.

Prof. Verma is a group leader of the Bioprocess and Bioenergy Laboratory at the Department of Microbiology, School of Life Sciences, CURAJ. His area of expertise involves microbial diversity, bioremediation, bioprocess development, and lignocellulosic and algal biomass-based biorefinery. He also holds 12 international patents in the field of microwave-assisted biomass pretreatment and bio-butanol production. He has more than 72+ research articles in peer-reviewed international journals and has contributed to several book chapters (36 published; ten in press) in different edited books. He has also edited 8 books for international publishers such as Springer, Taylor and Francis, CRC Press, and Elsevier. He is a guest editor for several journals such as *Biomass Conversion and Biorefinery* (Springer), *Frontiers in Nanotechnology* (Frontiers), and *International Journal of Environmental Research and Public Health* (mdpi). He is also an editorial board member for the journal *Current Nanomedicine* (Bentham Sciences). He is acting as reviewer for more than 40 journals in different publication houses such as Springer, Elsevier, RSC, ACS, Nature, Frontiers, and mdpi.

# Contributors

**Cecil Antony**
School of Biotechnology
National Institute of Technology Calicut
Kozhikode, Kerala, India

**S. Arisutha**
Energy Centre
Maulana Azad National Institute of Technology & Ecoscience and Technology
Bhopal, India

**Mamta Bhagat**
Department of Chemical Engineering
Deenbandhu Chottu Ram University of Science and Technology
Murthal, India

**Biswanath Bhunia**
Department of Bioengineering
National Institute of Technology Agartala
Jirania, India

**Manswama Boro**
Department of Microbiology
Sikkim University
Gangtok, Sikkim, India

**Dixita Chettri**
Department of Microbiology
Sikkim University
Gangtok, Sikkim, India

**Manali Das**
School of Bioscience
Indian Institute of Technology
Kharagpur, West Bengal, India

**Sushama U. Dessai**
Government College of Arts, Science, and Commerce
Khandola, Marcela, Goa, India

**Diksha**
Dr. S. S. Bhatnagar University Institute of Chemical Engineering and Technology
Panjab University
Chandigarh, India

**Irene J. Furtado**
Department of Microbiology
Goa University
Goa, India

**Sanket K. Gaonkar**
Department of Microbiology
Goa University
Goa, India

**Praveen Kumar Ghodke**
Department of Chemical Engineering
National Institute of Technology Calicut
Kozhikode, Kerala, India

**Amit Ghosh**
School of Energy Science and Engineering
Indian Institute of Technology Kharagpur
West Bengal, India

**Puja Ghosh**
Department of Biotechnology
School of Life Sciences
Pondicherry University
Pondicherry, India

**Anna Jose**
Discipline of Chemistry
Indian Institute of Technology Palakkad
Kerala, India

**S.K. Kansal**
Dr. S. S. Bhatnagar University Institute of Chemical Engineering and Technology
Panjab University
Chandigarh, India

**Ria Majumdar**
Department of Civil Engineering
National Institute of Technology Agartala
Jirania, India

**Bharat Manna**
School of Bioscience
Indian Institute of Technology Kharagpur
West Bengal, India

**Umesh Mishra**
Department of Civil Engineering
National Institute of Technology Agartala
Jirania, India

**Muthusivaramapandian Muthuraj**
Department of Bioengineering
National Institute of Technology Agartala
Jirania, India

**Pradipta Patra**
School of Energy Science and Engineering
Indian Institute of Technology Kharagpur
West Bengal, India

**Mintu Porel**
Discipline of Chemistry
Indian Institute of Technology Palakkad
Kerala, India

**Navnit Kumar Ramamoorthy**
Department of Biotechnology
School of Life Sciences
Pondicherry University
Kalapet, Pondicherry, India

**S. Renganathan**
Centre for Biotechnology
Anna University
Chennai, Tamil Nadu, India

**Mohd. Aseel Rizwan**
Dr. S. S. Bhatnagar University Institute of Chemical Engineering and Technology
Panjab University
Chandigarh, India

**Saravanakumar Thiyagarajan**
Department of Biochemistry and Molecular Biology
Michigan State University
East Lancing, Michigan, USA

**V. Venkateswara Sarma**
Department of Biotechnology
School of Life Sciences
Pondicherry University
Kalapet, Pondicherry, India

**P. K. Sinha**
Centre for Bioenergy and Renewables
Indian Institute of Technology Kharagpur
West Bengal, India

**Surinder Singh**
Dr. S. S. Bhatnagar University Institute of Chemical Engineering and Technology
Panjab University
Chandigarh, India

**S. Suresh**
Department of Chemical Engineering
Maulana Azad National Institute of Technology
Bhopal, M. P, India

**Liya Thurakkal**
Discipline of Chemistry
Indian Institute of Technology Palakkad
Kerala, India

**Anil Kumar Verma**
Department of Microbiology
Sikkim University
Gangtok, Sikkim, India

**Ashwani Kumar Verma**
Department of Microbiology
Sikkim University

**Sarika Verma**
CSIR-Advanced Materials and Processes Research Institute, Hoshangabad Road
Bhopal, India

# Abbreviations

| | |
|---|---|
| **AA** | auxiliary activities |
| **AC** | acid catalyst |
| **AFEX** | ammonia fiber explosion |
| **BG** | β-glucoside glucohydrolase |
| **CAGR** | compound annual growth rate |
| **CBD** | carbohydrate-binding domain |
| **CBH** | cellobiohydrolase |
| **CBMs** | carbohydrate binding module |
| **CBP** | consolidated bioprocessing |
| **CCR** | carbon catabolite repression |
| **CDs** | catalytic domains |
| **CE** | carbohydrate esterases |
| **CFD** | computational fluid dynamics |
| **$CH_4$** | methane |
| **CO** | carbon monoxide |
| **$CO_2$** | carbon dioxide |
| **CPW** | citrus processing waste |
| **CRISPR** | clustered regularly interspaced short palindromic repeats |
| **DES** | deep eutectic solvents |
| **DNA** | deoxy-ribonucleic acid |
| **DTT** | dithiothreitol |
| **EC** | European Union Regulation |
| **EG** | endoglucanase |
| **GH** | glycoside hydrolases |
| **GHG** | greenhouse gas |
| **GT** | glycosyl transferases |
| **HMF** | 5-hydroxy-methyl-furfural |
| **IEA** | International Energy Agency |
| **IUBMB** | International Union of Biochemistry and Molecular Biology |
| **LB** | lignocellulosic biomass |
| **LCB** | lignocellulosic biomass |
| **LiP** | lignin peroxidase |
| **LPMO** | lytic polysaccharide monooxygenase |
| **MnP** | manganese peroxidase |
| **MoEF** | Ministry of Environment and Forest |
| **MSW** | municipal solid wastes |
| **$N_2O$** | nitrous oxide |
| **NMMO** | N-methylmorpholine-N-oxide |
| **NREL** | National Renewable Energy Laboratory |
| **OTR** | oxygen transfer rate |
| **OUR** | oxygen uptake rate |
| **PL** | polysaccharide lyases |

**PSSSF** partial saccharification and simultaneous saccharification and fermentation
**SHF** separate hydrolysis and fermentation
**SLH** s-layer homology module
**SMF** submerged fermentation
**SPS** simultaneous pretreatment and saccharification
**SScF** semi-simultaneous saccharification and co-fermentation
**SSF** simultaneous saccharification and fermentation
**SSF** solid-state fermentation
**SSSF** semi-simultaneous saccharification and fermentation
**UN** United Nations
**USEIA** US Energy Information Administration

# 1 Bioprocessing Approaches for Enzyme-based Waste Biomass Saccharification

*Navnit Kumar Ramamoorthy and Puja Ghosh*
Pondicherry University, Pondicherry, India

*S. Renganathan*
Anna University, Chennai, India

*V. Venkateswara Sarma*
Pondicherry University, Pondicherry, India

## CONTENTS



DOI: 10.1201/9781003187684-1

## 1.1 INTRODUCTION

Biorefineries of the first generation of bioalcohol production employed edible grains as raw materials (Bertrand et al., 2016), which resulted in a food-fuel debate. Consequently, second generation (2G) biorefineries began to use lignocellulosic biomass as substrates (Muscat et al., 2020), where the cellulosic content in the biomass is fermented to the required bioalcohol. Majorly, the straw of food crops was recognized and used (Ben et al., 2016). According to the general definition from the US Energy Information Administration (USEIA), biomass is a renewable organic component obtained from animal and plant origin (source: "Biomass-renewable energy from plants and animals", (USEIA, 2021)). But, from the perspective of a biofuel researcher, and in the context of the present chapter, "biomass" refers to content/substrates obtained from plant origin.

Unicellular algal biomass, larger perennial trees (Kenechi, 2016), urban and semi-urban wastes (Ravi et al., 2018), residues from forests, and wastes from paper/paper-making industries (Ramamoorthy et al., 2020a), are all potential alternate lignocellulosic substrates. Researchers have thus begun focusing on such waste biomass as alternate sources of lignocelluloses for commercially important enzymes (such as cellulases) and biofuel production (Ramamoorthy et al., 2019b) Kumar and Verma, 2021 Furthermore, an anticipated competition between the conventional lignocelluloses' demand in the feed/fodder application and bioproduct application could be significantly reduced (Ramamoorthy et al., 2019b).

A second generation biorefinery (2G biorefinery) comprises the following stages: (i) Biomass choice –ones with higher cellulose contents are often preferred; (ii) Compositional analysis – quantification of the significant components of the biomass, such as cellulose, hemicellulose, and lignin (Ramamoorthy et al., 2020a); (iii) pretreatment, where the biomass is made free of lignin and hemicelluloses (Ramamoorthy et al., 2018c) (Loow et al., 2015); Chaturvedi and Verma, 2013, thereby preparing the cellulosic content to be enzymatically saccharified/hydrolyzed using cellulases (Falls et al., 2019); (iv) *in-house* cellulase production – with lignocellulosic substrates as the carbon source for fermentations performed – using cellulolytic microbes, which secrete sufficient quantities of cellulases; (v) enzymatic saccharification – the produced cellulases at optimum reaction conditions saccharify/hydrolyze the chosen pretreated biomass to simpler ß-D-glucose monomers (Li et al., 2019; Kumar et al., 2018); and (vi) fermentation – a micro-aerophilic/anaerobic process employing fermentative microbes to ferment the saccharified ß-D-glucose monomers to the respective bioalcohol (Mathew et al., 2015).

The present chapter discusses, in detail, the bioprocesses involved in cellulase production from waste/alternate lignocellulosic biomass. In general, cellulase production/procurement amounts to 40% of the total cost of a biorefinery process [Kuhad et al., 2011]. A detailed schematic representation signifies the need to opt for alternate lignocelluloses for the production of profitable bioproducts, such as enzymes and bioalcohols. Furthermore, it throws light on the impact of a pretreatment procedure on: the subsequent processing of a biomass material; the transcriptional regulatory mechanisms for cellulase synthesis in the major cellulase-producing bacteria and fungi; the bioprocess challenges to maintain efficient cellulase expressions in cellulolytic fungi while culturing them in a media with alternate biomass as the carbon source; cellulase-based saccharification's biochemical mechanisms and issues associated with the bioprocess; and improved combinatorial strategies reported in recent times by researchers.

## 1.2 THE BURGEONING DEMAND FOR CONVENTIONAL LIGNOCELLULOSIC BIOMASS

A forced demand for conventional lignocellulosic biomass is being created by the farmers as they speculate enormous revenue (Singh et al., 2016) by supplying crop residues (biomass) to biorefineries (Ganguli et al., 2018) and commercial enzyme producers. Prior to the advent of such biorefineries, these crop residues made their way to livestock fodder industries (Ben et al., 2016) at either an almost negligible, or certainly a much cheaper price.

Annually, the global production of lignocellulosic biomass is estimated to be 146 billion tonnes (source: Global Bioenergy Statistics, 2019, World Bioenergy Association). Wheat straw is reported to be the most highly used biomass (Fitria et al., 2019) and, in an agricultural country like India, roughly 40 crop residues can serve as potential alternatives, which could be obtained from 26 crop varieties cultivated across the nation (Hiloidhari et al., 2014). Worldwide, per annum, the biomass application for livestock fodder amounts to 2.3 billion tonnes (Herrero et al., 2013). In the year 2020, due to the COVID-19 pandemic-related lockdown and its adverse impact on the transportation sector, the worldwide fossil fuel demand dwindled to 10.8 million barrels per day (mb/d) (Source: Global Energy Review, 2020, "The Impacts of the Covid-19 Crisis on Global Energy Demand and $CO_2$ Emissions", Flagship report, April 2020). However, demand is expected to exponentially ascend to 103 million barrels per day (mb/d) by the year 2021 (Source: "Driving investment, Growth & Prosperity Across the Entire Guyana-Suriname Basins", Guyana Basins Summit, IAGP, 2021). Assuming a 20% (percentage volume basis) bioalcohol blending (roughly bioalcohol barrels of 20.6 million per day) of the commercially available fuels, the sole global dependence on fossil fuels could come down to 82.4 million barrels per day (mb/d). Based on our previous experiments, it could be inferred that, for cellulase production (Ramamoorthy et al., 2019b) and subsequent enzymatic saccharification of 1 kg of optimally pretreated biomass (Ramamoorthy et al., 2018c), the initial biomass requirement would be 2 kg (raw biomass), considering an enzyme loading of 15 FPU/g of pretreated biomass. In our work pertaining to 2G bioethanol (second generation bioethanol) production, we estimated a production of 180 mL ethanol from 1 kg of untreated raw biomass (Ramamoorthy et al., 2020b). In total, 3 kg biomass would be required to produce 180 mL of bioethanol; A ratio of ethanol: biomass requirement of 1:16.6 (v/w basis) could be arrived at.

The approximate cellulose compositions (in % w/w basis) of certain noted natural biomass are: 37% – cardoon; 30% – grapevine pruning; 29% – olive pruning; and 35% – riverbank residues (Takano and Hoshino, 2018). The average content of cellulose on a percentage weight basis is 35%. Roughly, 51 billion tonnes of cellulose from naturally available biomass sources, when subjected to complete saccharification (100% theoretical yield) and efficient fermentation, could produce 12.2 million barrels of bioethanol, annually. Assuming (on a volume basis) a 20% blending, it could be observed that the bioethanol requirement per day would be 20.6 million barrels per day (mb/d), which is several times less than the estimated bioethanol yield of 12.2 million barrels per year (mb/y). It is evident that biofuel industries are under insurmountable pressure to meet the bioalcohol demand. Hence, unconventional waste biomass sources are being investigated at an accelerated pace. Figure 1.1 presents a schematic representation of the competition that is bound to arise between biomass application for fuel/enzyme production and fodder purposes. Bioethanol's density has been considered as 789 kg/m$^3$, while the bioalcohol yield from cofermentation of hemicelluloses has not been considered for the present estimation.

Assumptions/considerations for the estimation:

1. The total biomass produced, 146 billion tonnes per year, is used for the production of bioethanol; no fraction of the biomass produced is utilized as livestock fodder.
2. The approximate (average) of the natural cellulose content (in % w/w) in various biomass sources is 35%.

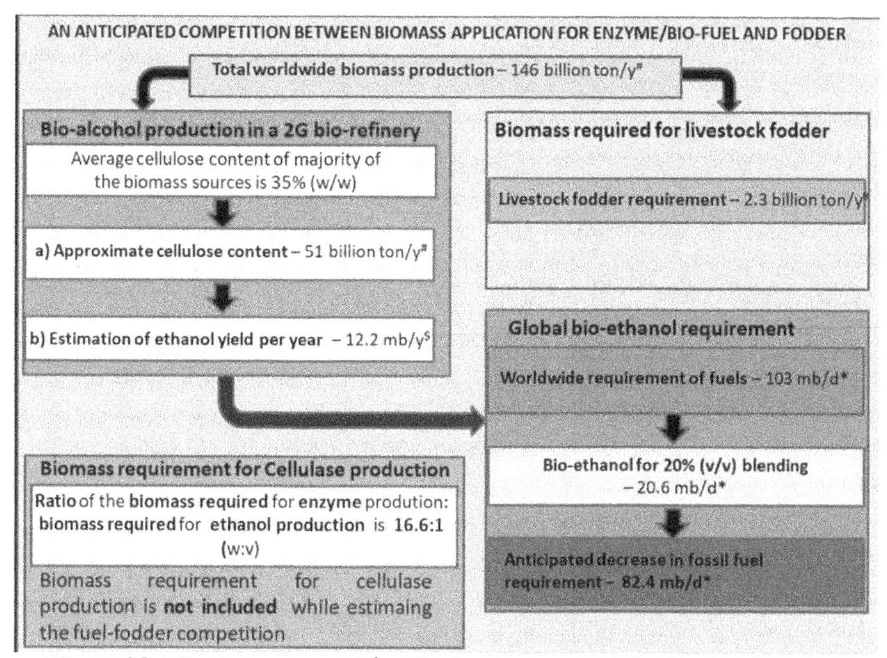


**FIGURE 1.1** A schematic representation of the anticipated demand for conventional lignocelluloses and the competition between its applications in the enzyme/fuel industries and livestock fodder.

3. It is assumed that from the cellulose content of 51 billion tonnes from various biomass sources, a 100% enzymatic saccharification is performed.
4. 0.5 g of ethanol is produced per gram of ß-D-glucose (0.5 g/g of sugar: maximum theoretical yield).

### 1.2.1 Municipal Wastes as Alternate Sources of Lignocellulosic Biomass

Considering the Indian scenario, we estimate that around 1% to 1.5% of solid municipal wastes are biomedical wastes (2 kg per hospital per day) (Tiwari and Kadu, 2013). Roughly 90% of wastes such as tissue paper, cotton gloves, aprons, dressing gauzes, and cotton are usually nonhazardous. Presently, these wastes (rich in lignocelluloses) are disposed of by incineration or burying in landfills, which results in environmental pollution (Tiwari and Kadu, 2013). Every city in the country records the generation of 700 tonnes of paper-based wastes; these are usually dumped in public dustbins to be later disposed of by the e municipality [Sarkar et al., 2014]. One-fourth of this enters the recycle chain, while the rest of the wastes are incinerated as recycling such wastes is generally considered a little expensive (Marousek et al., 2016). Interaction with the officials of a local municipality led to the conclusion that packaging wastes (cardboard, damaged cartons or boxes, papers, fibers) arising from e-commerce and supermarkets are buried in landfills (Marousek et al., 2016). Hence, such municipal wastes could be considered as potential sources of lignocelluloses (Maroušek et al., 2015).

## 1.3 PRETREATMENT: COMMONLY USED PROCEDURES, THEIR LIMITATIONS, AND RECENT ADVANCEMENTS

Pretreatment is a cost-incurring procedure in 2G bioalcohol refineries, which increases cellulose accessibility to the cellulose hydrolytic enzymes (cellulases) during the bioprocess termed, saccharification (Li et al., 2019). The process attenuates lignin's recalcitrance and enables lignin removal to a considerable extent (Loow et al., 2015); Kumar et al., 2020. When such pretreated biomass is subjected to the action of cellulases, the cellulose polymers in it are readily hydrolyzed to ß-D-glucose monomers. Commonly employed pretreatment methods include physicochemical agents, chemical, or biological techniques based on the use of various microbes (Yang et al., 2016).

Pretreatment procedures carried out at a reduced pH hydrolyze the hemicelluloses (Chen, 2015); while procedures which employ an increased pH result in significant lignin depolymerization (Rodríguez et al., 2017). Partial hydrolysis of hemicelluloses occurs at a neutral pH (Kucharska et al., 2018). Chemical pretreatment procedures may involve the usage of concentrated and dilute acids (sulphuric, phosphoric, and hydrochloric acids), sodium hydroxide, potassium hydroxide, ammonia, and calcium hydroxide (Guilliams et al., 2016). Acid pretreatments enable hemicellulose hydrolysis (predominantly xylose and arabinose) (Chen, 2015), while alkali pretreatments result in lignin and hemicellulose solubilization by catalyzing the cleavage of the intermolecular ester linkages between lignin and hemicelluloses. The degrees of polymerization and the crystallinity of celluloses are greatly altered (Ramamoorthy et al., 2018c).

The choice of a pretreatment temperature maximum of 160°C, and a pretreatment period of less than fou hours has been suggested to avoid the formation of inhibitors, such as acetic acid, p-coumaric acid, ferulic acid, furfurals, and 5-HMF (5-Hydroxy-Methyl-Furfural) (Wang et al., 2020a). Furfurals (furan aldehydes) are formed by the dehydration of pentose sugar monomers (Mariscal et al., 2016). While some yeasts like *P. stipitis and S. cerevisiae* can tolerate around 10–120 mM HMF (Liu et al., 2004), at higher titers, this compound is fatal to fermentative enzymes microbes (Iwaki et al., 2013). A list of various pretreatments along with their brief notes on their biochemical mechanisms, efficiencies, and improvements in the processes are shown in Table 1.1.

### 1.3.1 Bioprocess Inhibitors Generated in Pretreatment Processes

Mild pH variations at a maximum temperature of 180°C are to be followed to circumvent the formation of lignin breakdown products, which are known to inhibit cellulases, decrease their activities, and to be detrimental to the sugar yield in the enzymatic saccharification process. The aromatic, aldehyde-like, phenolic, and polyaromatic lignin breakdown products impart the fermentative yeasts with a

**TABLE 1.1**
**Various Pretreatments, Their Brief Biochemical Mechanisms, Efficiencies, and Improvements in the Processes**

| Pretreatment Technique | Process | Impact of the Process | Process Enhancements | References |
|---|---|---|---|---|
| Mechanical pretreatments: chipping, milling, grinding | Mechanical size reduction, abrasion, pulveriz-ation. | Reduction in the size and crystallinity of the biomass. | Improved processing equipment for higher biomass loading. | Ani (2016) |
| Hydrothermal pretreatment | Few min biomass treatments at high temperature saturated steam (150°C to 290°C) at high pressure of 10 to 50 atm. | 0.5% to 1% lignin depolymerization. Hydrolysis and removal of hemicellulose (predominantly pentose) | An acid treatment (1%–2%) before steam hydrolysis. Higher temperature steam treatment and stepwise release of the steam result in enhanced biomass deconstruction. | Zaafouri et al. (2017) |
| Ammonia Fibre Expansion (AFEX) | Biomass treatment using liquid ammonia at high tempera-tures of 60–180°C at pressures of 100–390 psi | Cellulose decrystallization; lignin depolymerization; and hemicellulose hydrolysis and removal. | 81% sugar yield and a 95% theoretical yield of bioethanol from sugarcane bagasse when a combination of sodium hydroxide, calcium hydroxide, and alkaline hydrogen peroxide is used with AFEX | Meléndez-Hernández et al. (2019) |
| Microwave-assisted pretreatment | Molecular collision initiated by dielectric polarization. | Disruption of biomass architecture. | Could be used in conjunction with certain chemical pretreatments such as $FeCl_3$ | Lu and Zhou (2015); Kumar and Verma (2019) |

(*Continued*)

**TABLE 1.1 (CONTINUED)**
**Various Pretreatments, Their Brief Biochemical Mechanisms, Efficiencies, and Improvements in the Processes**

| Pretreatment Technique | Process | Impact of the Process | Process Enhancements | References |
|---|---|---|---|---|
| Cold nonthermal plasma | Plasma radical and electron interaction results in dealkylation of the phenolic residues | Dielectric barrier discharge (DBD) enhances delignification and hemicellulose removal without the generation of inhibitors. | Better reactor volume to occupy larger quantities of biomass. | Ramamoorthy et al. (2020b) |
| Ultrasonic pretreatment | Shear forces created due to ultrasonic cavitation | Disintegration of biomass architecture | Sustained operational cycles for better biomass occupancy in the reactor. | Bundhoo and Mohee (2017) |
| Ionic liquid (IL) Pretreatment | IL ions from the solvents interrupt the hydrogen linkages in cellulose. Electron withdraw-ing groups in the alkyl chain facilitate the dissolution of cellulose. | Cellulose decrystallization and enhanced biomass deconstruction | Reduction of the ILs' toxicity toward fermenting microbes and opting for an economical IL source. | Greer et al. (2020) |
| Organosolv | Organic solvents disrupt the interconnect-ing bonds in hemicellu-lose and lignin. | Pure cellulose is obtained. | Acidic/basic catalysts reduce pretreatment temperature and enhance delignification rates. | Bajpai (2016) |
| Biological pretreatment | Enzyme complexes, such as laccases, lignin peroxidases, manganese peroxidases, versatile peroxidases in combination with feruloyl esterases, quinone reductases, catechol 2,3 dioxygenas-es, initiate an oxidative depolymerization of lignin mediated by hydrogen peroxide. | Lignin is selectively depolymerized without the generation of inhibitors. | Techniques to utilize cofunctioning microbes to efficiently deconstruct biomass in lesser time duration. | Kumar and Chandra (2020); Ramamoorthy et al. (2020c); Aftab et al. (2019) Bhardwaj et al. (2021a); Agrawal and Verma (2020) |

**TABLE 1.2**
**Significant Inhibitors Arising in the Pretreatment Processes, Their Sources, and Their Impact on Various Ensuing Bioprocesses**

| Inhibitor | Source of the Inhibitor | Impact on the Bioprocesses | References |
|---|---|---|---|
| 5 – Hydroxy-Methyl-Furfural (HMF) and furfurals | Hexose sugars | Hamper cell growth, respiration, inhibit of the action of the key fermentative enzymes, Alcohol Dehydrogenase, Aldehyde Dehydrogenase and Pyruvate decarboxylases in yeasts. Repress the translation activity by Stress Granules (SG) and cytoplasmic mRNP granules formation in *Saccharomyces cerevisiae*. | Iwaki et al., 2013; Olofsson et al., 2008; Liu et al., 2004; Field et al., 2015; Cheng et al., 2018 |
| Water, acetic acid, formic acid, xylose furfural. | Water – dehydration of hemicellulose<br>Acetic acid – elimination of acetyl group of xylose<br>Formic acid – carboxyl group of uronic acid<br>Xylose furfural – dehydration of xyloses | Water - Osmotic stress over microbes.<br>Acetic acid - lowers the fermentative microbes' cell pH resulting. Cause intracellular suppression of enolase and phosphoglyceromutase. | Jönsson and Martín, 2016; Casey et al., 2010; Caspeta et al., 2015 |

fragile cell membrane (Zeng et al., 2014). Cellulase's hydrolytic efficiency is greatly reduced in the presence of lignin as it non-specifically binds to the enzyme (Azar et al., 2020). Pseudolignin, which is a by-product formed during dilute acid pretreatments is known to nonproductively bind to cellulase and cause significant alteration in the saccharification yield (Wan et al., 2019). Pseudolignin formation occurs when lignin and carbohydrate degradation products interact and combine. Table 1.2 lists some important inhibitors from the pretreatment processes, along with their sources and their impact on various ensuing bioprocesses.

### 1.3.2 Techniques for Removal of Bioprocess Inhibitors

In activated carbon-based detoxification (w/v: AC-5%) (Li et al., 2020), there occurs adsorption of the inhibitors onto the surface of the AC (Freitas et al., 2019); the process has been recorded to cause 77.9% furan derivatives and 98.6% aromatic monomer removal (Kumar and Chandra, 2020). When calcium hydroxide is added to bring the pH of the hydrolysate close to 6 (overliming), 75% of the furan derivatives and 68% of the aromatic monomers are removed. At higher pH ranges (8–10), complete degradation of the inhibitors has been recorded (Ria et al., 2002). When AC and overliming procedures were used in tandem, a significant bioalcohol titer (0.22 g/g of sugar) was reported in a fermentation performed using *C. saccharobutylicum*

(Kumar and Chandra, 2020). In another procedure, which functions through hydrophobic interaction, organic solvents are used to remove the inhibitors (Qi et al., 2011). Around 90% of the aromatic inhibitory compounds have been effectively removed using a surfactant-based extraction procedure; the hydrophobic-hydrophilic interactions of the inhibitor and the surfactant employed serve as the basis for this procedure (Dhamole et al., 2013). However, the procedure is relatively expensive and biologically unsafe, too (Dhamole et al., 2013). Though nanofiltration and microfiltration techniques are effective with respect to molecular-sized-based separation (Li et al., 2020) of the sugars and the inhibitors, the continuous operation of electric pumps, and the excessive usage of water (Qi et al., 2011) make the procedure expensive. NF-270 and NF-245 membranes have recorded a glucose rejection of 80% and an inhibitor removal of 94% (Nguyen et al., 2015). Membrane fouling is cited as a major shortcoming while such membrane-based techniques are used (Li et al., 2020). In bioabatement/biological abatement, a technique that employs microbes such as *Coniochaeta ligniaria* for inhibitor removal, furfural, phenolics, and HMF removal of 50% has been reported (Cao et al., 2013).

## 1.4 CELLULASES

Cellulases, apart from their significance in biorefineries, are used for cotton biopolishing in textile industries (Ali et al., 2012); and as an anti-bacterial agent for certain bacterial strains, which are antibiotic-resistant (Kuhad et al., 2011). In other industries, such as coffee making; wine brewing; detergent manufacturing; and pulp and paper making, cellulase finds application (Kuhad et al., 2011).

### 1.4.1 Enzyme-based Deconstruction of Lignocellulosic Biomass

Cellulose is degraded into cellobiose units (repeating homo-disaccharide (β-D-glucose) units (Lakhundi et al., 2015) or monomeric β-D-glucose units by a complex of enzymes termed cellulases (Lakhundi et al., 2015). Three major enzymes namely: endoglucanases/CMCases (EG) (EC 3.2.1.4); cellobiohydrolases (CBH) (EC 3.2.1.91); and β-glucosidases/cellobiases (BGL) (EC3.2.1.21) comprise the cellulase complex (Ekundayo et al., 2017). As shown in Figure 1.2, the synergistic actions of these three enzymes help in the complete hydrolysis of cellulose (Lakhundi et al., 2015). The predominant mechanism of the action of cellulases is an exo-endo synergism. A random disruption of the β-1,4 glycosidic bonds in a cellulose polymer (de-polymerization) is catalyzed by the action of endoglucanase/CMCases. The fragmented chain thus resulting, has additional segments, which can be further acted upon by exoglucanase/FPases; the subsequent combinatorial hydrolytic activities of CBH–I and II (Cello-Bio Hydrolases I and II) commence from the reducing and the nonreducing ends of the cellulose fragments, respectively, thereby improving the exo-exo synergistic action. Exoglucanase/FPases usually begin their catalytic action (cleavage) from the reducing terminal of the fragmented cellulose chain. The resulting end products, cellobioses, are further catalytically cleaved by cellobiases/β-glucosidases to β-D-glucose monomers (Thoresen et al., 2021).

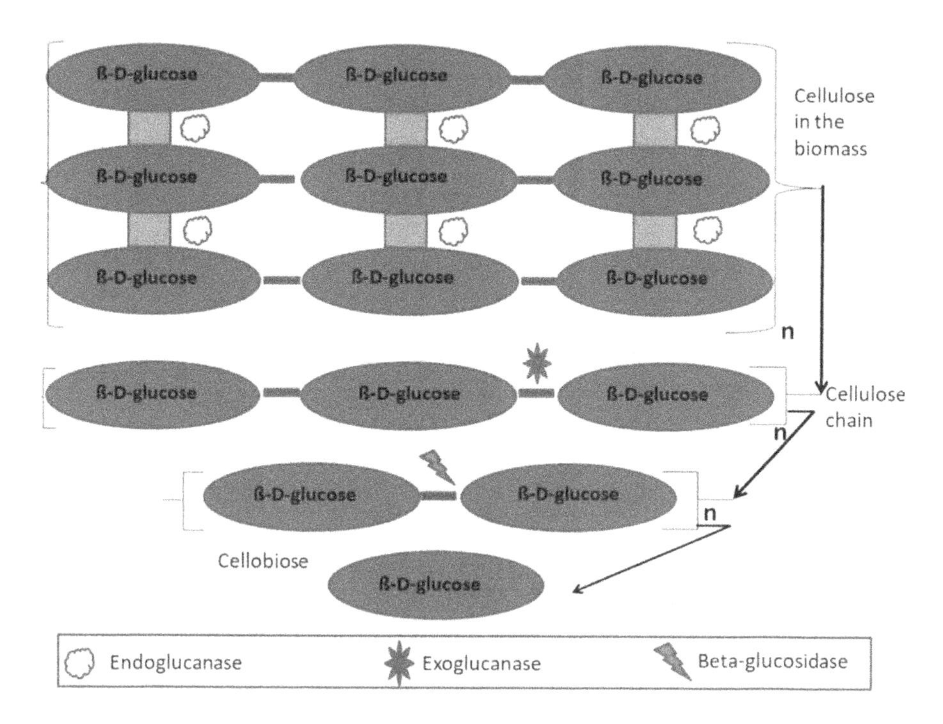


**FIGURE 1.2** A schematic representation of the cellulolytic saccharification of lignocellulosic biomass.

Certain fungal species produce accessory cellulase-degrading enzymes, such as cellodextrinase (EC 3.2.1.74); cellodextrin phosphorylase (EC 2.4.1.49); cellobiose phosphorylase (EC 2.4.1.20); cellobiose epimerase (EC 5.1.3.11); and lytic polysaccharide mono-oxygenases (LPMOs), which have specific actions to enhance the effective deconstruction of cellulose (Muller et al., 2018). Apart from the major cellulolytic enzymes and the accessory enzymes, a collection of some noncellulolytic proteins, namely swollenins (fungal origin) (Zhou et al., 2011), which are analogous in action to the plant expansions (Zhou et al., 2011), relieve the tension (by loosening) of the tight architecture in the cellulose macrofibrils, thereby improving the accessibility of the cellulolytic enzymes into the interiors of the cellulose fibrils (Zhou et al., 2011).

### 1.4.2 Cellulolytic Microbes

#### 1.4.2.1 Major Cellulolytic Fungi

Microorganisms, which help in degrading/decomposing plant remains, are significant contributors to the maintenance of the ecosystem's carbon cycle (Bani et al., 2018). A plethora of wood-rotting white and brown rot fungi, basidiomycetes, certain pathogens of plants, and several novel isolates, which degrade cellulose has been reported (Wang et al., 2020a); Bhardwaj et al., 2021b. Basidiomycetes, due to their saprophytic nature of growing on dead plant material, wood and waste, have been extensively studied (Thorn et al., 1996).

*Trichoderma reesei*, a well-studied ascomycete, and a filamentous fungus was originally isolated from the South Pacific and is a hypercellulase producer (Wang et al., 2020a). *T. reesei*, which inhabits the soil, and multiplies fast on substrates rich in cellulose (Wang et al., 2020a). For large-scale and industrial applications, several synthetic biology approaches have been used to harvest the complete cellulolytic potential of *T. reesei* (Paloheimo et al., 2016). Techniques, such as blending enzymes from various cellulolytic species (Obeng et al., 2018), and cocultivation of cellulase producers (Karuppiah et al., 2021), have been reported to exhibit enhanced cellulase synthesis and increased cellulose degradation.

#### 1.4.2.1.1 The Substrate Carbon Source's Impact on Fungal Cellulase Production

The regulation of cellulase production (cellulase and xylanase) in the transcriptional level in fungi is based on the external carbon source it grows over, or the carbon source it is provided with (Ilmén et al., 2021). The expression of cellulases is possibly controlled by certain signal transduction pathways, which are activated in the presence of inducers (glucose/cellobiose) in stipulated quantities (Salmon et al., 2014). Initially, a minor quantity of the soluble sugars (glucose/cellobiose) in the cellulosic substrate used for fungal growth induces the cellulase/hemicellulase genes in the fungal system (Salmon et al., 2014). These soluble sugars are the primary product of cellulose deconstruction. However, the presence of such easily metabolizable carbon sources as glucose or cellobiose in excess (inhibitory quantity – 2g/L for the genus *Trichoderma* (Ramamoorthy et al., 2019a), causes inhibition of cellulase production (Amore et al., 2013). The initial cellulose breakdown product, cellobiose, is further hydrolyzed into β-D-glucose monomers by the action of β-glucosidases. Since β-D-glucose is much easier to metabolize than cellobiose and since it is available in excess, the fungal system prefers to consume the excess β-D-glucose present in the culture environment (Amore et al., 2013). Until the complete consumption of β-D-glucose occurs, cellulase production isstalled (Amore et al., 2013). This is due to a phenomenon called carbon catabolite repression (CCR). Recent studies have been focusing on the removal of β-glucosidase encoding genes from the genomes of the cellulolytic fungi, thereby anticipating a reduction in cellulase inhibition or an enhancement in cellulase production (Fowler and Brown, 1992).

In the cellulolytic fungal systems, the sensors of the carbon sources, their transport, and their cellular signaling pathways are still not completely understood (Wang et al., 2020a). The cAMP/PKA signaling, as a response to the presence of glucose, is efficiently regulated by a heteromeric G-protein Gan B (alpha)-Sfa D (beta)- Gpg A (gamma) (Lafon et al., 2005). Gan B has been proposed to be effectively involved in sensing various carbon sources and initiating the signal transduction, downstream (Lafon et al., 2005). Furthermore, glucose and a xylose transporter, HxtB, are localized to the plasma membrane and function in the metabolism and signaling of glucose (Dos Reis et al., 2017). A protein kinase, PskA plays a significant role in controlling the sugar flux and its metabolism (Wang et al., 2020a).

#### *1.4.2.1.2 The Molecular Basis for Enzyme Production in Major Cellulolytic Fungi*

The transcriptional factors involved in cellulase regulation fall under the family of Zn2Cys6 zinc binuclear cluster. These transcription factors bind to specific target sequences in the promoter region of the genes and control the expression of the enzymes (Transcription Factors, Reference Module in Biomedical Sciences). Researchers have reported several transcriptional regulators after thoroughly studying the cellulolytic enzyme-producing genes and their promoter regions (Wang et al., 2020a). XlnR (in *A. niger*) and Xyr1 (in *T. reesei*), which are positive regulators of the Zn2Cys6 type, involve themselves in regulating cellulolytic and xylanolytic genes (Benocci et al., 2017). Additionally, it has been reported that despite the presence of a cellulase induction mechanism, the deletion of Xyr1/XlnR results in an attenuated expression of the key cellulolytic and xylanolytic genes (Benocci et al., 2017). cbh1 and cbh2 are the genes responsible for the production of exoglucanases/FPases; egl1 contributes to Endoglucanases/CMCases expression, and bgl1 participates in the production of beta-glucosidases/cellobiases; and xyn1, xyn2 are involved in the production of xylanases (Amore et al., 2013). The presence of easily metabolizable sugars (glucose/cellobiose/hemicelluloses) results in the up-regulation of the transcription factors, which are involved in the production of xylanases (XyR1) and beta-glucosidases (BglR) in higher quantities (Benocci et al., 2017). This, in turn, would inhibit cellulase synthesis (Wang et al., 2020a). Clr1, Clr2, and Clr3, the three transcriptional factors reported to be present in the cellulolytic fungus *Neurospora crassa*, function depending on the regulatory activities of each other (Craig et al., 2015). In the presence of cellobiose (or any other inducing carbon source), Clr1 activates the beta-glucosidase-producing genes (Craig et al., 2015), and it also activates another transcriptional regulator called Clr2 (Craig et al., 2015), which further activates the major cellulolytic genes. Clr2 has been found to activate the cellulase genes even in the absence of an inducer source (Craig et al., 2015). The deletion of the factors Clr1 and Clr2 leads to reduced cellulase expression in the cellulolytic fungus (Beier et al., 2020). In a mutant species of *N. crassa*, another factor called Clr3 inhibits the function of Clr1. The deletion of Clr3 showed that Clr1 could be functional in the absence of an inducer too (Huberman et al., 2017). In certain pathway-specific cases, the transcription factors may remain clustered in locations, which may be the same as the target genes (Wang et al., 2020a). In some peculiar cases, certain other transcription factors may be activated based on their substrate specificities or remain regulated by the phenomenon CCR (Amore et al., 2013). Cre transcription factor mediates CCR, and it inhibits the expression of cellulases by binding to specific locations on the target gene's promoter region (Amore et al., 2013). CreA, a transcriptional regulator identified first in the *Aspergillus* species, inhibits the commencement of transcription of the cellulose hydrolyzing genes when there are simple sugars (monomeric sugars) like glucose/fructose/xylose in its vicinity (Reijngoud et al., 2021). Reports have shown that the CreA repressor binds specifically to the sequence region, SYGGRG located on the promoter sites of the genes, which are responsible for the production of cellulose hydrolyzing enzymes, thereby inhibiting their expressions (Reijngoud et al., 2021). The above said inhibitory mechanisms could be dependent on a post-translational modification of the CreA protein or interaction between a few proteins (Reijngoud et al., 2021). Additionally, the end product of the cellulolytic action of the cellulase complex, glucose, is a known suppressor (Amore et al., 2013).

In *T. reesei*, CreI, has been identified to participate in CCR during hemicellulase and cellulase expressions, which was further confirmed while analyzing CreI mutants of *T. reesei* RUT C-30 (Zhang et al., 2018b). The expression of xyn1 and cbh1 is regulated by Cre1, while it does not directly get involved in the expression of xyn2 and cbh2 (Zhang et al., 2018b). Apart from Cre1, another regulator termed Cre2 has been reported in *T. reesei*, which is known to take part in CCR (Zhang et al., 2018b). Cre2, a ubiquitin c-terminal hydrolase, involves itself in the de-ubiquitination of Cre1 (Zhang et al., 2018b). Under both CCR and nonCCR conditions, Cre2 is proved to have an association with Cre3 WD-40 repeating protein, which in turn could stabilize Cre2 and prevent its proteolytic degradation (Zhang et al., 2018b). CreD, another member of the Cre protein group, acts in an opposing manner to Cre2 and Cre3 and inhibits Cre1 (Han et al., 2020).

In *T. reesei*, AceI and AceII have been reported to be the genes responsible for encoding cellulase regulators (Zou et al., 2012). AceI, an orthologue, found in *Aspergillus* species too, inhibits the expression of cellulases and xylanases (Zou et al., 2012). AceII, a transcriptional activator, found only in *Trichoderma* species, activates the key cellulose hydrolyzing enzymes namely cbh1, cbh2, egl1, egl2, and xyn2 (xylanolytic gene) (Payne et al., 2015). Another transcriptional regulator termed Ace3 has been identified, the over-expression of which markedly increases cellulase production (Payne et al., 2015). The deletion of Ace3 reduced cellulase expression (Zou et al., 2012). In solid-state and submerged fermentations (SSF and SMF), the growth of filamentous fungi and the secretion of enzymes is dependent on the ambient pH too (Ramamoorthy et al., 2019a). A pH-responsive transcription factor, PacC/Pac1 regulator, found in *Aspergillus* species and *T. reesei*, activates alkali-expressed genes and suppresses acid-expressed genes at high ambient pH (Barda et al., 2020). pac1's deletion causes an enhanced activity of Xyr1, while its influence on cellulase production is often impacted by the presence of certain other regulatory mechanisms (Barda et al., 2020).

#### 1.4.2.2 Major Cellulolytic Bacteria

Fungal cellulases are preferred due to their higher specific yield and enzyme activities in comparison to bacterial cellulases (Ramamoorthy et al., 2019a). However, features such as ease of heterologous expression; enhanced specific activity (IU/mg); and nonstringent requirements of reactor conditions (pH, temperature, oxygen requirement), make bacterial cellulases, a possible alternative. Species of bacteria, such as *Cellulomonas, Bacillus, Streptomyces*, and *Pseudomonas* have been reported to produce cellulases (Obeng et al., 2017a, 2017b). Bacteria are known to produce extracellular enzymes/proteins, which remain attached to their cell surfaces (Desvaux et al., 2006). Though *Bacillus subtilis* produces enhanced quantities of endoglucanases, *Clostridium* and *Theromotoga* species have been reported to produce thermostable cellulases, which are produced and remain functional over a wide range of pH and temperatures (60–125°C) (Liu et al., 2012). Additionally, the multimodular enzyme system, cellulosome, assists in the efficient degradation of the complex biomass architecture (Obeng et al., 2017a). The temperature-stable cellulase production/saccharification process contributes to the better techno-economic feasibility of the processes. Such an operation at a higher temperature may result in: fewer chances of contamination; less viscosity, thereby reducing foaming and antifoam additions, resulting in better oxygen mass transfer coefficient ($K_La$) (Routledge, 2012); and enhanced substrate solubility. *C. bescii*, a gram-positive, nonspore-forming bacteria, produces cellulosome-independent cellulases/free cellulases (75°C- tolerant) (Brunecky et al., 2017). Owing to the several advantages of bacterial cellulases, the cell-free system and the whole-cell system have been investigated for cellulase production/saccharification. Additionally, cell surface display of the cellulolytic proteins (Inokuma et al., 2016) and engineering bacterial cellulase-producing genes into recombinant expression hosts (Vadala et al., 2021), have been performed.

#### *1.4.2.2.1 The Cellulosomal System*

With the discovery of *Clostridium thermocellum*'s cellulolytic activity, the cellulosomal multienzyme system came to light. Cellulosomes are multienzyme complexes comprising carbohydrate-binding modules (CBMs); catalytic domains (CDs); and scaffoldin subunits, which bring together various CDs of the enzymes functioning together utilizing a dockerin-cohesin interaction (Matte et al., 2009). The complexed enzyme system works in unison to depolymerize cellulose (Matte et al., 2009). The architecture of a cellulosome comprises a fundamental unit of scaffoldin, which hosts continuously repeating cohesion (type I) units (Matte et al., 2009). This has a high specificity to an enzyme-containing type I dockerin, thereby enabling the organization of several enzymes in a systematic order (Matte et al., 2009). Usually, the scaffoldins have six to nine distinct cohesins, which could efficiently bring together/bind close to 26 distinct enzymes making up the cellulosome (Matte et al., 2009). A type II cohesion-dockerin binding to an anchoring scaffoldin facilitates the attachment of cellulosome to the surface of the cell using an s-layer homology module (SLH) (Matte et al., 2009). The expression of various cellulosomes in a single organism becomes possible as the assembly is entirely dependent on an intermodular cohesion-dockerin interaction, which is based on the subunit composition of the enzymes (Arora et al., 2015). In an attempt to engineer or mimic the natural cellulosomes, artificial cellulosomal constructs/chimeras have been developed by researchers (Vazana et al., 2013). However, owing to changes at the microenvironmental level, the artificial cellulosomes show slower rates of hydrolysis in comparison to their natural counterparts (Vazana et al., 2013).

## 1.5 BIOPROCESS TECHNIQUES FOR MICROBIAL CELLULASE PRODUCTION

Cellulase is produced in the laboratory/biofuel industries in bioreactors under controlled conditions using cellulase producers, majorly fungus (Kuhad et al., 2011). In-house cellulases are significantly inexpensive; this makes biofuel researchers opt for in-house cellulases instead of the expensive commercially marketed cellulases (Ramamoorthy et al., 2019a). In our earlier works, we observed that the production cost of in-house cellulases, obtained using a mixture of the peels of vegetables and fruits, was $8.5 (159 CMCase/100 mL) (Ravi et al., 2018), and the cost was $8 (20 FPase/100mL) when we attempted cellulase production from a mixture of hospital wastes (Ramamoorthy et al., 2019a).

### 1.5.1 Solid-state Fermentation – SSF

Solid-state fermentation (SSF), a commonly used and relevantly simple technique of microbial cultivation, has microorganisms growing on a bed of substrates, which are periodically moistened appropriately using minimal essential media (MEM) components. Such a timed moisture/media addition regulates the moisture and also replenishes the nutrients in the cultivation media. Microbial cultivation in an SSF may be aerobic or anaerobic, and there is a limited free water requirement (60–90% w/w). The pH of the cellulase production system is maintained at an optimum of 4–6 pH, which is optimal to facilitate efficient biosynthesis of cellulase and fungal biomass formation (Hmad and Gargouri, 2017). Cellulases are reported to be functional and efficient in saccharification at a decreased pH (4–6) (Hmad and Gargouri, 2017) A higher pH than the stipulated range may result in growth retardation and reduced enzyme synthesis (Gorems and Tenkegna, 2014). The carbon source used in the media bed is usually conventional lignocellulosic substrates such as the straw of edible crops, which mainly contains cellulose to initiate (induce) cellulase production in the fungus, to obtain easily metabolizable ß-D-glucose. However, researchers have attempted the usage of municipal waste in an SSF system (Abdullah et al., 2016); coir in an SSF and a submerged fermentation (SMF) system (Mrudula and Murugammal, 2011); and a mixture of waste paper and soybean hull in an SSF mode (Julia et al., 2016). Our research group has used a mixture of waste surgical cotton and waste packaging cardboard for cellulase production in an SSF and an SMF system (Ramamoorthy et al., 2019a). Similarly, our research group has also performed cellulase production from a mixture of the peels of sapodilla, kiwi, potato, and coir in an SSF system (Ravi et al., 2018). The SSF system has been shown to circumvent issues related to contamination (Ramamoorthy et al., 2019a). Essentially, the substrate/cultivation media should possess and provide the vital carbon sources, nitrogen sources, and trace amounts of the minimal salts required for the growth and metabolism of the required organism. The maintenance of an optimum ratio of C/N is another critical factor, which is essential for efficient fungal growth. As SSF progresses, the fungus produces mycelia and grows aerially and on the surface. In SSF, the fungus may propagate in the aerial direction, and on the surface.

#### 1.5.1.1 Common Issues Pertaining to an SSF for Cellulase Production

In an SSF for cellulase production from waste biomass substrates, researchers (Ramamoorthy et al., 2019a) have reported issues related to:

a. The maintenance of temperature throughout the process and within the various layers of the fermentation media.

b. The maintenance and regulation of the optimum pH, pertinent to the production of cellulases.
c. The regulation and periodical maintenance of moisture in the media.
d. An even mass transfer and sufficient oxygen supply.
e. Uneven/nonuniform growth of the fungus in various layers of the substrate bed.
f. Lack of an appropriate technique/methodology to assess the fungal growth/fungal biomass concentration during the fermentation.

#### 1.5.1.2 Process Improvements in an SSF for Cellulase Production

##### *1.5.1.2.1 The Impact of Pretreated Substrates in the SSF*

In a previous work of ours, in a nine-day SSF using a pulverized mixture of surgical waste cotton and waste packaging carboard (1:1 ratio), following a preliminary phase of adaptation of 48 hours, the mycelial growth intensified, and after 72 hours it was observed that the growth was uniform in various layers of the fermentation bed. Hemicelluloses could cause catabolite repression and delay the production of cellulases (Amore et al., 2013). Hence, a 5% (v/v) acid pretreated waste substrate mixture (devoid of hemicelluloses) was used in a modified batch SSF process (Ramamoorthy et al., 2019a). During an acid pretreatment, since hemicelluloses are removed, it was observed that the pretreated substrate in an SSF produced 1.3 fold higher cellulase activity than the SSF using an untreated substrate mixture (Ramamoorthy et al., 2019a). A pretreatment step may expose the cellulosic content in the substrate to the fungus, which could use it for its nutritional requirements. For the solid-state batch operation using a hospital waste mixture, our group reported: 6.46 FPU/g.ds, and a CMCase activity of 41.86 CMCase/g.ds. Mrudula and Murugammal, 2011 carried out a solid-state fermentation employing a coir waste substrate and recorded activities of 8.89 FPase/g.ds FPase, and 3.562 CMCase/g.ds. Stepwise technoeconomic analyses of the SSF and SMF cellulase-production processes showed that an SSF is 19.3% lower in cost than an SMF (Ramamoorthy et al., 2019a).

##### *1.5.1.2.2 Fixed Volume Cyclic Fed-batch SSF*

In another earlier work of ours, potato, sapodilla, and kiwi peels, and coir (in a weight-based ratio 1:1:1:1) (untreated) were used for cellulase production in a self-designed aerated tray reactor for ten days (Ravi et al., 2018). An SSF is a considerably time-consuming process; a method to initiate quicker growth in an SSF avoiding the lag/adaptation phase is a fed-batch operation. Hence, in our first-of-its-kind constant volume cyclic fed-batch SSF, the volume of the media in the reactor was maintained constant during each cycle of the fed-batch (Ravi et al., 2018). The process began with a batch-culture operation, which lasted ten days. The fed-batch cycles involved the removal of a stipulated quantity of the biomass mixture and replacement with a similar quantity of the fresh biomass mixture. Two cycles were carried out and the durations of the fermentations were days per cycle. Such a cyclic fed-batch operation decreased the overall production time by 1.5 fold and produced cellulases, which showed a saccharification efficiency of 45% (Ravi et al., 2018).

### 1.5.2 Submerged Fermentation (SMF)

In an SMF, a primary culture/starter inoculum is required to commence the growth of the fungus within the bioreactor. Once the starter inoculum is introduced into the reactor, factors such as temperature, pH, aeration, agitation, and foaming are regulated throughout the fermentation. Efficient heat and mass transfers have been recorded in SMF systems in comparison to SSF systems. Computational fluid dynamics (CFD) studies were used by our research group (Ramamoorthy et al., 2020a) to arrive at the anchor-type impeller, which causes optimal mixing of the media components, as well as providing efficient gas and heat transfers in a variable viscosity-possessing cellulase production media. The lesser amount of mycelia formed in an SMF media and constant breakage of mature mycelia (Ramamoorthy et al., 2019b) by the hydrodynamic shear caused due to the constantly rotating impeller (Muhammad et al., 2016), and the currents produced due to agitation results in a lesser amount of cellulase being secreted into the production media. Secondary mycelia are more responsible for cellulase production than primary mycelia (Ma et al., 2013). Primary mycelia are characterized by structures, which are less dense and less branched than secondary mycelia. The mature secondary mycelia are a highly branched network of porous hyphae, which lodge the cellulases in them before they are secreted into the culture media (He et al., 2016). Contrary to this fact, certain researchers have postulated that the maintenance of a higher rate of agitation in the reactor lets the mycelia divide in a compact, hairy, messy network (pelleted appearance), which could improve cellulase productivity (Xia et al., 2014). Opposing this view, (Veiter et al., 2018) recorded that such a pelleted form could result in a decreased nutrient and oxygen transfer, which may hamper their productivity.

In a finding reported for the first time, our research group (Ramamoorthy et al., 2019a) put forth that the fungi (after an initial 36–72 hours of growth in accelerated phase) was anchoring to the agitator blades, baffles, shaft of the impeller before it continued to increase in proportion and started to produce cellulase in substantial quantities. This was indicative of a partial solid-state growth in a submerged operation. The fungus anchoring to the solid components of the reactor was producing cellulase along with the dense fungal population which was observed in the media (Ramamoorthy et al., 2019a). This phenomenon was recorded despite the addition of a marginal quantity of surfactant, which was anticipated to prevent the adherence of the fungus to the reactor's mechanical parts.

The estimation of parameters such as the oxygen uptake rate per hour or day (OUR); the biomass yield coefficients on the basis of substrate consumption and oxygen consumption ($Y_{x/s}$, $Y_{x/O2}$); and the threshold values of viscosity and accumulation of sugars, would maintain and regulate sufficient cellulase production. The maintenance of coefficients on the basis of oxygen uptake and substrate uptake ($mO_{2x}$ and $ms_x$); the mass transfer coefficient ($K_La$); and the oxygen transfer rate (OTR), in an SMF system are essential to scale-up a laboratory scale process to a pilot scale or commercial production scale (Ramamoorthy et al., 2019a).

#### 1.5.2.1 Major Shortcomings of an SMF for Cellulase Production

An SMF for cellulase production from a waste biomass mixture usually suffers from the following drawbacks:

a. As a result of an increased viscosity in the culture broth due to the produced cellulases, there is a considerable decrease in the distribution of oxygen (Clarke, 2013) within the culture broth. Raising the temperature to counteract the rise in viscosity (Wang et al., 2014) may be detrimental to the growth of the fungus as *Trichoderma* spp., (the major cellulase producer) grows at an optimum temperature of 25–28°C (Qiu et al., 2017). An increase in the speed of agitation to circumvent the rise in viscosity (up to 3.2 cP Ramamoorthy et al., 2020a) may result in breakage of the mature mycelia, which harbor the cellulases in their tips. This breakage and the secretion of cellulases into the media may result in a further increase in viscosity (Blunt et al., 2019). As a result, there would be uncontrollable foaming (Stoyanova et al., 2014). The foam may rise to the flange plate of the reactor, thereby increasing the chances of contamination of the culture media (Hill, 2003). Furthermore, as the mature mycelia are broken, the primary mycelia would have to commence growth from the start to attain maturity to secrete cellulases (Ramamoorthy et al., 2019b).
b. The cellulases produced by the fungus could commence saccharification/hydrolysis of the waste biomass substrate and accumulate the detrimental monomeric sugars (glucose and xylose) beyond the threshold level (1.8–2 g/L), and reduce the further secretion of cellulase by the fungus. The carbon catabolite repression (CCR) phenomenon is known to be the cause (Amore et al., 2013).
c. The growth of the fungus within the reactor results in a rapid and enhanced uptake of dissolved oxygen (Xia et al., 2014). A decreased oxygen hold-up is observed in the broth, which is caused by an increased viscosity due to the accumulated cellulases. This may result in the nonavailability of oxygen to the growing fungus and a sudden stall in the growth of the fungus is observed (Ramamoorthy et al., 2019a). Hence, a significant reduction in the oxygen transfer rate (OTR), oxygen uptake rate (OUR), and mass transfer coefficient ($K_La$) have been recorded.

#### 1.5.2.2 A Sequential Cellulase Induction-partial Saccharification-Catabolite Repression in an SMF for Cellulase Production

In an SMF bioreactor, as the glucose and cellobiose start accumulating, a portion of the cellulose in the biomass is left unconsumed. But, the fungal biomass concentration increases as it consumes the reducing sugars in the cultivation broth (Shahzadi et al., 2013). In our earlier works (Ramamoorthy et al., 2019a, 2019b), we hypothesized and proved that cellulase production is impacted by three major phases. The cellulose in the starter inoculum behaves as an inducer to initiate cellulase secretion (Salmon et al., 2014). The secreted cellulases hydrolyze the cellulose component of the media releasing the easily digestible, soluble oligosaccharides (xylose, cellobiose, and glucose) into the production media (Ramamoorthy et al., 2019a). The starter inoculum, along with these sugars, when transferred to the bioreactor, induces cellulase production (Jourdier et al., 2012). The sharp rise in enzyme production (0.3 FPase/mL) within the reactor during a batch operation using a waste hospital mixture could be attributed to this induction mechanism (Ramamoorthy et al., 2019b).

As the fermentation progresses, the produced cellulase hydrolyzes the waste biomass mixture and releases sugars causing catabolite repression (Antoniêto et al., 2015).

#### 1.5.2.3 Bottlenecks in the SMF Operation for Cellulase Production

In order to improve the utilization of cellulose in the substrate by the fungus, the stress level/nutritional limitation needs to be kept high (Shahzadi et al., 2013). But, care has to be taken to avoid too much starvation as it may result in autophagy, which in turn could result in growth retardation of the mycelia due to starvation (Pollock et al., 2013) and anaerobic stress. The secreted cellulase in the bioreactor may partially saccharify the biomass substrate mixture (as optimum saccharification occurs at 50°C) (Sartori et al., 2015). This may result in glucose and cellobiose accumulation, and nonconsumption of cellulose. Hence, based on the measurement of the glucose/cellobiose accumulation, timely removal of the media (Ramamoorthy et al., 2019b) along with cellulase, and the sugars, may provide the required uninterrupted stress, which could facilitate enhanced cellulase yields (Shahzadi et al., 2013).

The pH of the fermentation media decreases to 3–3.5, which indicates that the nitrogen source of the media component is consumed prior to the consumption of the carbon source (Boberg et al., 2014). It is further evident as a higher C/N ratio is observed in the media composition (Yu et al., 1998). Free $H^+$ are hence liberated in the broth, thereby decreasing the fermentation pH (Garcia-Ochoa and Gomez, 2009). An automated cascaded operation between the dissolved oxygen percentage (DO%) and the agitation (RPM) at optimum minimum and maximum levels help maintain a sustained production of the enzyme (Khdhiri et al., 2014). However, the cascade could fail after an excessive production of the viscous, extracellular proteins (of which cellulases are a part), which increases the broth viscosity to intolerable/unregulatable levels (Zhao et al., 2021). In a batch SMF process of cellulase production of ours (Ramamoorthy et al., 2019a), the volumetric oxygen mass transfer coefficient, ($K_La$), reached 19/hour from an initially recorded value of 498/hour as the fungal biomass increased in proportions also followed by an increase in the viscosity of the culture broth (Cascaval et al., 2003). Modified SMF operational modes, such as fed-batch, cyclic-fed batch, and chemostat modes could be possible alternatives (Yang and Sha, 2016). Prior to performing a stepwise nutrient feeding process, a yeast coculture in a batch cellulase production using *T. harzanium* was carried out. As anticipated, the yeast biomass started consuming the accumulated oligosaccharides while maintaining the nutrition stress levels high for the fungus, thereby improving cellulase production by 15% in comparison to a regular batch process (Charles et al., 2018).

#### 1.5.2.4 A Fed-batch SMF Operation for Improved Cellulase Yield

A fed-batch operation possesses certain advantages over a batch operation as it regulates the growth rate (Assawajaruwan et al., 2018) of the fungus and controls the production of certain significant metabolites (Rodrigues et al., 2019). An effective expression of the transcriptional promoters is observed in the fed-batch nutrient feeding. Some recent techniques being employed in a fed-batch operation are: an open-loop operation/feeding on the basis of predicted or analyzed information (Assawajaruwan et al., 2018); maintenance of a constant specific growth rate

(Morcelli et al., 2018); maintenance of a stipulated pH range (Wang et al., 2011); maintenance of a constant respiratory quotient (Xia et al., 2014); maintenance of a constant dissolved oxygen concentration (Chitra et al., 2018); maintenance of a constant substrate concentration (as in a pulse fed-batch) (Sawatzki et al., 2018); and constant maintenance of the concentrations of certain notable metabolites (Morcelli et al., 2018).

Additionally, a fed-batch SMF has the following advantages:

1. As fresh media is fed into the reactor, a constant dilution of the broth occurs, thereby reducing the broth's viscosity (Faheina et al., 2015) and diluting the sugar concentration in the broth (Sanchez et al., 2014). This, in turn, would relieve the cellulase production operation from catabolite repression (Li et al., 2018).
2. A fall in viscosity may occur following the addition of the fresh feed (Bhargava et al., 2005) facilitating an increase in the OUR, OTR, and ($K_La$) of the process (Chitra et al., 2018).
3. An improved product yield and an increased microbial biomass formation could be obtained (Roeva et al., 2010).
4. A significant reduction in the manpower and time could be recorded (Abdel-Rahman et al., 2014).
5. A lag/adaptation phase of growth is absent during a fed-batch operation as the microbial culture is maintained at an accelerated and a higher rate of growth within the bioreactor prior to feeding. This could save considerable time.
6. The costs and the energy involved in autoclaving the media, starter inocula preparation (Abdel-Rahman et al., 2014), and so forth, are considerably reduced (Pollock et al., 2013).

#### *1.5.2.4.1 Shortcomings of the Fed-batch SMF for Cellulase Production*

In a fed-batch operation, care should be taken while the nutrients are being fed. A higher concentration of the substrate source may result in stunted fungal growth and stalled cellulase production in a phenomenon called, substrate inhibition (Väljamäe et al., 2001). Typically, a hypertonic environment may result in osmotic stress on the growing mycelia, thereby adversely affecting fungal metabolism and microbial biomass formation (Zhang et al., 2015). During fed-batch feeding, autophagy may occur and slow down growth while a sharp decrease in the glucose concentration during dilution (fed-batch feeding) may occur, too. This could be attributed to a delay/lag in the increase in the proportions of the biomass before the fungus could adapt to the usage of cellulose from the fed substrate (Pollock et al., 2013). This nutrient adaptation could take time as vacuolation may have occurred and the mycelia could have started utilizing the endogenous carbon. Enzyme production is higher in an SSF in comparison to an SMF, as fungus are known to grow uninhibited, adhered to a moistened solid substrate, which is unlike the environment that is seen in a liquid media-based SMF (Barrios and Tarragó-Castellanos, 2015). There is a marked difference between the fundamental physiology of the mycelia formation in an SMF and an SSF as elucidated in a microscopic analysis by (Barrios and Tarragó-Castellanos, 2015).

At the end of a 120 hour SMF, using a waste cotton-cardboard mixture and the fungus *T. harzanium*, our group (Ramamoorthy et al., 2019a Process Biochem) recorded 1.9 FPase/mL and 2.47 CMCase/mL. Cellulase produced using an SSF is 23% higher in enzyme activity than the cellulase produced using an SMF (Ramamoorthy et al., 2019a). In an exponential fed-batch operation using the exactly same substrate, our group recorded 2.7% higher enzyme activity than a batch SMF for cellulase production (Ramamoorthy et al., 2019b). (Cunha et al., 2012) recorded a 0.05 CMCase/mL while using sugarcane bagasse as the biomass substrate and the cellulolytic fungus *A. niger* A12. Using coir as the substrate and *A. niger* as the cellulolytic fungus, (Mrudula and Murugammal, 2011) obtained 2.3 Fpase/mL, and 3.29 CMCase/mL. Jourdier et al. (2012) reported a 10% reduction in the protein yield in an exponentially fed-batch process using *T. reesei* for cellulase production. Masampally et al. (2018) employed a gaussian process regression (GPR) algorithm and performed a fed-batch operation using the design and emerged successfully. Li et al. (2017) reported 13 FPase/mL while using lactose pulses in a fed-batch operation for cultivating recombinant *T. reesei*. In an approach, which could employ relatively lesser fermentation maintenance challenges, our group heterologously expressed the FPase/exoglucanase gene (*Cel 7A* of *T. reesei*) in a recombinant host, *E. coli* shuffle. An exoglucanase/FPase yield of 2.5 IU/mL (protein yield: 0.7 g/L and molecular weight 58 kDa) was observed (Ramamoorthy et al., 2018a).

## 1.6 INTEGRATED BIOPROCESSES FOR IMPROVED SACCHARIFICATION YIELD

### 1.6.1 The Biochemical Basis of an Enzymatic Saccharification Process

Glycosyl hydrolases either function through an invertion or a retention mechanism of hydrolysis (Elferink et al., 2020). A two-stage, double displacement in the nucleophile residue and the acid or base has been reported to occur in a hydrolytic mechanism, which involves retention. In the first stage of glycosylation, there occurs a transfer of a proton from the acid or base residue to the glycosidic oxygen. Following this, there is a glycosyl enzyme intermediate complex formation (GEI), when an attack on the anomeric carbon in the carbohydrate in the −1 site of binding is initiated by the nucleophile. When there is an attack on the anomeric carbon by a water molecule, the bonding in the GEI complex is broken; this is the second stage, called deglycosylation. These processes facilitate the transfer of a proton to the acid or base. As a result of these series of reactions, the nucleophile and the acid or base residues return to their native states, thereby indicating the end of the catalytic process (Ribeiro et al., 2019). The enzymatic saccharification process is hampered/inhibited by cellobioses (Chen, 2015). A residue of tryptophan located close to the cellobiohydrolase's active site has a binding initiated by cellobiose molecules, which in turn results in a stearic hindrance. Subsequent saccharification is inhibited as the cellulose molecule being saccharified cannot diffuse itself into the cellbiohydrolase's site of activity. Furthermore, cellobiohydrolase's conformation at the molecular level is altered due to the binding of cellobiose; this results in cellulose adsorbing to the enzyme in a nonproductive manner. This phenomenon of inhibition by the end product (cellobiose, in this case) is detrimental and, at times, stalls complete saccharification too, as it poses great stress to the microfibril removal from the cellulose polymer. (Yue et al., 2004).

#### 1.6.1.1 Shortcomings of Traditional Enzymatic Saccharification

End product inhibition as a result of the accumulation of cellobiose/glucose in an enzymatic saccharification process has been reported in a separate hydrolysis and fermentation (SHF) procedure (Zhang et al., 2018a). In the age-old SHF process, two bioprocesses, namely saccharification and fermentation, are carried out in two separate reactors (Ishizaki and Hasumi 2014). The highly viscous end products (sugars) bind to the cellulases at the air-liquid interface (Bhagia et al., 2019) and result in incomplete biomass saccharification (da Silva et al., 2020). Surfactant addition was found to circumvent the inhibitors' activity in small-scale operations, while for the process involving higher enzyme loadings, the technique of surfactant addition was deemed to be a failure (Bhagia et al., 2019). A step of presaccharification of the biomass has been reported to decrease the enzyme loading from 33 to 22 mg protein per gram of glucan (Shi et al., 2019), which makes it evident that a presaccharification could improve the recovery of sugars from the biomass.

### 1.6.2 Simultaneous Saccharification and Fermentation (SSF): The Process, Its Drawbacks, and Advancements

In an SHF, the end product inhibition attenuates the fermentable ß-D-glucose yields. If the accumulated sugars resulting in such an end product inhibition are recovered/removed from the reactor, periodically, (calculated on the basis saccharification rate), it could circumvent the product inhibition issue. The simultaneous saccharification of the biomass and fermentation of the saccharified monomers (β-D-glucose) to the bioalcohol occurs in a single vessel in a process termed SSF (Ramamoorthy et al., 2018c). As there is a continuous fermentation of the accumulated sugars, they do not involve themselves in the end product inhibition (Charles et al., 2018). In a previous work of ours, we observed that a coculture of *Saccharomyces cerevisiae* RW 143 (5% w/v) and *Trichoderma harzanium* ATCC® 20846™ (15% v/v), enhanced the production of cellulase (2.035 FPase/mL). A miniscule quantity of ethanol was recorded (0.09 g/L). This process was performed with an aim to circumvent carbon catabolite repression, which may have otherwise occurred due to the accumulated sugar monomers (Charles et al., 2018). You et al., 2017 reported SSF as an efficient method for producing bioalcohol from paper sludge. The combination of two major bioprocesses in a single stage makes this an economical and attractive alternative to the traditional cellulase synthesis/bioalcohol production processes (Olofsson et al., 2008). Furthermore, as there is a miniscule quantity of bioethanol accumulation in the reactor, a significant reduction in the contamination could be observed (Olofsson et al., 2008).

In such combinatorial processes, optimal pH and temperature maintenance is a major challenge. The optimal enzymatic saccharification temperature is 50°C, while fermentation has been reported to occur at a temperature range of 25–30°C (Ramamoorthy et al., 2020c). However, researchers have recorded efficient enzymatic saccharifications at temperatures close to 37°C (Charles et al., 2018). Wang et al.

2020b, in a recently advanced procedure, reported that a temperature modification, every six hours, during the process of a fed-batch, nonisothermal-SSF resulted in an ethanol yield of 30.1%. In an attempt to reduce the production of inhibitors, Mendes and his research group chose to perform a batch/fed-batch SSF and reported a bioethanol yield of 22.7 g/L while using *Saccharomyces cerevisiae;* the biomass was sludge (containing unbleached pulp and 60% carbohydrates) collected from the effluent stream of paper manufacturing industries (Mendes et al., 2017).

### 1.6.3 Partial Saccharification and Simultaneous Saccharification and Fermentation (PSSSF)

A PSSSF integrates SSF and SHF, where a step of partial presaccharification (50°C) is performed prior to an SSF. Pratto et al. 2019 reported 290 L of ethanol production while performing a PSSSF for sugarcane straw biomass.

### 1.6.4 Semi-simultaneous Saccharification and Fermentation (SSSF)

In semi-simultaneous saccharification and fermentation (SSSF), a step of presaccharification, for eight hours at a temperature of 50°C, is performed prior to the SSF. The SSF is performed by lowering the reactor temperature to reach the temperature required for fermentation (Zeng et al., 2020). An SSSF for a biomass mixture of coconut waste and cactus, yielded 89% v/w ethanol (Goncalves et al., 2014). Elliston et al. (2013) performed a fed-batch SSSF with a calculated substrate feeding and reported 11.6% (v/v) ethanol yield and a cumulative substrate concentration of 65% (w/v).

### 1.6.5 Semi-simultaneous Saccharification and Cofermentation (SScF)

In an SScF, within a single reactor, an SSF is performed, wherein a microbe capable of cofermenting the biomass (i.e., having the ability to ferment both pentose and hexose sugars) is used (Lee et al., 2017). During pretreatments, when there is a significant quantity of hemicelluloses recorded, an SScF is used (Lee et al., 2017). The biomass could naturally contain a higher quantity of pentose sugars while in certain cases, the pretreatment techniques employed could falter in efficiently solubilizing and removing the pentose sugar components of the biomass. The use of engineered *Saccharomyces cerevisiae* XUSE (containing the pathway genes of the pentose phosphate pathway), and an enhanced version of the engineered *S. cerevisiae* XUSEA, expressing the genes *xylA*3* and $RPE_1$ with an ability to coferment xylose along with glucose has been reported (Tran et al., 2020). A research group reported a yield of 0.32 g ethanol per g of sugar while they used a xylose-glucose cofermenting microbe *Saccharomyces cerevisiae* (KE6-12) in an SScF of wheat straw biomass (Bondesson and Galbe, 2016).

### 1.6.6 Consolidated Bioprocessing (CBP)

A Consolidated Bioprocessing (CBP) combines the three significant biorefinery stages: cellulase production; enzymatic saccharification of the biomass; and fermentation of the sugars in the saccharified hydrolysate (Cunha et al., 2020). In one of our earlier works, a CBP was attempted by using chemically fused (PEG and calcium chloride mediated) protoplasts of *T. harzanium* ATCC® 20846™ and *Saccharomyces cerevisiae* RW 143 and surgical waste cotton and waste packaging cardboard biomass. An ethanol concentration of 0.04 g/L and a cellulase activity of 0.9 FPase/mL were reported. We hypothesized that such an integrated process would significantly mitigate the detrimental catabolite repression (Ramamoorthy et al., 2019c). In a first-of-its-kind solid-state CBP, our research group reported 11.2 g/L ethanol yield while using a hospital waste biomass mixture as the substrate in a self-designed aerated tray reactor (Ramamoorthy et al., 2018b).

## 1.7 CONCLUSION

The usage of waste, alternate lignocellulosic biomass substrates for enzyme production, or biofuel application, apart from mitigating environmental pollution occurring due to their improper disposal methodologies, could also serve as a potential technique to recycle such inexpensive wastes into commercially valuable bioproducts. However, while using alternate or waste lignocellulosic substrates of different origins, issues pertaining to: finalizing the optimal pretreatment techniques; batch-to-batch variation of the substrate compositions; and modifications in bioreactor operations may arise. Compositional analysis of each batch of alternate lignocelluloses would help in deducing the quantities of various fundamental components of the biomass, thereby enabling the process engineer to formulate subsequent bioprocesses while also working out techno-economic aspects beforehand. Globally, during cellulase production processes, several research groups have been facing an ordeal in the attainment of a standard procedure to perform an online/inline estimation of the increase in cellulolytic fungal biomass. Furthermore, although the NREL's recommended IUPAC's cellulase complex's activity measurement procedure has been used by a majority of researchers, a few research groups resort to the use of alternate procedures to report cellulase activities; such practices result in ambiguity while reporting and comparing the efficiencies of cellulase production strategies practiced among various research groups. Although scaled-up bioprocesses have been innovated in recent times, these are often found to be microbe-specific or substrate biomass-specific. Future researches in this domain could aim to design a larger spectrum of microbial bioprocesses and offer generalized techniques for processing a range of biomass substrates.

## ACKNOWLEDGMENTS

We would like to thank the head of the Department of Biotechnology, School of Life Sciences, Pondicherry University, India for the facilities provided.

## REFERENCES

Abdel-Rahman, M. A., Xiao, Y., Tashiro, Y., Wang, Y., Zendo, T., Sakai, K., Sonomoto, K., 2014. Fed-batch fermentation for enhanced lactic acid production from glucose/xylose mixture without carbon catabolite repression, *Journal of Bioscience and Bioengineering*, 119. https://doi.org/10.1016/j.jbiosc.2014.07.007

Abdullah, J. J., Greetham, D., Pensupa, N., Tucker, G. A., Du, C., 2016. Optimizing cellulase production from Municipal Solid Waste (MSW) using Solid State Fermentation (SSF), *Journal of Fundamentals of Renewable Energy and Applications*, 6: 206. https://doi.org/10.4172/2090-4541.1000206

Aftab, N., Iqbal, I., Riaz, F., Karadağ, A., Tabatabaei, M. et al., 2019. Different pretreatment methods of lignocellulosic biomass for use in biofuel production, *Biomass for Bioenergy-Recent Trend and Future Challenges*. https://doi.org/10.5772/intechopen.84995

Agrawal, K., Verma, P., 2020. Laccase-mediated synthesis of bio-material using agro-residues. In: *Biotechnological Applications in Human Health*. Springer, Singapore, pp. 87–93.

Ali, H., Mohamed, M., Shaker, N., Ramadan, M., El-Sadek, B., Hady, M. A., 2012. Cellulase enzyme in bio-finishing of cotton-based fabrics: effects of process parameters, *Research Journal of Textile and Apparel*, 16, 57–65. https://doi.org/10.1108/RJTA-16-03-2012-B006

Amore, A., Giacobbe, S., Faraco, V., 2013. Regulation of cellulase and hemicellulase gene expression in fungi, *Current Genomics*, 14(4). https://doi.org/10.2174/1389202911314040002

Ani, F., 2016. Utilization of bioresources as fuels and energy generation, *Electric Renewable Energy Systems*, 140–155. https://doi.org/10.1016/B978-0-12804448-3.00008-6

Antoniêto, A. C. C., De Paula, R. G., Castro, L. D. S., Silva-Rocha, R., Persinoti, G. F., Silva, R. N., 2015. *Trichoderma reesei* CRE1-Mediated carbon catabolite repression in response to sophorose through RNA sequencing analysis, *Current Genomics*, 17, 1–1. https://doi.org/10.2174/1389202917666151116212901

Arora, R., Behera, S., Sharma, N. K. 2015. Bioprospecting thermostable cellulosomes for efficient biofuel production from lignocellulosic biomass, *Bioresourse and Bioprocesing*, 2, 38. https://doi.org/10.1186/s40643-015-0066-4

Assawajaruwan, S., Kuon, F., Funke, M., Hitzmann, B., 2018. Feedback control based on NADH fluorescence intensity for *Saccharomyces cerevisiae* cultivations, *Bioresour Bioprocessing*, 5, 1–9. https://doi.org/10.1186/s40643-018-0210-z

Azar, R. L., Emilio, S., Junior, B., Craig, L., Jordan, S., Drew, F., Daehwan, K., 2020. Effect of lignin content on cellulolytic saccharification of liquid hot water pretreated sugarcane bagasse, *Molecules*, 25, 623. https://doi.org/10.3390/molecules25030623

Bajpai, P., 2016. Pretreatment of lignocellulosic biomass, *Springer Briefs in Green Chemistry for Sustainability*, 111–144. https://doi.org/10.1007/978-981-10-0687-6_4

Bani, A., Pioli, S., Ventura, M., Panzacchi, P., Borruso, L., Tognetti, R., Tonon, G., Brusetti, L., 2018. The role of microbial community in the decomposition of leaf litter and deadwood, *Applied Soil Ecology*, 126. https://doi.org/10.1016/j.apsoil.2018.02.017

Barda, O., Maor, U., Sadhasivam, S., Bi, Y., Zakin, V., Prusky, D., Sionov, E., 2020. The pH-responsive transcription factor PacC governs pathogenicity and ochratoxin a biosynthesis in *Aspergillus carbonarius*, *Frontier Microbiology*, 11, 210. https://doi.org/10.3389/fmicb.2020.00210

Barrios, G. J., Tarragó-Castellanos, M. R., 2015. Solid-state fermentation: Special physiology of fungi, *Fungal Metabolites*, 319–347. https://doi.org/10.1007/978-3-319-25001-4_6

Beier, S., Hinterdobler, W., Bazafkan, H., Schillinger, L., Schmoll, M., 2020. CLR1 and CLR2 are light dependent regulators of xylanase and pectinase genes in *Trichoderma reesei*, *Fungal Genetics and Biology*, 136, 103315. https://doi.org/10.1016/j.fgb.2019.103315

Ben, F. N., Jayet, P. A., Darzi, P. L., 2016. Competition between food, feed, and (bio)fuel: A supply-side model based assessment at the European scale, *Land Use Policy*, 52, 195–205. https://doi.org/10.1016/j.landusepol.2015.12.027

Benocci, T., Aguilar-Pontes, M. V., Zhou, M., Seiboth, B., de Vries, R. P. (2017). Regulators of plant biomass degradation in ascomycetous fungi, *Biotechnology for Biofuels*, 10, 152. https://doi.org/10.1186/s13068-017-0841-x

Bertrand, E., Vandenberghe, L. P. S., Soccol, C. R., Sigoillot, J. C., Faulds, C., 2016. First generation bioethanol. In: Soccol, C., Brar, S., Faulds, C., Ramos, L. (eds.) *Green Fuels Technology, Green Energy and Technology*. Springer, Cham. https://doi.org/10.1007/978-3-319-30205-8_8

Bhagia, S., Wyman, C. E., Kumar, R., 2019. Impacts of cellulase deactivation at the moving air–liquid interface on cellulose conversions at low enzyme loadings, *Biotechnology for Biofuels*, 12, 96. https://doi.org/10.1186/s13068-019-1439-2

Bhardwaj, N., Agrawal, K., Kumar, B., Verma, P., 2021a. Role of enzymes in deconstruction of waste biomass for sustainable generation of value-added products. In: Thatoi, H., Mohapatra, S., Das, S. K. (eds.) *Bioprospecting of Enzymes in Industry, Healthcare and Sustainable Environment*. Springer, Singapore, pp. 219–250.

Bhardwaj, N., Kumar, B., Agrawal, K., Verma, P., 2021b. Current perspective on production and applications of microbial cellulases: A review, *Bioresources and Bioprocessing*, 8, 95. https://doi.org/10.1186/s40643-021-00447-6

Bhargava, S., Wenger, K., Rane, K., Rising, V., Marten, M., 2005. Effect of cycle time on fungal morphology, broth rheology, and recombinant enzyme productivity during pulsed addition of limiting carbon source, *Biotechnology and Bioengineering*, 89, 524–529. https://doi.org/10.1002/bit.20355

Biomass-renewable energy from plants and animals, USEIA. https://www.eia.gov/energy explained/biomass/ (accessed on 6 April 2021).

Blunt, W., Gaugler, M., Collet, C., Sparling, R., Gapes, D. J., Levin, D. B., Cicek, N., 2019. Rheological behavior of high cell density *Pseudomonas putida* LS46 cultures during production of medium chain length polyhydroxyalkanoate (PHA) polymers, *Bioengineering* (Basel, Switzerland), 6(4), 93. https://doi.org/10.3390/bioengineering6040093

Boberg, J. B., Finlay, R. D., Stenlid, J., Ekblad, A., Lindahli, B. D., 2014. Nitrogen and carbon reallocation in fungal mycelia during decomposition of boreal forest litter, *Plos One*, 9(3), e92897. https://doi.org/10.1371/journal.pone.0092897

Bondesson, P. M., Galbe, M., 2016. Process design of SSCF for ethanol production from steam-pretreated, acetic-acid-impregnated wheat straw, *Biotechnol Biofuels*, 9, 222. https://doi.org/10.1186/s13068-016-0635-6

Brunecky, R., Donohoe, B. S., Yarbrough, J. M. et al., 2017. The multi domain *Caldicellulosiruptor bescii* CelA cellulase excels at the hydrolysis of crystalline cellulose, *Scientific Report*, 7, 9622. https://doi.org/10.1038/s41598-017-08985-w

Bundhoo, Z., Mohee, R., (2017). Ultrasound-assisted biological conversion of biomass and waste materials to biofuels: A review, *Ultrasonics Sonochemistry*, 40. https://doi.org/10.1016/j.ultsonch.2017.07.025

Cao, G., Ximenes, E., Nichols, N. N., Zhang, L., Ladisch, M., 2013. Biological abatement of cellulase inhibitors, *Bioresource Technology*, 146, 604–610. https://doi.org/10.1016/j.biortech.2013.07.112

Cascaval, D., Oniscu, C., Galaction, A. I., 2003. Rheology of fermentation broths. 2 - Influence of the rheological behavior on biotechnological processes, *Revue Roumaine de Chimie*, 48, 339–356.

Casey, E., Sedlak, M., Ho, N. W., Mosier, N. S., 2010. Effect of acetic acid and pH on the cofermentation of glucose and xylose to ethanol by a genetically engineered strain of *Saccharomyces cerevisiae*, *FEMS Yeast Research*, 10(4), 385–393. https://doi.org/10.1111/j.1567-1364.2010.00623.x

Caspeta, L., Castillo, T., Nielsen, J., 2015. Modifying yeast tolerance to inhibitory conditions of ethanol production processes, *Frontiers in Bioengineering and Biotechnology*, 3, 184. https://doi.org/10.3389/fbioe.2015.00184

Charles, S., Ramamoorthy, N., Sambavi, T., Sahadevan, R., 2018. Yeast co-culture with *Trichoderma harzanium* ATCC® 20846™ in submerged fermentation enhances cellulase production from a novel mixture of surgical waste cotton and waste card board, *International Journal of Mordern Science and Technology*, 3, 117–125.

Chaturvedi, V., Verma, P., 2013. An overview of key pretreatment processes employed for bioconversion of lignocellulosic biomass into biofuels and value added products. *3 Biotech*, 3(5), 415–431. https://doi.org/10.1007/s13205-013-0167-8

Chen, H., 2015. Lignocellulose biorefinery feedstock engineering, *Lignocellulose Biorefinery Engineering*, 37–86. https//doi.org/10.1016/B978-0-08-100135-6.00003-X

Cheng, C., Tang, R. Q., Xiong, L. et al., 2018. Association of improved oxidative stress tolerance and alleviation of glucose repression with superior xylose-utilization capability by a natural isolate of *Saccharomyces cerevisiae*, *Biotechnol Biofuels*, 11, 28. https://doi.org/10.1186/s13068-018-1018-y

Chitra, M., Pappa, N., Abraham, A., 2018. Dissolved oxygen control of batch bioreactor using model reference Adaptive control scheme, *IFAC-Papers OnLine*, 51, 13–18. https://doi.org/10.1016/j.ifacol.2018.06.008

Clarke, K. G., 2013. Bioprocess scale up, *Bioprocess Engineering*, Woodhead Publishing, 171–188. https://doi.org/10.1533/9781782421689.171

Craig, J. P., Coradetti, S. T., Starr, T. L., Glass, N. L., 2015. Direct target network of the *Neurospora crassa* plant cell wall deconstruction regulators CLR-1, CLR-2, and XLR-1, *mBio*, 6(5), e01452-15. https://doi.org/10.1128/mBio.01452-15

Cunha, F. M., Esperança, M. N., Zangirolami, T. C., Badino, A. C., Farinas, C. S., 2012. Sequential solid-state and submerged cultivation of *Aspergillus niger* on sugarcane bagasse for the production of cellulase, *Bioresource Technology*, 112, 270–274. https://doi.org/10.1016/j.biortech.2012.02.082

Cunha, J. T., Romaní, A., Inokuma, K. et al., 2020. Consolidated bioprocessing of corn cob-derived hemicellulose: engineered industrial *Saccharomyces cerevisiae* as efficient whole cell biocatalysts, *Biotechnol Biofuels*, 13, 138. https://doi.org/10.1186/s13068-020-01780-2

da Silva, A. S., Espinheira, R. P., Teixeira, R., de Souza, M. F., Ferreira-Leitão, V., Bon, E., 2020. Constraints and advances in high-solids enzymatic hydrolysis of lignocellulosic biomass: a critical review, *Biotechnology for Biofuels*, 13, 58. https://doi.org/10.1186/s13068-020-01697-w

Desvaux, M., Dumas, E., Chafsey, I., Hébraud, M., 2006. Protein cell surface display in Gram-positive bacteria: from single protein to macromolecular protein structure, *FEMS Microbiology Letters*, 256(1), 1–15. https://doi.org/10.1111/j.1574-6968.2006.00122.x

Dhamole, P., Wang, B., Feng, H., 2013. Detoxification of corn stover hydrolysate using surfactant based aqueous two phase system, *Journal of Chemical Technology & Biotechnology*, 80, 1744. https://doi.org/10.1002/jctb.4032

Dos Reis, T. F., Nitsche, B. M., de Lima, P. B., de Assis, L. J., Mellado, L., Harris, S. D., Meyer, V., Dos Santos, R. A., Riaño-Pachón, D. M., Ries, L. N., Goldman, G. H., 2017. The low affinity glucose transporter HxtB is also involved in glucose signalling and metabolism in *Aspergillus nidulans*, *Scientific Reports*, 7, 45073. https://doi.org/10.1038/srep45073

Driving Investment, growth & prosperity across the entire guyana-surinamebasins, Guyana Basins Summit, IAGP. https://www.iogp.org/event/guyana-basin-summit-2021/, 2021 (accessed on 6 April 2021).

Ekundayo, F. O., Ekundayo, E. A., Ayodele, B. B., 2017. Comparative studies on glucanases and β-glucosidase activities of *Pleurotus ostreatus* and *P. pulmonarius* in solid state fermentation, *Mycosphere*, 8(8), 1051–1059. https://doi.org/10.5943/mycosphere/8/8/16

Elferink, H., Bruekers, J., Veeneman, G. H., and Boltje, T. J., 2020. A comprehensive overview of substrate specificity of glycoside hydrolases and transporters in the small intestine: "A gut feeling", *Cellular and Molecular Life Sciences: CMLS*, 77(23), 4799–4826. https://doi.org/10.1007/s00018-020-03564-1

Elliston, A., Collins, S. R. A., Wilson, D. R., Roberts, I. N., Waldron, K. W., 2013. High concentrations of cellulosic ethanol achieved by fed batch semi simultaneous saccharification and fermentation of waste-paper, *Bioresource Technology*, 134, 117–126. https://doi.org/10.1016/j.biortech.2013.01.084

Faheina, G., Faheina, S., Vieira, M., Fonseca, D., Amorim, S., Souza, C., Menezes, D., Sousa, D., Alves, S. K., Adolfo, G. P. S., 2015. Strategies to increase cellulase production with submerged fermentation using fungi isolated from the Brazilian biome, *Acta Scientiarum Biological Science*, 37, 15–22. https://doi.org/10.4025/actascibiolsci.v37i1.23483

Falls, M., Madison, M., Liang, C., Karim, M. Sierra, R., Holtzapple, M. et al., 2019. Mechanical pretreatment of biomass – Part II: Shock treatment, *Biomass and Bioenergy*, 126, 47–56. https://doi.org/10.1016/j.biombioe.2019.04.016

Field, S. J., Ryden, P., Wilson, D. et al., 2015. Identification of furfural resistant strains of *Saccharomyces cerevisiae* and *Saccharomyces paradoxus* from a collection of environmental and industrial isolates, *Biotechnol Biofuels*, 8, 33. https://doi.org/10.1186/s13068-015-0217-z

Fitria, F., Ruan, H., Fransen, S., Carter, A., Tao, H., Yang, B. et al., 2019. Selecting winter wheat straw for cellulosic ethanol production in the Pacific Northwest, U.S.A, *Biomass and Bioenergy*, 123, 59–69. https://doi.org/10.1016/j.biombioe.2019.02.012

Fowler, T., Brown, R. D., Jr, 1992. The bgl1 gene encoding extracellular beta-glucosidase from *Trichoderma reesei* is required for rapid induction of the cellulase complex, *Molecular Microbiology*, 6(21), 3225–3235. https://doi.org/10.1111/j.1365-2958.1992.tb01777.x

Freitas, J., Nogueira, F., Farinas, C., 2019. Coconut shell activated carbon as an alternative adsorbent of inhibitors from lignocellulosic biomass pretreatment, *Industrial Crops and Products*, 137, 16–23. https://doi.org/10.1016/j.indcrop.2019.05.018

Ganguli, S., Somani, A., Motkuri, R. K., Bloyd, C. N., 2018. India Alternative Fuel Infrastructure: The Potential for Second-generation Biofuel Technology. OSTI.SOV, Technical Report.

Garcia-Ochoa, F., Gomez, E., 2009. Bioreactor scale-up and oxygen transfer rate in microbial processes: an overview, *Biotechnology Advances*, 27, 153–176. https://doi.org/10.1016/j.biotechadv.2008.10.006

Global Bioenergy Statistics 2019. World Bioenergy Association.

Global Energy Review 2020. The impacts of the Covid-19 crisis on global energy demand and CO2 emissions, Flagship report — April 2020.

Gonçalves, F. A., Ruiz, H. A., Nogueira, C. C., Santos, E. S., Teixeira, J., Macedo, G. R., 2014. Comparison of delignified coconuts waste and cactus for fuel-ethanol production by the simultaneous and semi-simultaneous saccharification and fermentation strategies, *Fuel*, 131, 66–76. https://doi.org/10.1016/j.fuel.2014.04.021

Gorems, W., Tenkegna, T. A., 2014. Production and optimization of cellulase from *Trichodrma* isolates under liquid state fermentation (LSF), *SINET: Ethiopian Journal of Science*, 37, 131–142.

Greer, A. J., Jacquemin, J., Hardacre, C., 2020. Industrial applications of ionic liquids, *Molecules*, 25, 5207. https://doi.org/10.3390/molecules25215207

Guilliams, A., Pattathil, S., Willies, D. et al., 2016. Physical and chemical differences between one-stage and two-stage hydrothermal pretreated hardwood substrates for use in cellulosic ethanol production, *Biotechnology for Biofuels*, 9, 30. https://doi.org/10.1186/s13068-016-0446-9

Han, L., Tan, Y., Ma, W., Niu, K., Hou, S., Guo, W., Liu, Y., Fang, X., 2020. Precision engineering of the transcription factor Cre1 in Hypocrea jecorina (*Trichoderma reesei*) for efficient cellulase production in the presence of glucose, *Frontiers Bioengineering and Biotechnology*, 8, 852. https://doi.org/10.3389/fbioe.2020.00852

He, R., Li, C., Ma, L., Dongyuan, Z., Shulin, C., 2016. Effect of highly branched hyphal morphology on the enhanced production of cellulase in *Trichoderma reesei* DES-15, *3 Biotech*, 6, 1–10. https://doi.org/10.1007/s13205-016-0516-5

Herrero, M., Havlík, P., Valin, H., Notenbaert, A., Rufino, M., Thornton, P., Blümmel, M., Weiss, F., Grace, D., Obersteiner, M. et al., 2013. Biomass use, production, feed efficiencies, and greenhouse gas emissions from global livestock systems, *Proceedings of the National Academy of Sciences of the United States of America*, 110. https://doi.org/10.1073/pnas.1308149110

Hill, R. M., 2003. *Silicone (Siloxane) Surfactants, Encyclopedia of Physical Science and Technology* (Third Edition), Academic Press, 793–804. https://doi.org/10.1016/B0-12-227410-5/00690-6

Hiloidhari, M., Das, D., Baruah, D., 2014. Bioenergy potential from crop residue biomass in India, *Renewable and Sustainable Energy Reviews*, 32, 504–512. https://doi.org/10.1016/j.rser.2014.01.025

Hmad, I., Gargouri, A., 2017. Neutral and alkaline cellulases: Production, engineering, and applications, *Journal of Basic Microbiology*, 57, 653–658. https://doi.org/10.1002/jobm.201700111

Huberman, L. B., Coradetti, S. T., Glass, N. L., 2017. Network of nutrient-sensing pathways and a conserved kinase cascade integrate osmolarity and carbon sensing in *Neurospora crassa*, *Proceedings of the National Academy of Sciences of the United States of America*, 114(41), E8665–E8674. https://doi.org/10.1073/pnas.1707713114

Ilmén, M., Saloheimo, A., Onnela, M. L., Penttilä, M. E., 2021. Regulation of cellulase gene expression in the filamentous fungus *Trichoderma reesei*, *Applied and Environmental Microbiology*, 63(4), 1298–1306. https://doi.org/10.1128/aem.63.4.1298-1306.1997

Inokuma, K., Bamba, T., Ishii, J., Ito, Y., Hasunuma, T., Kondo, A., 2016. Enhanced cell-surface display and secretory production of cellulolytic enzymes with *Saccharomyces cerevisiae* Sed1 signal peptide, *Biotechnology and Bioengineering*, 113(11), 2358–2366. https://doi.org/10.1002/bit.26008

Ishizaki, H., Hasumi, K., 2014. Chapter 10 - Ethanol production from Â biomass. In: *Research Approaches to Sustainable Biomass Systems*. Academic Press, pp. 243–258. https://doi.org/10.1016/B978-0-12-404609-2.00010-6

Iwaki, A., Kawai, T., Yamamoto, Y., Izawa, S., 2013. Biomass conversion inhibitors furfural and 5-hydroxymethylfurfural induce formation of messenger RNP granules and attenuate translation activity in *Saccharomyces cerevisiae*, *Applied and Environmental Microbiology*, 79(5), 1661–1667. https://doi.org/10.1128/AEM.02797-12

Jönsson, L. J., Martín, C., 2016. Pretreatment of lignocellulose: Formation of inhibitory by-products and strategies for minimizing their effects, *Bioresource Technology*, 199, 103–112. https://doi.org/10.1016/j.biortech.2015.10.009

Jourdier, E., Chaabane, F., Poughon, L., Larroche, C., Monot, F., 2012. Simple kinetic model of cellulase production by *Trichoderma reesei* for productivity or yield maximization, *Chemical Engineering Transactions*, 27, 313–318. https://doi.org/10.3303/CET1227053

Julia, B. M., Belén, A. M., Georgina, B., Beatriz, F., 2016. Potential use of soybean hulls and waste paper as supports in SSF for cellulase production by *Aspergillus niger*, *Biocatalysis and Agricultural Biotechnology*, 6, 1–6. https://doi.org/10.1016/j.bcab.2016.02.003

Karuppiah, V., Zhixiang, L., Liu, H. et al., 2021. Co-culture of Vel1-overexpressed *Trichoderma asperellum* and *Bacillus amyloliquefaciens*: An eco-friendly strategy to hydrolyze the lignocellulose biomass in soil to enrich the soil fertility, plant growth and disease resistance, *Microbial Cell Factories*, 20, 57. https://doi.org/10.1186/s12934-021-01540-3

Kenechi, N. O., 2016. Utilization of agricultural waste for bioethanol production-a review, *International Journal of Current Research and Review*, 8(19), 1–5.

Khdhiri, H., Potier, O., Leclerc, J. P., 2014. Aeration efficiency over stepped cascades: better predictions from flow regimes, *Water Research*, 55, 194–202. https://doi.org/10.1016/j.watres.2014.02.022

Kucharska, K., Rybarczyk, P., Hołowacz, I., Łukajtis, R., Glinka, M., Kamiński, M. et al., 2018. Pretreatment of lignocellulosic materials as substrates for fermentation processes, *Molecules*, 23(11), 2937. https://doi.org/10.3390/molecules23112937

Kuhad, R. C., Gupta, R., Singh, A., 2011. Microbial cellulases and their industrial applications, *Enzyme Research*, 2, 1–10. https://doi.org/10.4061/2011/280696

Kumar, A., Chandra, R., 2020. Ligninolytic enzymes and its mechanisms for degradation of lignocellulosic waste in environment, *Heliyon*, 6(2). https://doi.org/10.1016/j.heliyon.2020.e03170

Kumar, B., Bhardwaj, N., Alam, A., Agrawal, K., Prasad, H., Verma, P., 2018. Production, purification and characterization of an acid/alkali and thermo tolerant cellulase from *Schizophyllum commune* NAIMCC-F-03379 and its application in hydrolysis of lignocellulosic wastes, *AMB Express*, 8(1), 173, 1–16. https://doi.org/10.1186/s13568-018-0696-y

Kumar, B., Bhardwaj, N., Agrawal, K., Chaturvedi, V., Verma, P., 2020. Current perspective on pretreatment technologies using lignocellulosic biomass: an emerging biorefinery concept, *Fuel Processing Technology*, 199, 106244. https://doi.org/10.1016/j.fuproc.2019.106244

Kumar, B., Verma, P., 2019. Optimization of microwave-assisted pretreatment of rice straw with FeCl3 in combination with $H_3PO_4$ for improving enzymatic hydrolysis. In: Kundu, Rita, Narula, Rajiv (eds.) *Advances in Plant & Microbial Biotechnology*. Springer, pp. 41–48.

Kumar, B., Verma, P., 2021. Biomass-based biorefineries: An important architype towards a circular economy. *Fuel*, 288, 119622.

Lafon, A., Seo, J. A., Han, K. H., Yu, J. H., d'Enfert, C., 2005. The heterotrimeric G-protein GanB(alpha)-SfaD(beta)-GpgA(gamma) is a carbon source sensor involved in early cAMP-dependent germination in *Aspergillus nidulans*, *Genetics*, 171(1), 71–80. https://doi.org/10.1534/genetics.105.040584

Lakhundi, S., Siddiqui, R., Khan, N. A. 2015. Cellulose degradation: a therapeutic strategy in the improved treatment of Acanthamoeba infections, *Parasites Vectors*, 8, 23. https://doi.org/10.1186/s13071-015-0642-7

Lee, C. R., Sung, B., Lim, K. M. et al., 2017. Co-fermentation using recombinant *Saccharomyces cerevisiae* yeast strains hyper-secreting different cellulases for the production of cellulosic bioethanol, *Scientific Report*, 7, 4428. https://doi.org/10.1038/s41598-017-04815-1

Li, C. Lin, F., Zhou, L., Qin, L., Li, B. Z., Zhou, Z., Jin, M., Chen, Z., 2017. Cellulase hyper-production by *Trichoderma reesei* mutant SEU-7 on lactose, *Biotechnology for Biofuels*, 10, 228. https://doi.org/10.1186/s13068-017-0915-9

Li, Y., Qi, B., Yinhua, W., 2020. Separation of monosaccharides from pretreatment inhibitors by nanofiltration in lignocellulosic hydrolysate: Fouling mitigation by activated carbon adsorption, *Biomass and Bioenergy*, 136, 105527. https://doi.org/10.1016/j.biombioe.2020.105527

Li, Y., Zhai, R., Jiang, X., Chen, X., Yuan, X., Liu, Z., Jin, M. et al., 2019. Boosting ethanol productivity of *Zymomonas mobilis* 8b in enzymatic hydrolysate of dilute acid and ammonia pretreated corn stover through medium optimization, high cell density fermentation and cell recycling, *Frontiers in Microbiology*, 10, 2316. https://doi.org/10.3389/fmicb.2019.02316

Li, Z., Zhou, Y., Yang, H., Zhang, D., Wang, C., Liu, H., Li, X., Zhao, J, Wei, C., 2018. A novel strategy and kinetics analysis of half-fractional high cell density fed-batch cultivation of *Zygosaccharomyces rouxii*, *Food Science and Nutrition*, 6, 1162–1169. https://doi.org/10.1002/fsn3.666

Lu, J. L., Zhou, P. J., 2015. Ethanol production from Microwave-assisted $FeC_{13}$ pretreated rice straw, *Energy Sources Part A: Recovery Utilization, and Environmental Effects*, 37(21), 2367–2374. https://doi.org/10.1080/15567036.2011.649335

Liu, J. M., Xin, X. J., Li, C. X., Xu, J. H., Bao, J., 2012. Cloning of thermostable cellulase genes of *Clostridium thermocellum* and their secretive expression in *Bacillus subtilis*, *Applied Biochemistry and Biotechnology*, 166(3), 652–662. https://doi.org/10.1007/s12010-011-9456-z

Liu, Z. L., Slininger, P. J., Dien, B. S., Berhow, M. A., Kurtzman, C. P., Gorsich, S. W. et al., 2004. Adaptive response of yeasts to furfural and 5-hydroxymethylfurfural and new chemical evidence for HMF conversion to 2,5-bis-hydroxymethylfuran, *Journal of Industrial Microbiology & Biotechnology*, 31(8), 345–352. https://doi.org/10.1007/s10295-004-0148-3

Loow, Y. L., Wu, T. Y., Tan, K. A., Lim, Y. S., Siow, L. F., Jahim, J. M., Mohammad, A. W., Teoh, W. H. et al., 2015. Recent advances in the application of inorganic salt pretreatment for transforming lignocellulosic biomass into reducing sugars, *Journal of Agricultural and Food Chemistry*, 63(38), 8349–8363. https://doi.org/10.1021/acs.jafc.5b01813

Ma, L., Li, C., Yang, Z., Jia, W., Dongyuan, Z., Shulin, C., 2013. Kinetic studies on batch cultivation of *Trichoderma reesei* and application to enhance cellulase production by fed-batch fermentation, *Journal of Biotechnology*, 166, 192–197. https://doi.org/10.1016/j.jbiotec.2013.04.023

Mariscal, R., Maireles-Torres, P., Ojeda, M., Sadaba, I., Granados, M. et al., 2016. Furfural: A renewable and versatile platform molecule for the synthesis of chemicals and fuels, *Energy & Environmental Science*, 9. https://doi.org/10.1039/C5EE02666K

Marousek, J., Haskova, S., Zeman, R., Zak, J., Vanickova, R., Marouskova, A., Vachal, J., Myskova, K., 2016. Polemics on ethical aspects in the compost business, *Science and Engineering Ethics*, 22, 581–590. https://doi.org/10.1007/s11948-015-9664-y

Maroušek, J., Maroušková, A., Myšková, K., 2015. Techno-economic assessment of collagen casings waste management, *International Journal of Environmental Science and Technology*, 12, 3385–3390. https://doi.org/10.1007/s13762-015-0840-z

Masampally, V. S., Pareek, A., Runkana, V., 2018. Cascade Gaussian process regression framework for biomass prediction in a fed-batch reactor, 18417394. https://doi.org/10.1109/SSCI.2018.8628937

Mathew, A., Wang, J., Luo, J. et al., 2015. Enhanced ethanol production via electrostatically accelerated fermentation of glucose using *Saccharomyces cerevisiae*, *Scientific Report*, 5, 15713. https://doi.org/10.1038/srep15713

Matte, A., Kozlov, G., Trempe, J. F., Currie, M. A., Burk, D., Jia, Z., Gehring, K., Ekiel, I., Berguis, A. M., Cygler, M., 2009. Preparation and characterization of bacterial protein complexes for structural analysis, *Advances in Protein Chemistry and Structural Biology*, 76, 1–42. https://doi.org/10.1016/S18761623(08)76001-2

Meléndez-Hernández, P. A., Hernández-Beltrán, J. U., Hernández-Guzmán, A. et al., 2019. Comparative of alkaline hydrogen peroxide pretreatment using NaOH and $Ca(OH)_2$ and their effects on enzymatic hydrolysis and fermentation steps, *Biomass Conservation and Biorefinery*. https://doi.org/10.1007/s13399-019-00574-3

Mendes, C., Rocha, J., de Menezes, F. F., Carvalho, M., 2017. Batch and fed-batch simultaneous saccharification and fermentation of primary sludge from pulp and paper mills, *Environmental Technology*, 38(12), 1498–1506. https://doi.org/10.1080/09593330.2016.1235230

Morcelli, A., Rech, R., Klafke, A., Pelegrini, R., Ayub, M., 2018. Exponential fed-batch cultures of *Klebsiella pneumoniae* under anaerobiosis using raw glycerol as a substrate to obtain value-added bioproducts, *Journal of Brazilian Chemistry Society*, 29, 2278–2286. https://doi.org/10.21577/0103-5053.20180104

Mrudula, S., Murugammal, R., 2011. Production of cellulose by *Aspergillus niger* under submerged and solid state fermentation using coir waste as a substrate, *Brazilian Journal of Microbiology: [Publication of the Brazilian Society for Microbiology]*, 42(3), 1119–1127. https://doi.org/10.1590/S1517-83822011000300033

Muhammad, N., Martua, M., Joko, S., Endar, M., 2016. Effect of agitation speed and cultivation time on the production of the emestrin produced by *Emericella nidulans* Marine Fungal, *Squalen Bulletin of Marine and Fisheries Postharvest and Biotechnology*, 10, 73–78. https://doi.org/10.15578/squalen.v10i2.122

Müller, G., Chylenski, P., Bissaro, B., Eijsink, V. G. H., Horn, S. J., 2018. The impact of hydrogen peroxide supply on LPMO activity and overall saccharification efficiency of a commercial cellulase cocktail, *Biotechnolohy for Biofuels* 11, 209. https://doi.org/10.1186/s13068-018-1199-4

Muscat, A., de Olde, E. M., de Boer, I. J. M., Bosch, R. R., 2020. The battle for biomass: A systematic review of food-feed-fuel competition, *Global Food Security*, 25, 100–330. https://doi.org/10.1016/j.gfs.2019.100330

Nguyen, D. T. N. N., Fargues, C., Guiga, W., Lameloise, M.-L., 2015. Assessing nanofiltration and reverse osmosis for the detoxification of lignocellulosic hydrolysates, *Journal of Membrane Science*, 487, 40–50. https://doi.org/10.1016/j.memsci.2015.03.072

Obeng, E. M., Adam, S. N. N., Budiman, C. et al. 2017a. Lignocellulases: a review of emerging and developing enzymes, systems, and practices, *Bioresources and Bioprocessing*, 4, 16. https://doi.org/10.1186/s40643-017-0146-8

Obeng, E. M., Budiman, C., Ongkudon, C. M. 2017b. Identifying additives for cellulase enhancement—a systematic approach, *Biocatalysis and Agricultural Biotechnology*, 11, 67–74. https://doi.org/10.1016/j.bcab.2017.06.006

Obeng, E., Ongkudon, C., Budiman, C., Maas, R., Jose, J., 2018. An optimal blend of single autodisplayed cellulases for cellulose saccharification – a proof of concept, *Journal of Chemical Technology & Biotechnology*, 93(9), 2719–2728. https://doi.org/10.1080/15567036.2019.1602208-10.1002/jctb.5628

Olofsson, K., Bertilsson, M., Lidén, G., 2008. A short review on SSF - an interesting process option for ethanol production from lignocellulosic feedstocks, *Biotechnology for Biofuels*, 1(1), 7. https://doi.org/10.1186/1754-6834-1-7

Paloheimo, M., Haarmann, T., Mäkinen, S., Vehmaanperä, J., 2016. Production of industrial enzymes in *Trichoderma reesei*. In: Schmoll, M., Dattenböck, C. (eds.) *Gene Expression Systems in Fungi: Advancements and Applications, Fungal Biology*. Springer, Cham. https://doi.org/10.1007/978-3-319-27951-0_2

Payne, C. M., Knott, B. C., Mayes, H. B., Hansson, H., Himmel, M. E., Sandgren, M., Ståhlberg, J., Beckham, G. T., 2015. Fungal cellulases, *Chemical Reviews*, 115(3), 1308–1448. https://doi.org/10.1021/cr500351c

Pollock, J., Ho, S., Farid, S., 2013. Fed-batch and perfusion culture processes: Economic, environmental, and operational feasibility under uncertainty, *Biotechnology and Bioengineering*, 110(1), 206–219. https://doi.org/10.1002/bit.24608

Pratto, B., Santos-Rocha, M., Longati, A., Sousa, R., Cruz, A., 2019. Experimental optimization and techno-economic analysis of Bio-ethanol production by simultaneous saccharification and fermentation process using sugarcane straw, *Bioresource Technology*, 297, 122494. https://doi.org/10.1016/j.biortech.2019.122494

Qi, B., Luo, J., Chen, X., Hang, X., Wan, Y., 2011. Separation of furfural from monosaccharides by nanofiltration, *Bioresource Technology*, 102(14), 7111–7118. https://doi.org/10.1016/j.biortech.2011.04.041

Qiu, Z., Wu, X., Zhang, J., Huang, C., 2017. High temperature enhances the ability of *Trichoderma asperellum* to infect *Pleurotus ostreatus* mycelia, *PloS One*, 12(10), e0187055. https://doi.org/10.1371/journal.pone.0187055

Ramamoorthy, N., Ravi, S., Renganathan, S., 2018a. Heterologous expression of exoglucanase from *Trichoderma reesei* in *E. coli*, *International Journal of Modern Science and Technology*, 3, 65–71.

Ramamoorthy, N., Ravi, S., Sahadevan, R., 2018b. Consolidated bioprocessing in solid state fermentation for the production of bioethanol from a novel mixture of surgical waste cotton and waste cardboard, *International Journal of Modern Science and Technology*, 3. 173–180.

Ramamoorthy, N. K., Ravi, S., Sahadevan R., 2018c. Production of bio-ethanol from an innovative mixture of surgical waste cotton and waste card board after ammonia pretreatment, *Energy Sources, Part A: Recovery, Utilization, and Environmental Effects*, 2451–2457. https://doi.org/10.1080/15567036.2018.1502843

Ramamoorthy, N., Sambavi, T. R., Renganathan, S., 2019a. A study on Cellulase production from a mixture of lignocellulosic wastes, *Process Biochemistry*, 83, 148–158. https://doi.org/10.1016/j.procbio.2019.05.006

Ramamoorthy, N., Ravi, S., Sahadevan, R., 2019b. Assessment of fed-batch strategies for enhanced Cellulase production from a waste lignocellulosic mixture, *Biochemical Engineering Journal*, 152, 107–387. https://doi.org/10.1016/j.bej.2019.107387

Ramamoorthy, N., Ravi, S., Sahadevan, R., 2019c. A novel strain developed through protoplast fusion for consolidated bioprocessing of lignocellulosic waste mixture, *International Journal of Modern Science and Technology*, 4, 128–137.

Ramamoorthy, N., Sambavi, T., Baskar, G., Sahadevan, R., 2020a. Experimental validation of optimization by statistical and CFD simulation methods for Cellulase production from waste lignocellulosic mixture, *International Journal of Modern Science and Technology*, 5, 45–58.

Ramamoorthy, N. K., Nagarajan, R., Ravi, S., Sahadevan, R., 2020b. An innovative plasma pretreatment process for lignocellulosic bio-ethanol production, *Energy Sources, Part A: Recovery, Utilization, and Environmental Effects*, 10, 1080. https://doi.org/10.1080/15567036.2020.1815900

Ramamoorthy, N. K., T r, T. R., Sahadevan, R., 2020c. Production of bioethanol by an innovative biological pretreatment of a novel mixture of surgical waste cotton and waste card board, *Energy Sources, Part A: Recovery, Utilization, and Environmental Effects*, 42(8), 942–953. https://doi.org/10.1080/15567036.2019.1602208

Ravi, S., Ramamoorthy, N., Sahadevan, R., 2018. Mixture of potato, sapodilla, kiwi peels and coir as a substrate for the production of cellulases using *Trichoderma atroviride* ATCC® 28043™ by a solid state cyclic fed-batch strategy and evaluation of its saccharification efficiency, *Iferp*, 4(6), 130–133.

Reijngoud, J., Arentshorst, M., Ruijmbeek, C., Reid, I., Alazi, E. D., Punt, P. J., Tsang, A., Ram, A., 2021. Loss of function of the carbon catabolite repressor CreA leads to low but inducer-independent expression from the feruloyl esterase B promoter in *Aspergillus niger*, *Biotechnology Letters, Advance Online Publication*. https://doi.org/10.1007/s10529-021-03104-2

Ria, M., Niklasson, C., Taherzadeh, M., 2002. Effect of pH, time and temperature of overliming on detoxification of dilute-acid hydrolyzates for fermentation by *Saccharomyces cerevisiae*, *Process Biochemistry*, 38, 515–522. https://doi.org/10.1016/S0032-9592(02)00176-0

Ribeiro, A., Tyzack, J., Borkakoti, N., Holliday, G., Thornton, J., 2019. A global analysis of function and conservation of catalytic residues in enzymes, *Journal of Biological Chemistry*, 295. https://doi.org/10.1074/jbc.REV119.006289

Rodrigues, D., Pillaca-Pullo, O., Torres-Obreque, K., Flores-Santos, J., Sánchez-Moguel, I., Pimenta, M. V., Basi, T., Converti, A., Lopes, A., Monteiro, de S. G., Fonseca, L. P., Pessoa, A., 2019. Fed-batch production of *Saccharomyces cerevisiae* L-asparaginase II by recombinant *Pichia pastoris* MUT strain, *Frontiers of Bioengineering and Biotechnology*, 7, 1–12. https://doi.org/10.3389/fbioe.2019.00016

Rodríguez, F. J., Erdocia, X., Alriols, M., Labidi, J. et al., 2017. Lignin depolymerization for phenolic monomers production by sustainable processes, *Journal of Energy Chemistry*, 26. https://doi.org/10.1016/j.jechem.2017.02.007

Roeva, O., Tzonkov, S., Hitzmann, B., 2010. Optimal feeding trajectories design for E. coli fed-batch fermentations, *International Journal of Bioautomation*, 14, 89–98.

Routledge, S., 2012. Beyond de-foaming: The effects of antifoams on bioprocess productivity, *Computational and Structural Biotechnology Journal*, 3, e201210014. https://doi.org/10.5936/csbj.201210014

Salmon, D. N. X., Spier, M. G., Soccol, C. R., Vandenberghe, L. P. S., Montibeller, V. W., Bier, M. C. J., Faraco, V., 2014. Analysis of inducers of xylanase and cellulase activities production by *Ganoderma applanatum* LPB MR-56, *Fungal Biology*, 118, 655–662. https://doi.org/10.1016/j.funbio.2014.04.003

Sánchez, O. J., Montoya, S., Vargas, L. M., 2014. *Polysaccharide Production by Submerged Fermentation, Polysaccharides*. Springer, Cham, pp. 451–473.

Sarkar, P., Devdutt, U., Tasadoq, J., 2014. Estimation of paper waste generated at few of the lottery centres in and around Pune, Maharashtra, India, *Journal of Environmental Research and Development*, 9, 435–445.

Sartori, T., Tibolla, H., Prigol, E., Colla, L. M., Costa, J. A., Bertolin, T. E., 2015. Enzymatic saccharification of lignocellulosic residues by cellulases obtained from solid state fermentation using *Trichoderma viride*, *BioMed Research International*, 2015, 342716. https://doi.org/10.1155/2015/342716

Sawatzki, A., Hans, S., Narayanan, H., Haby, B., Krausch, N., Sokolov, M., Glauche, F., Riedel, S., Neubauer, P., Cruz, B. M., 2018. Accelerated bioprocess development of endopolygalacturonase-production with *Saccharomyces cerevisiae* using multivariate prediction in a 48 mini-bioreactor automated platform, *Bioengineering*, 5, 101. https://doi.org/10.3390/bioengineering5040101

Shahzadi, T., Ikram, N., Afroz, A., But, H., Anwar, Z., Irshad, M., Mahmood, A., 2013. Optimization of physical and nutritional factors for induced production of cellulase by co-culture solid-state bio-processing of corn stover, *WSEAS Transactions on Environment and Development*, 9, 263–267.

Shi, X., Liu, Y., Dai, J., Liu, X., Dou, S., Teng, L., Meng, Q., Lu, J., Ren, X., Wang, R., 2019. A novel integrated process of high cell-density culture combined with simultaneous saccharification and fermentation for ethanol production, *Biomass and Bioenergy*, 121, 115–121. https://doi.org/10.1016/j.biombioe.2018.12.020

Singh, R., Srivastava, M., Shukla, A., 2016. Environmental sustainability of bioethanol production from rice straw in India: A review, *Renewable and Sustainable Energy Reviews*, 54, 202–216. https://doi.org/10.1016/j.rser.2015.10.005

Stoyanova, E., Forsthuber, B., Pohn, S., Fuchs, W., Bochmann, G., 2014. Reducing the risk of foaming and decreasing viscosity by two-stage anaerobic digestion of sugar beet pressed pulp, *Biodegradation*, 25, 277–289. https://doi.org/10.1007/s10532-013-9659-9

Takano, M., Hoshino, K., 2018. Bioethanol production from rice straw by simultaneous saccharification and fermentation with statistical optimized cellulase cocktail and fermenting fungus, *Bioresource and Bioprocessing*, 5, 16. https://doi.org/10.1186/s40643-018-0203-y

Thoresen, M., Malgas, S., Mafa, M. S., Pletschke, B. I., 2021. Revisiting the phenomenon of cellulase action: Not all endo- and exo-cellulase interactions are synergistic, *Catalysts*, 11, 170. https://doi.org/10.3390/catal11020170

Thorn, R. G., Reddy, C., Harris, D., Paul, E., 1996. Isolation of saprophytic basidiomycetes from soil, *Applied and Environmental Microbiology*, 62, 4288–4292. https://doi.org/10.1128/AEM.62.11.4288-4292.1996

Tiwari, A. V., Kadu, P. A., 2013. Biomedical waste management practices in India-A review, *International Journal of Current Engineering and Technology*, 3, 2030–2033.

Tran, H. N. P., Ko, J. K., Gong, G., Um, Y., Lee, S. M., 2020. Improved simultaneous co-fermentation of glucose and xylose by *Saccharomyces cerevisiae* for efficient ligno cellulosic biorefinery, *Biotechnology for Biofuels*, 13, 12. https://doi.org/10.1186/s13068-019-1641-2

Transcription Factors, Reference Module in Biomedical Sciences, Elsevier, 2014. https://doi.org/10.1016/B978-0-12-801238-3.05466-0

Vadala, B. S., Deshpande, S., Apte-Deshpande, A., 2021. Soluble expression of recombinant active cellulase in *E.coli* using *B.subtilis* (natto strain) cellulase gene, *Journal, Genetic Engineering & Biotechnology*, 19(1), 7. https://doi.org/10.1186/s43141-020-00103-0

Väljamäe, P., Pettersson, G., Johansson, G. (2001). Mechanism of substrate inhibition in cellulose synergistic degradation, *European Journal of Biochemistry*, 268(16), 4520–4526. https://doi.org/10.1046/j.1432-1327.2001.02377.x

Vazana, Y., Barak, Y., Unger, T. et al., 2013. A synthetic biology approach for evaluating the functional contribution of designer cellulosome components to deconstruction of cellulosic substrates, *Biotechnology for Biofuels*, 6, 182. https://doi.org/10.1186/1754-6834-6-182

Veiter, L., Rajamanickam, V., Herwig, C., 2018. The filamentous fungal pellet—relationship between morphology and productivity, *Applied Microbiology Biotechnology*, 102, 2997–3006. https://doi.org/10.1007/s00253-018-8818-7

Wan, G., Zhang, Q., Li, M., Jia, Z., Guo, C., Luo, B., Wang, S., Min, D., 2019. How pseudo-lignin is generated during dilute sulfuric acid pretreatment, *Journal of Agricultural and Food Chemistry*, 67(36), 10116–10125. https://doi.org/10.1021/acs.jafc.9b02851

Wang, B. T., Hu, S., Yu, X. Y., Jin, L., Zhu, Y. J., Jin, F. J. (2020a). Studies of cellulose and starch utilization and the regulatory mechanisms of related enzymes in fungi, *Polymers*, 12(3), 530. https://doi.org/10.3390/polym12030530

Wang, C., Shengju, L., Wu, J., Li, Z., 2014. Effects of temperature-dependent viscosity on fluid flow and heat transfer in a helical rectangular duct with a finite pitch, *Brazilian Journal of Chemical Engineering*, 31, 787–797. https://doi.org/10.1590/0104-6632.20140313s00002676

Wang, J., Ye, J., Feng, E., Xiu, Z., 2011. Modeling and parameter estimation of a nonlinear switching system in fed-batch culture with pH feedback, *Application of Math Model*, 36, 4887–4897. https://doi.org/10.1016/j.apm.2011.12.025

Wang, Z., Ning, P., Hu, L., Nie, Q., Liu, Y., Zhou, Y., Jianming, Y., 2020b. Efficient ethanol production from paper mulberry pretreated at high solid loading in Fed-nonisothermal-simultaneous saccharification and fermentation, *Renewable Energy*, 160. https://doi.org/10.1016/j.renene.2020.06.128

Xia, X., Lin, S., Xia, X., Cong, F. S., Zhong, J. J., 2014. Significance of agitation-induced shear stress on mycelium morphology and lavendamycin production by engineered *Streptomyces flocculus*, *Applied Microbiology and Biotechnology*, 98, 4399–4407. https://doi.org/10.1007/s00253-014-5555-4

Yang, S., Zhang, Y., Yue, W., Wang, W., Wang, Y. Y., Yuan, T. Q., Sun, R. C. et al., 2016. Valorization of lignin and cellulose in acid-steam-exploded corn stover by a moderate alkaline ethanol post-treatment based on an integrated biorefinery concept, *Biotechnology for Biofuels*, 9, 238. https://doi.org/10.1186/s13068-016-0656-1

Yang, Y., Sha, M., 2016. A beginner's guide to bioprocess modes – batch, fed-batch, and continuous fermentation. Application Note No. 408, *Engineering, Biology*, 1–16.

You, Y., Liu, S., Wu, B., Wang, Y., Zhu, Q., Qin, H., Tan, F., Ruan, Z., Ma, K., Dai, L., Zhang, M., Hu, G., Mingxiong, H., 2017. Bio-ethanol production by *Zymomonas mobilis* using pretreated dairy manure as a carbon and nitrogen source, *RSC Advances*, 7, 3768–3779. https://doi.org/10.1039/C6RA26288K

Yu, X. B., Nam, J. H., Yun, H. K., Koo, Y. M., 1998. Optimization of cellulase production in batch fermentation by Trichoderma reesei, *Biotechnology and Bioprocess Engineering*, 3, 44–47. https://doi.org/10.1007/BF02932483

Yue, Z., Bin, W., Baixu, Y., Peiji, G., 2004. Mechanism of cellobiose inhibition in cellulose hydrolysis by cellobiohydrolase, *Science in China, Series C, Life Sciences*, 47(1), 18–24. https://doi.org/10.1360/02yc0163

Zaafouri, K., Ziadi, M., Ben, H. T. A., Mekni, S., Aïssi, B., Alaya, M., Bergaoui, L., Hamdi, M. et al., 2017. Optimization of hydrothermal and diluted acid pretreatments of Tunisian *Luffa cylindrica* (L.) fibers for 2G bioethanol production through the cubic central composite experimental design CCD: Response surface methodology, *BioMed Research International*, 9524521. https://doi.org/10.1155/2017/9524521

Zeng, G., You, H., Wang, K. et al., 2020. Semi-simultaneous saccharification and fermentation of ethanol production from *Sargassum horneri* and biosorbent production from fermentation residues, *Waste and Biomass Valorization*, 11, 4743–4755. https://doi.org/10.1007/s12649-019-00748-0

Zeng, Y., Zhao, S., Yang, S., Ding, S. Y., 2014. Lignin plays a negative role in the biochemical process for producing lignocellulosic biofuels, *Current Opinion in Biotechnology*, 27, 38–45. https://doi.org/10.1016/j.copbio.2013.09.008

Zhang, H., Wei, W., Zhang, J. et al., 2018a. Enhancing enzymatic saccharification of sugarcane bagasse by combinatorial pretreatment and Tween 80, *Biotechnology for Biofuels*, 11, 309. https://doi.org/10.1186/s13068-018-1313-7

Zhang, J., Zhang, G., Wang, W. et al. 2018b. Enhanced cellulase production in *Trichoderma reesei* RUT C30 via constitution of minimal transcriptional activators, *Microbial Cell Factories*, 17, 75. https://doi.org/10.1186/s12934-018-0926-7

Zhang, Q., Wu, D., Wang, X., Kong, H., Tanaka, S., Lin, Y., 2015. Substrate and product inhibition on yeast performance in ethanol fermentation, *Energy Fuels*, 29, 1019–1027. https://doi.org/10.1021/ef502349v

Zhao, Q., Liu, Q., Wang, Q., Qin, Y., Zhong, Y., Gao, L., Liu, G., Qu, Y., 2021. Disruption of the *Trichoderma reesei* gul1 gene stimulates hyphal branching and reduces broth viscosity in cellulase production, *Journal of Industrial Microbiology and Biotechnology*, 48, 1–2. https://doi.org/10.1093/jimb/kuab012

Zhou, Q., Lv, X., Zhang, X. et al., 2011. Evaluation of swollenin from *Trichoderma pseudokoningii* as a potential synergistic factor in the enzymatic hydrolysis of cellulose with low cellulase loadings, *World Journal of Microbiology and Biotechnology*, 27, 1905–1910. https://doi.org/10.1007/s11274-011-0650-5

Zou, G., Shi, S., Jiang, Y., van den Brink, J., de Vries, R. P., Chen, L., Zhang, J., Ma, L., Wang, C., Zhou, Z., 2012. Construction of a cellulase hyper-expression system in *Trichoderma reesei* by promoter and enzyme engineering, *Microbial Cell Factories*, 11, 21. https://doi.org/10.1186/1475-2859-11-21

# 2 Developments in Hydrolysis Processes Toward Enzymatic Hydrolysis in Biorefinery

*Liya Thurakkal, Anna Jose, and Mintu Porel*
Indian Institute of Technology Palakkad, Palakkad, India

## CONTENTS



DOI: 10.1201/9781003187684-2

## 2.1 INTRODUCTION

Woeful climatic changes have compelled the world to seek valuable replacement of fossil fuel with low-carbon energy and fuel. The utilization of biofuels is a good solution on account of the potential reduction of the largest source of greenhouse gases and carbon emissions. The production of biofuels from lignocellulosic mass feedstock is evolving as the significant mode of operation toward this state of affairs. Lignocellulose refers to cellulose, hemicellulose, and lignin which are crystalline glucose polymer, an amorphous polymer of xylose and other units, and branched, substituted, aromatic polymer respectively. The main sources of this lignocellulosic biomass are landfill waste; crop stubble (which is otherwise burned); and wood and forest wastes. Strategic lignocellulosic biomass conversion generally comprises three steps (see Figure 2.1) which are: 1) conversion of recalcitrant lignocellulosic structure into reactive form by pretreatment; 2) hydrolysis of these intermediates to fermentable sugar by various methods; and 3) conversion of these sugars to other value-added products (Khare et al., 2015; Chaturvedi & Verma, 2013).

Pretreatment is an essential step in biowaste valorization as the plant cell wall is very difficult to hydrolyze. Therefore, the pretreatment process reduces the crystallinity of the cellulose and removes hemicellulose (Mosier et al., 2005). Different methods have been employed for the pretreatment process including physical (milling, microwave irradiation, extrusion, ultrasonication); chemical (acidic or alkaline

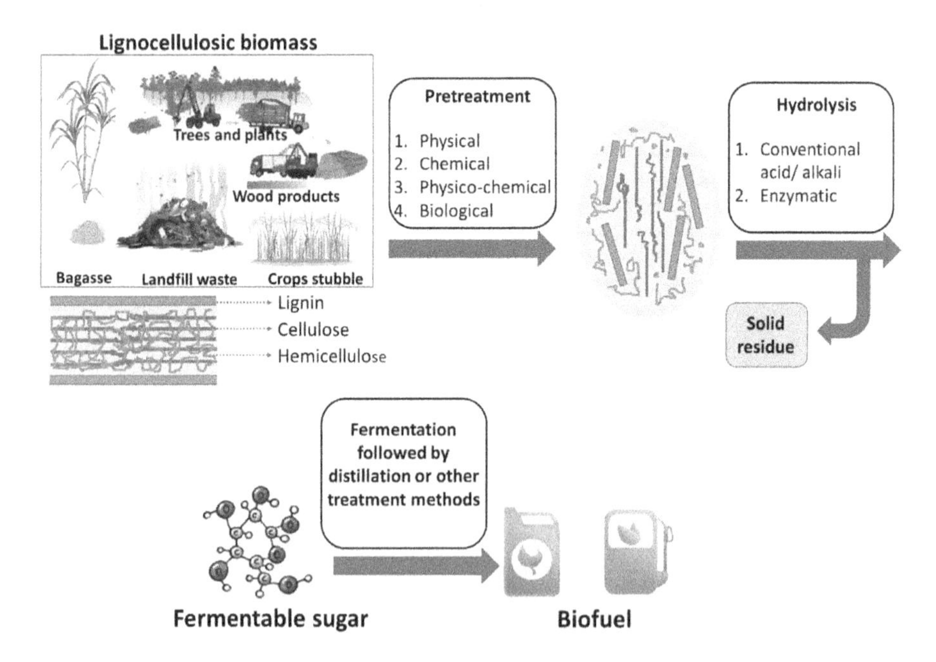


**FIGURE 2.1** Stepwise formation of value-added products by strategic treatment of lignocellulosic biomass.

hydrolysis, ionic liquids, organosolv process, deep eutectic solvents); physicochemical (steam explosion, ammonia fiber explosion (AFEX), $CO_2$ explosion, liquid hot water); and biological (enzymatic) methods (Rocha-Meneses et al., 2017; Kumar et al., 2020). To select one of the methods as the best would not be conclusive, but there are certain desired elements for the pretreatment method. A few of them are: low cost, minimum waste, fast reaction, and high product yield. The product obtained after the pretreatment process is subjected to hydrolysis.

Hydrolysis, otherwise called saccharification, is the process of producing fermentable sugar from the pretreated lignocellulosic biomass. This step is a crucial one as it determines the efficiency of the overall process of valorization. Out of the various methods of hydrolysis, chemical and biological processes of hydrolysis have emerged as the better ones. Environmental hazards and other drawbacks of conventional hydrolysis processes like strong acid hydrolysis have made the community consider more efficient and environmentally friendly methods. Lately, a highly efficient method has been developed, enzymatic hydrolysis, which utilizes the action of enzymes from various microorganisms for the conversion of biomass to fermentable sugar. The sugar obtained is subjected to fermentation to get useful products like bioethanol (Kumar & Verma, 2020).

Methods of directly converting lignocellulosic biomass to the desired products in a single step are yet to be developed and are a significant focus area for research. As of now, the biomass is converted to smaller units by hydrolysis and then fermented. Subsequently, the efficacy and practicability of a process are determined mainly by the efficiency of the hydrolysis step. In this chapter, we discuss the development of various types of hydrolysis practices from conventional methods to competent enzymatic methods.

## 2.2 HISTORICAL BACKGROUND OF THE HYDROLYSIS PROCESS

The production of ethanol has been one of the primary areas of research for a very long time. Before the Second World War, efforts were taken to produce sugars for the production of ethanol in some industrial plants in the United States by adopting the method of acid hydrolysis (Sherrard & Kressman, 1945). Sulfurous acids of different strengths were used to convert wood wastes to fermentable sugars. Although companies could obtain the desired product, small profits meant that economic success was difficult to achieve. The main reason was small product yield which subsequently increased the cost of production (Katzen & Othmer, 1942).

In Germany, similar methods for the hydrolysis of lignocellulosic materials were tried. A German scientist, Friedrich Bergius, first produced sugar from wood and received a Nobel Prize in 1931. This method was modified for better productivity by different groups, using concentrated hydrochloric acid (HCl). This method, later known as the Bergius-Rheinau process, used 41% HCl with the acid-wood ratio of 3:1. However, the use of acid in high concentration in this ratio could not be escalated to an industrial scale synthesis due to high cost and successive tedious procedures (Bergius, 1937). Consequently, another method called the Udic-Rheinau process was instigated which consisted of three hydrolysis steps (Wright et al., 1985). First, 35%

hydrochloric acid was used for prehydrolysis, then 41% HCl for the main hydrolysis, and finally dilute HCl for the posthydrolysis step. Chemical methods like distillation at elevated temperature were used for acid recovery, and other physical methods were also used for purification of the product obtained. The Scholler process was a subsequent attempt by German scientists to produce fermentable sugar xylose (wood sugar). Dilute sulfuric acid of 0.8% was used at elevated temperature and 10 atm. A yield of about 50% of the reducible sugar was obtained through the process which made this particular technique the frontrunner among the competent techniques. But, the drawback was that the resultant sugar solution was very dilute having only about 4% sugar in it (Faith, 1945). The historical view of the development of acid hydrolysis of cellulose is given in Table 2.1.

During the Second World War, there was a demand for more and cheaper fuel. The high cost of biofuel limited its usage and, therefore, its large-scale production. At the same time, petroleum products were available at much cheaper rates than products made from lignocellulosic biomass. Many industrial plants worldwide were forced to shut down, unable to meet the expense. Then, a milestone in the field of biomass conversion was reached when the US military in the South Pacific detected some microorganisms that were destroying their uniforms. In the 1950s, a scientist named Reese from the Natick Massachusetts Laboratory of US Army Materials Command analyzed the cloth samples and identified the fungus (Katzen & Schell, 2006). He named it *Trichoderma viridae* and this fungus made an incredible difference to the field of biomass hydrolysis. The fungus is now known as *T. reesei* QM6a and all currently used strains are derived from that isolate. In 1970, Mandels et al. worked on the strain to exploit it for biomass degradation. For this purpose, they developed an improved strain, *T. reesei* QM6a and later many scientists around the world attempted to mutate the strain for enriched cellulase synthesis (Mandels et al., 1974; Persson et al., 1991).

Now, enzymatic hydrolysis has become a widely used technique for hydrolytic breakdown of the biomass because of the advantages of high product yield, low operational cost, and higher selectivity when compared to the other conventional methods. Advancements in biotechnology can further improve the present situation and provide products at even more economical rates.

**TABLE 2.1**
**The Historic View on the Development of Acid Hydrolysis of Cellulose (Ragg & Fields, 1987)**

| Year | Process | Reagent | Temperature |
|---|---|---|---|
| 1923 | Scholler | Dil. $H_2SO_4$ | 170°C |
| 1937 | Bergius | Con. HCl | Ambient |
| 1945 | Madison | Dil. $H_2SO_4$ | 180°C |
| 1960 | Noguchi | Gaseous HCl | Ambient |
| 1978 | Grethlein | Dil. $H_2SO_4$ | 240°C |
| 1981 | Houchst | Anhydrous HF | 40°C |

## 2.3 CONVENTIONAL METHODS OF HYDROLYSIS AND THEIR DRAWBACKS

### 2.3.1 Acid-Catalyzed Hydrolysis

Acid hydrolysis is a hydrolyzation procedure with wide usage and acceptance for the conversion of cellulose and other macromolecular structures into simpler units. Concentrated and dilute sulfuric acid and hydrochloric acid-like protic acids have been used for the purpose. The concentrated acid breaks the intermolecular hydrogen bonding between the cellulose units and breaks its crystallinity. Once the crystallinity is lost and it becomes an amorphous state, cellulose forms a gelatinous homogenous form with the acid. Hydrolysis at this stage is uncomplicated, resulting in the generation of glucose (LaForge & Hudson, 1918).

The mode of action of acid in the hydrolysis process is due to the Bronsted acid property of hydrochloric acid, or sulfuric acid. They donate a proton in the presence of water, and the oxygen atom in the glycosidic linkage is protonated. This charged hydroxyl group leaves the polymer chain and this site is occupied by the hydroxyl group from the water leaving the proton. This step is continued till all the glycosidic linkage is broken down to result in simpler monomers such as glucose, xylose, arabinose (Harmsen et al., 2010) (Figure 2.2).

The conversion is enhanced when the samples are pretreated to remove hemicellulose. Even though the concentrated acid hydrolysis resulted in a moderate yield of the sugar, the after-effects made by the process were penitent due to corrosiveness, toxicity, and hazard. The use of dilute acid was augmented for this reason, but it resulted in lower yield at moderate temperatures. The dilute acid hydrolysis at higher temperatures resulted in good yield and went great guns after much research and development. But the cost of operation limited its commodious usage.

### 2.3.2 Base-Catalyzed Hydrolysis

Although not as widely accepted as acid hydrolysis, hydrolysis under alkaline conditions has also been used for the production of fermentable sugars by strategic conversion of lignocellulosic biomass. A base like NaOH saponifies the ester cross-linking bond between hemicellulose and lignin which, in turn, increases the porosity (Tarkow & Feist, 1969). Dilute NaOH and $NH_3$ arere also employed for sugar production from different sources of biomass. Accordingly, the base-catalyzed hydrolysis is divided mainly into two groups, one based on sodium, potassium, and calcium hydroxide, and another based on ammonia. The alkaline treatment is executed by

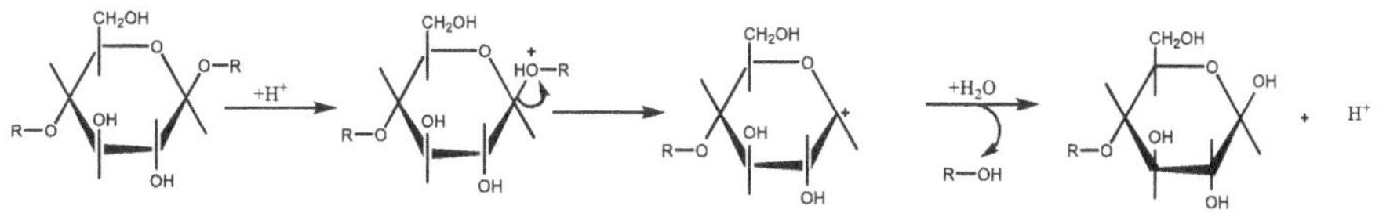

**FIGURE 2.2** Acid-catalyzed hydrolysis of hemicellulose by glycosidic bond cleavage (Loow et al., 2016).

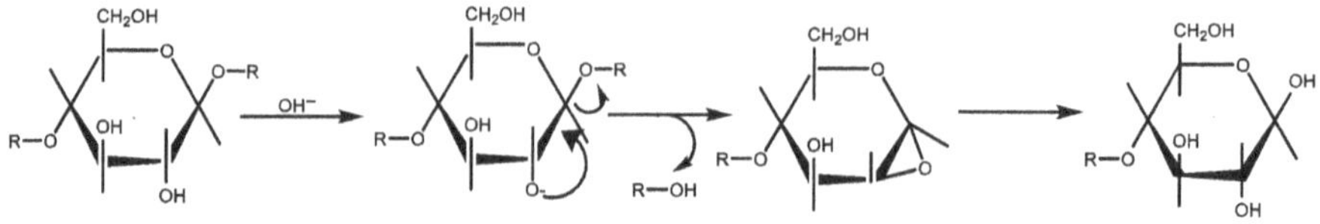

FIGURE 2.3 Alkaline hydrolysis of cellulose.

making a slurry of the base with water which is sprayed onto the biomass and kept for a long time. NaOH is the most widely used due to its low cost, higher basicity, and fractionation capacity for a wide range of biomass samples (Bensah et al., 2011). As shown in Figure 2.3, an ether bond is hydrolyzed to produce monomeric sugar units. The base abstracts the proton from the hydroxyl group which results in the formation of an epoxide kind of intermediate. This intermediate is subjected to a nucleophilic substitution reaction with hydrogen to result in the breakdown product (Krazzig & Schurz, 2008).

The reaction works only in the presence of a strong base and at a minimum temperature of 150°C. However, the product yield is not satisfactory and the process works better as a pretreatment step to remove lignin which has less recalcitrance.

### 2.3.3 Hydrolysis in Ionic Liquid

This is a milder technique to increase the digestibility of less recalcitrant lignin from the biomass. A nonvolatile solvent with low melting and boiling points and consisting of ions held together by noncovalent interaction is called an ionic liquid. A judicious selection of ionic liquid helps the cellulose in the biomass to get solubilized and is beneficial for the next step as well (Swatloski et al., 2002). It is more convenient to combine this process with other hydrolysis techniques, like enzymatic or acid hydrolysis.

### 2.3.4 Drawbacks of Conventional Methods of Hydrolysis

Conventional methods of hydrolysis were prevalent for a long time and are still practiced in rare cases. But there are certain points where research would be beneficial for improved results. The major drawbacks of chemical hydrolysis are as follows:

- Corrosion issues: The use of strong acids or bases corrodes the machinery used. This demands expensive materials in the construction of the machinery to withstand the harsh conditions which increases production costs. (Mosier et al., 2005).
- Degradation products: The action of chemicals produces some unwanted compounds as a result of degradation of the biomass. Furfural, acetic acid and 5 hydroxymethylfurfural (HMF), levulinic acid, and formic acid are a few of such formed compounds, which also affect the successive fermentation step (Jennings & Schell, 2011). Therefore, washing off the solid residue from the mixture is mandatory for better hydrolysis. This is achieved

by detoxification agents such as gypsum, use of membrane, or ammonia explosion among others.

- Subsequent to cellulose, hemicellulose hydrolysis, the formed monomer or pentose/hexose sugar also undergo degradation, sometimes producing furfural or HMF when subjected to a higher temperature and higher residence time (McKillip & Collin, 2002).
- Oligomers: The formation of smaller units of cellulose or hemicellulose other than its monomer results in making the purification process tedious.
- Acid recovery: The economic acceptance of acid-catalyzed hydrolysis is extremely dependent on the efficiency of recovery of acid from the reaction mass after completion of the reaction. Recovery is important to proceed with the treatment procedures toward fermentation; at the same time, the reusability of the acid is also crucial to make the whole strategy economical.
- Environmental hazards: The prolonged use of concentrated acid and its disposal present environmental risks as the acid destroys the beneficial microorganism community, and harms air and soil.

## 2.4 ENZYMATIC HYDROLYSIS

Enzymatic hydrolysis is the crucial process of converting cellulosic biomass to sugar products, facilitated by enzymes produced by bacteria, fungus, or other microorganisms. This biological conversion of biomass to economically important chemicals, fuels, and energy resources in high yield with comparatively lower costs is developing as a powerful competent technique. The first indication of the capability of enzymes for hydrolysis was obtained during the Second World War. A potential *T. reseei* Rut C30 strain was developed after mutagenesis and wise strain selection from the wild strain at Rutgers University. Although the cellulosic biomass conversion process with the aid of enzymes resulted in higher yield, lower cost, very high selectivity, lower energy cost, and mild reaction conditions than the prevailing chemical processes, the execution of this procedure was considered to be risky and challenging at that time. However, subsequent developments in biotechnology offered dramatic changes in operational cost and simplicity

Many microorganisms, like bacteria and fungi, have the potential to produce enzymes that can break down the polymeric structure of lignocellulose in the biomass. The microorganisms are aerobic, anaerobic, thermophilic, or mesophilic. The enzymes that hydrolyze cellulose are called cellulase. Cellulases are produced mainly by bacteria belonging to the group *Clostridium, Bacillus, Cellulomonas*, bacteriodes, *Streptomyces, Microbispora*, and *Acetovibrio* (Bisaria et al., 1991; Bhardwaj et al., 2021). Out of the bacteria showing excellent cellulolytic capacity (for example, *Clostridium thermocellum* and *Bacteroides cellulosolvens)*, many are anaerobic bacteria. The anaerobic bacteria require anaerobic conditions and their growth conditions are stringent which increase the chances of contamination. To combat the living conditions of bacteria, an advanced sophisticated setup should be assembled. For this reason, fungi are also used for enzyme retrieval for hydrolysis purposes (Sun & Cheng, 2002). Species of *Penicillium, Trichoderma, Aspergillus, Schizophyllum* are a few of the fungi used.

Enzymatic hydrolysis proceeds mainly via two processes: 1) physical disruption of the crystallinity of cellulose; and 2) hydrolysis of disrupted cellulose to sugar by glycosidic bond cleavage. In contrast to chemical hydrolysis, enzymatic hydrolysis is executed in milder conditions with pH in the range of 4–5 and temperature in the range of 45–50°C. The efficiency of the process can be improved by changing aspects of various factors including enzymes, substrate, and also pretreatment conditions. The pretreatment process separates the lignin part from the mass. Enzymes include cellulase, xylanase, peroxidase, and laccase, among others (Bhardwaj & Verma, 2021; Agrawal & Verma, 2020).

### 2.4.1 Cellulase

Cellulase is the O-glycoside oxidase enzyme that hydrolyzes the glycoside bond to produce glucose from cellulose with the synergistic action of three different enzymes: endoglucanase (EG), cellobiohydrolase (CBH) andβ-glucosidase (BG) (Table 2.2). Each microorganism has a distinct number of these cellulase enzymes and most of the time they act synergistically. The most extensively studied fungi, *Trichoderma reesei* produces five EGs, two CBHs, and two BGs.

The action of cellulase hydrolysis commences by the initiation of endoglucanases which dismantle the cellulose construct. Both the reducible and nonreducible terminals are uncovered upon which CBH acts. Cello-oligosaccharides and cellobiose units are released by the action of CBH. In succession to that, the cellobiose thus formed is cleaved by BG to form individual glucose units. At the same time, EG acts on the amorphous part of the cellulose and this triggers the action of CBH to cleave β-1,4-glucosidic bonds from the chain terminals to form oligosaccharides. BG hydrolyzes these oligosaccharides and releases monomeric sugar (Figure 2.4) (Du et al., 2013). CBH and EG undertake by the action of a dual-domain: catalytic domain, and cellulose-binding domain. The enzyme catalytic module is brought into the close vicinity of the substrate in optimum by the cellulose-binding domain. The involvement of this domain in the cellulase action is significant because the absence of this adversely affects crystalline cellulose hydrolysis. The catalytic domain is connected with the cellulose-binding domain with a linker of flexible glycosylate.

The synergistic action of three enzymes is carried out in three steps: 1) adsorption of cellulase onto the surface of cellulose; 2) conversion of cellulose to glucose; and 3)

**TABLE 2.2**
**Cellulase Enzymes and Their Functions**

| Enzyme | Functions |
|---|---|
| Endoglucanase (EG) | Hydrolyzes internal β-1,4-glycosidic linkages of cellulose chain randomly. |
| Cellobiohydrolase (CBH) or exoglucanase | Hydrolyzes the cellobiose units of the cellulose from the ends. |
| β-glucosidase (BG) or β-glucoside glucohydrolase | Hydrolyzes the cellobiose to glucose and also cleaves out the simple glucose units from the cello-oligosaccharides. |

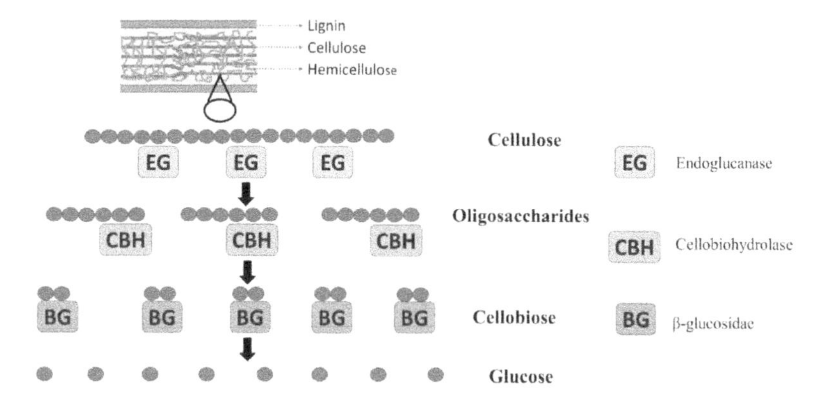


**FIGURE 2.4** Mode of action of cellulase enzyme on cellulose.

desorption of cellulase enzyme from the cellulose structure. In some cases, the cellulase gets adsorbed on the cellulose surface reversibly and it causes deactivation (Converse et al., 1988). This phenomenon is avoided by the addition of surfactants which modify the surface of cellulose. Various surfactants added for this purpose in the enzyme hydrolysis are polyoxyethylene glycol, emulgen 147, sophorolipid, rhamnolipid, bacitracin, and others. (Sun & Cheng, 2002). The recovery of the enzyme after hydrolysis is crucial and is one of the prime aspects which regulates the overall efficiency of the process. The recovery is rather difficult due to the homogenous nature of the enzyme in solution (Wahlström & Suurnäkki, 2015). One of the strategies developed to tackle this limitation is immobilization of the enzymes by trapping them into a solid inert material. Polymeric kinds of materials are extensively used for this purpose as they seldom interfere in the processes. It has been reported that enzyme turnover is higher when they are immobilized on the polymer polystyrene and calcium alginate gel particles (Tsai & Meyer, 2014).

### 2.4.2 Hemicellulases

Hemicelluloses are another major component present in the lignocellulosic biomass. Lignocellulose biomass contains orderly packed cellulose with a major portion of crystalline structure and hemicellulose is arranged in a disordered fashion around the cellulose array. Hence, the removal of hemicellulose or xylan using xylanase opens up the cellulose and makes it vulnerable to enzymatic hydrolysis. Hemicelluloses are made up of the carbohydrate xylan which is the polysaccharide of a pentose sugar called xylose. As the prime function of hemicellulose is to hydrolyze xylene, they are often also called xylanase. But hemicellulose hydrolysis is performed by various other enzymes named xylosidase, glucuronidase, ferulic and coumaric acid esterase, acetylgalactan esterase, arabinofuranosidase, acetylxylan esterase, among others. A summary of the action of various hemicellulose hydrolyzing enzymes is tabulated in Table 2.3. (Charles et al., 2004). As hemicellulose holds a complex structure, hydrolysis is comparatively difficult and it needs multiple isozymes of xylanase for its effective hydrolysis. The capability of xylanases is limited in hydrolyzing

**TABLE 2.3**
**The Action of Various Hemicellulase Enzymes**

| S. No. | Enzyme | Break Down Unit and Other Functions |
|---|---|---|
| 1 | Xylosidase | β-1,4 linkages of unreduced terminals and deliver xylose |
| 2 | Xylanase | β-1,4 linkages of unsubstituted unreduced terminals |
| 3 | Coumaric acid and ferulic acid esterase | Coumaric acid and ferulic acid linked with arabinose |
| 4 | Arabinanase | α-1,5 linkages |
| 5 | Arabinosidase | β, α-1,2 and α-1,3 linkages |
| 6 | Esterases of acetylgalactan, acetylmannan and acetylxylan | Ester linkages and eliminates acetyl functional from arabinogalactan, galactoglucomannan and xylan |
| 7 | Glucuronidase | Side chain substituted 4-*O*-methylglucopyranosuric-acid and α-1,2 linked glucopyranosuric-acid |

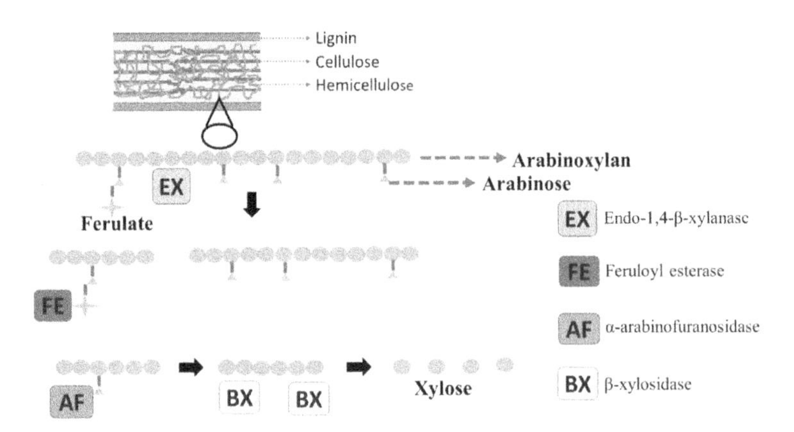


**FIGURE 2.5** Mode of action of hemicellulase enzymes.

glycosidic bonds in between the substituted xylans. This effect is attributed to the high selectivity nature of the enzymes. To overcome this effect, other enzymes are used to hydrolyze the substituents on the xylan first, then these treated xylans are hydrolyzed by the enzyme xylanase to produce the sugar. The synergistic action of various hemicellulase enzymes is given in Figure 2.5 depicting an example of the hydrolysis of ferulate substituted arabinoxylan. Endo 1,4- β-xylanase breaks the linkages from the nonsubstituted part and depolymerizes it. The substituted ferulic acid or ferulate is hydrolyzed by feruloyl esterase. Arabinofuranosidase is acted on the other part cleaving and the resulting chain is hydrolyzed by xylosidase to give the product xylose sugar. *T. reesei, Humicola insolens, Aspergillus niger*, and *Bacillus* are a few of the commercial xylanases and they function at the optimum temperature of 40 to 60°C.

### 2.4.3 PEROXIDASE

Lignin is that part of biomass that encrusts the embedded cellulose by hindering its hydrolysis. This lignin is hydrolyzed by a group of enzymes that require peroxides as oxidants and are called peroxidases. The typical action of peroxidases is given in the scheme:

$$2S + 2H_2O_2 + 2e^- \longrightarrow 2S_{ox} + 2H_2O + O_2 \quad (2.1)$$

(S: electron donor substrate, $S_{ox}$: oxidized substrate)

The class of peroxidases is mainly classified into two: heme peroxidase and non-heme peroxidase. The heme peroxidase is further classified as peroxidase cyclooxygenase superfamily and peroxide catalase superfamily. Lignin peroxidase (LiP) is the heme peroxidase of the peroxide catalase superfamily. LiP is otherwise called diaryl propane oxygenase which catalyzes oxidative degradation of lignin in presence of hydrogen peroxide. LiP oxidizes arylglycerol-aryl ethers with β-O-4 linkage of different non-phenolic lignin model compounds (Kirk et al., 1986). The action of this enzyme involves one-electron oxidation and radical cations formation. Apart from the oxidation of non-phenolic compounds several phenolic compounds with ring and N-substituted amines are also oxidized (Baciocchi et al., 2001).

The mechanism of the catalyst lignin peroxidase enzyme proceeds through three steps. First, hydrogen peroxidase accepts one electron from the ferric enzyme (III) which turns to Fe (IV) oxo-ferryl intermediate (compound I). The next step is initiated by the transfer of one electron to compound (I) which is $2e^-$ deficient by the non-phenolic aromatic substrate (S). The second intermediate compound II is thus formed which has a deficiency of $2e^-$. The catalytic cycle is completed by the reduced substrate when one more electron is transferred to compound II leading itself to the preliminary stage (Figure 2.6) (Abdel-Hamid et al., 2013).

Manganese peroxidase (MnP) is a heme-containing enzyme which belongs to the oxidoreductase family and is used extensively for lignin degradation. MnP oxidizes the phenolic lignin structures by a redox process exploiting the oxidation capacity of Mn (II) to Mn (III). The mechanism of action (Figure 2.7) resembles the action of LiP. The catalytic cycle is initiated by the hydrogen peroxide cleavage which is complexed with heme resulting in a Fe (IV) radical complex (MnP compound I). $Mn^{2+}$ is

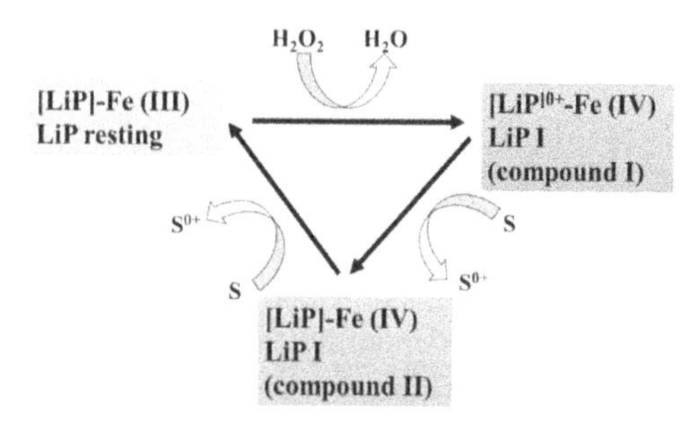


FIGURE 2.6 Catalytic loop of LiP enzyme.

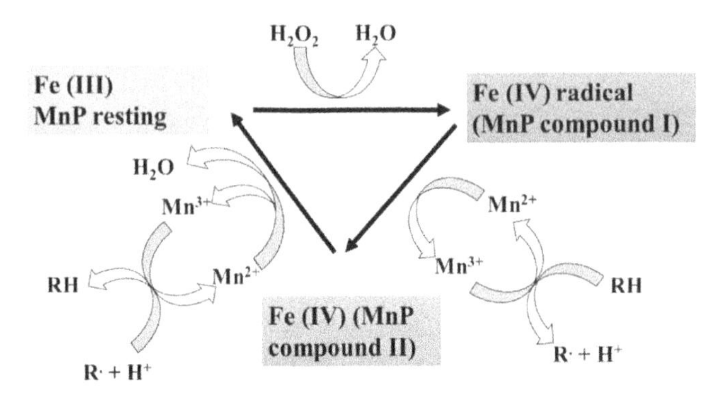


**FIGURE 2.7** Catalytic loop of MnP enzyme.

converted to $Mn^{3+}$ in the next step by donating an electron to Fe (IV) radical and at the same time, the phenolic lignin substrate is converted to phenoxy radicals and other products. The formed MnP compound II release is finally returned to the initial stage by oxidation of $Mn^{2+}$ followed by the evolution of one more molecule of water (Kumar & Chandra, 2020).

### 2.4.4 Laccase

Laccase is a blue copper oxygen oxidoreductase enzyme that contains copper as the central metal. It is capable of oxidizing a wide range of organic compounds like aniline, polyphenol, and phenol by a single electron transfer mechanism (Agrawal et al., 2018). This enzyme is capable of the degradation of lignin, so it is used for the hydrolysis of lignocellulosic biomass. Apart from this, laccase removes the inhibitors produced and also improves the xylanase activity during hydrolysis by phenolic tolerance. The recombinant laccase shows extra thermostability and other properties which make it capable of using for delignification (Rai et al., 2019).

## 2.5 FACTORS AFFECTING ENZYMATIC HYDROLYSIS

The nature, yield, and quality of the sugar obtained after the enzymatic hydrolysis are greatly dependent on the chemical and structural pattern of the lignocellulosic biomass and the pretreatment conditions followed. During enzymatic hydrolysis, cellulase activity is sometimes low due to the irreversible binding of cellulase with the lignin. Therefore, the amount of lignin present in the biomass has a significant role in the efficiency of the enzymatic hydrolysis process. The amount of lignin in the substrate mass is contingent on the pretreatment process employed, the dosage of the enzyme, and the physical conditions specified during the time. Even though the impact of each individual factor on the efficiency is arduous to determine and is not fully established, many of the factors are fulfilledin the whole enzyme-assisted saccharification method. The factors which control the efficiency of enzymatic hydrolysis are broadly classified as enzyme-related factors, substrate-related factors, and a few others.

### 2.5.1 Enzyme-related Factors

There are various factors, bracketed with the nature of the enzyme, which influence the hydrolysis process that involves insoluble reactant or the substrate and also the soluble enzyme catalyst part. A few of the factors are: enzyme adsorption, enzyme concentration, end-product inhibition, synergism, thermal inactivation, nonspecific binding, enzyme source, enzyme processibility, enzyme incompatibility, specific activity, and irreversible binding with lignin. Many of the factors like the nature of the enzyme, stereochemical mechanism of hydrolysis, mode of action, and so forth are interconnected. Temperature, pH, and other physiological conditions of the reactor also affect the yield of the saccharified product. When the conditions are not optimum, the disadvantage of higher enzyme loading could arise in order to get the desired yield.

#### 2.5.1.1 Enzyme Loading

The total expense of enzymatic hydrolysis relies on the amount of enzyme used up for the maximum conversion of cellulose to respective sugar. This, in turn, is dependent on the efficiency of the strategy used for pretreatment. Alkaline pretreatment requires more enzyme loading than acid hydrolysis for the same level of saccharification (Wyman et al., 2011). If the lignin persists after the pretreatment the loaded enzymes are vulnerable to nonproductive adsorption. This requires higher loading of the enzyme for the desired hydrolysis rate (Alvira et al., 2010). Combinations of enzymes such as cellulase in presence of xylanase and others will reduce the enzyme loading.

#### 2.5.1.2 Synergy of the Enzymes

Synergy of many enzymes is required for the effectual working of the complex lignocellulosic substrate. If the pretreatment is not particularly effective, or if the substrate is in varied form, the complexity is higher and needs more rigorous enzymatic conditions (Van Dyk & Pletschke, 2012). A synergy of the enzyme is obligatory in such instances. A few of the examples of the synergy of hemicellulase, cellulase interaction on complex structures are endoxylanase and cellulase on corn membrane (Murashima et al., 2003), endoxylanase and endomannose present in the pretreated cane mass (Beukes & Pletschke, 2011), endoxylanase, cellobiohydrolase, ferulic acid esterase and acetyl xylan esterase on pretreated corn wastes (M. J. Selig et al., 2008a) arabinofuranosidase and endomannanase on nontreated bagasse of sugarcane (Beukes & Pletschke, 2010).

#### 2.5.1.3 Recycling of Enzymes

Reusing the enzyme over several batches increases the catalytic productivity and thus reduces the total cost of the hydrolysis step. The cellulase enzyme will be present in the residual substrate in the bound form or the supernatant in the free form. Recovering the cellulase or other enzyme from both the solid and liquid phases is the prime intent behind recycling. Recycling cellulases extracted from the liquid fraction of the hydrolysis crude has been reported widely, and enzymes adsorbed on the residual solid are also equally important (Tu et al., 2007), because cellulase-degrading enzymes have a high affinity toward cellulose. Removing unabsorbed enzymes is also necessary even though it has only a negligible effect on the yield of hydrolysis (Hu et al., 2018).

### 2.5.2 pH, Temperature, and Mixing

Faulds et al. studied the influence of pH on the activity of enzymatic hydrolysis considering different strains of microorganisms on brewers' stem grain over a range of 3.2–11.2 (Faulds et al., 2008). It was observed that pH has an adequate influence on the enzymatic action and it is distinct for different microorganisms. Trichoderma-derived enzyme showed efficiency at lower acidic pH while the enzyme mixture from *Humicola* was well capable for action in the entire range but optimum at 6–8 and with maximum solubilization at pH = 9. The synergistic action of the enzyme is also affected in suboptimal conditions. Reid et al. remarked on the effect of temperature on the hydrolysis process to remove lignin and noted that a temperature of 15° to 35°C is optimum for the delignification process (Reid, 1989). The temperature should be controlled rigorously as an increase would lead to degradation of the product as well as fungal growth. Han et al. reported that the best temperature range for cellulase activity is from 45° to 55°C for *T. reesei* (Han et al., 2012). In large-scale reactions, mixing of the substrate in the reactor or assay is important especially when the substrate is insoluble. Changes have been reported even when the type of shaking was different, that is, carried out in an orbital shaker and mixed by tumbling, in which the latter showed better conversion (Merino & Cherry, 2007). Even though there are disagreements between scientists regarding the extent of the effect of mixing on the whole process, there are many reports which found similar results (Kristensen et al., 2009).

### 2.5.3 Effect of Surfactants

The addition of surfactants during the enzymatic hydrolysis process was reported to be a prolific strategy to increase the conversion rate and yield. The mechanism behind the action of surfactants was undiscoveredfor many years and numerous efforts and experiments were accomplished to analyze this. One among the many proposed mechanistic actions is that surfactants disrupt or modify the rigid structure of lignocellulose rendering the accessibility of the enzyme to the cellulose. Alteration in enzyme-substrate reaction in presence of surfactants could be the cause of less nonproductive enzyme adsorption, thus allowing higher yield in the presence of surfactants. Another proposed mechanism is that surfactants play the role of stabilizers for the enzymes which adsorb at the surface of liquid resulting in the denaturation in the course of various mechanical processes like agitation. (Kim et al., 1982). Studies have suggested that when screening different surfactants to improve hydrolysis on steam-pretreated spruce, nonionic surfactants showed better productivity. Erricson et al. explained that during the mechanism of surfactant activity, the reason for the improved effect could be due to the elimination and reduction of unproductive enzyme adsorption to lignin of the substrate. It was stated that the nonspecifically bound enzyme is released by the hydrophobic interaction of surfactant with lignin on the lignocellulose surface. Mixed charged and anionic surfactants have also proved to be good candidates for improving enzymatic hydrolysis (Eriksson et al., 2002). Fatty acid esters of sorbitan polyethylene glycol, and polyethoxylates (tween80, tween20) were reported to be the most effective surfactants (Kaar & Holtzapple, 1998).

In 2020, Wang et al. reported that the surfactant effect on enzyme hydrolysis on various substrates such as sugarcane bagasse, cypress, and *Pterocarpus soyauxii* (Wang et al., 2020). Tween 20 was appended to the acid- and alkali-treated substrate. They observed differences in the activity enhancement in ground and unground samples, acid- and alkali-treated samples, and also different for different substrates. The activity was found to increase even when the adsorption of the enzyme was increased. From the observations, it was inferred that the promotion effect upon the addition of surfactant is not solely because of the hindrance against adsorption, but it is also dependent on the lignocellulose features, its impurity, and also the surface morphological aspect. Consequently, the surfactant addition effect has become a substrate-related factor as well.

### 2.5.4 Substrate-related Factors

One of the main challenges to confront for the better performance of enzymatic hydrolysis is to cast the substrate into the downright form by removing other stuff like lignin and hemicellulose. The proficiency of the enzymatic hydrolysis of lignocellulosic biomass intensely relies on the characteristics of the substrate used. The characteristics include the degree of polymerization, crystallinity, size, structure, and the surface area of the cellulose which can be accessed by the enzyme and this, in turn, affects the synergy of the enzyme, cellulase adsorption, and processivity.

The degree of polymerization of cellulose and its effect on hydrolysis is still a matter of study and exact findings on the same are not corroborated. A lesser degree of polymerization of the substrate is meant to be favorable for hydrolysis as it may increase the adsorption of cellulose, thus raising the catalytic action. Zhang et al. reported that the degree of polymerization has less impact on hydrolysis than increasing the accessibility toward glycosidic bonds (Zhang & Lynd, 2006).

Crystallinity is dependent on the hydrolysis rate and yield such that an increased hydrolysis rate was applied to cellulose with a more amorphous structure than the crystalline (Pinto et al., 2006). The enzyme adsorption declined when crystallinity increased in the glycosyl hydrolase system, individual enzyme components, and cellulose-binding module. Crystallinity also affects the synergism of the various enzymes and it alters the efficiency of the adsorbed cellulase compounds (Hall et al., 2010).

The accessibility of the surface of cellulose for the enzyme is a major criterion for enzymatic hydrolysis to move forward smoothly because enzyme hydrolysis of cellulose is a surface phenomenon. The accessibility is reduced by nanometer-sized microfibrils and cellulose chain crosslinking present on cellulose, even though it gives strength and protection to the plant. The limitations due to low surface accessibility can be overcome by the pretreatment technique to an extent. The National Renewable Energy Laboratory (NREL) has recommended that the biomass substrate should not be dried after aqueous pretreatment because this leads to irreversible pore collapse (Selig et al., 2008b). Even though drying increases the accessible surface area, it is more important to conserve the pores for the active enzymes.

### 2.5.5 Recent Advances in Enzymatic Hydrolysis

Progressions in enzymatic hydrolysis techniques have helped to elevate the whole process of waste valorization to a more economic and adaptable practice. Extensive research has been carried out for the development of the absolute enzyme for the hydrolysis of lignocellulosic biomass. Instead of encountering new enzymes, or genetically modifying the existing ones, interest has recently been focused on enzymatic consortium or enzymatic cocktails. These are a pool of various enzymes which work for a specific function like cell-wall destruction by synergistic action. They are effective in the way that the product formed by the action of a single enzyme is further broken down by other enzymes in the enzyme cocktail (Lopes et al., 2018). However, the optimization of such cocktails is obligatory to modulate the type of function, the extent, and their concentration for lucrative hydrolysis of the substrate.

Recently, works on lytic polysaccharide monooxygenase (LPMO) enzymes have entered the limelight. LPMO can be used to improve the capability of the cellulolytic process by removing a small amount of hemicellulose which is otherwise recalcitrant (Østby et al., 2020). The lignocellulosic conversion in presence of LPMO has recently been reported to be enhanced upon the addition of hydrogen peroxide (Costa et al., 2020). Thus, the potential of LPMO for efficient saccharification has the capacity for further exploration, and it might help in suppressing the cost curve for the process.

Interdisciplinary approaches are also testified to be a worthy strategy as a defect in one field could be rectified by another. For example, lignocellulosic biomass degradation by combining the hydrothermal and biological techniques has been reported to improve enzymatic hydrolysis to a profitable level (Song et al., 2021). Research studies investigating the optimum solvent for the process are also gaining ground, and ionic liquids are currently considered the best option. Pretreatment procedures have a significant role in the successive hydrolysis steps; thus, better hydrolysis can be achieved by precise and cautious pretreatment.

## 2.6 FUTURE OUTLOOK

Advancements in biotechnology take a pivotal role in commercializing enzymatic hydrolysis for the making of biofuel and other value-added products from lignocellulosic biomass waste products. Even though the enzymatic hydrolysis processes have many advantages, a highly economical strategy that could compete with the cost and capability of existing fossil fuels has not yet proved economically viable. The basic requirement for the competent and productive success of the hydrolysis technique is in-depth knowledge about the structure and mechanism of the substrate and enzyme respectively. Better characterization techniques and analytical methods are also essential for the understanding of such processes. Research on the same could predict the exact requisite to enrich the value of each step toward product formation.

The simultaneous saccharification and fermentation (SSF) process has been a breakthrough in the field of bioethanol production from lignocellulosic biomass.

In this process, glucose formed after hydrolysis is directly taken up by microorganisms for fermentation. This is advantageous because inhibition to the enzyme caused by various products formed is minimized as there is a minimum concentration of sugar in the media. Further advancement was observed when nonisothermal SSF prevailed. Consolidated bioprocessing (also called direct microbial conversion) is another technique for the production of not only ethanol, but also the enzymes required for the conversion of cellulose to ethanol. Subsequently, an assembly of mechanical and biotechnological machineries for biomass conversion directly to the value-added preferred product, with in situ enzyme generation, high yield, and low cost, would be an advantage in this field.

Identification of a unique enzyme that can collectively achieve the properties of various hydrolytic enzymes such as cellulase, xylanase, and peroxidase is a highly promising solution to reduce the challenges caused by enzymatic hydrolysis today. Emergent DNA recombinant technology produces cost-effective potential enzymes by strategical cloning and expressing the recombinant strains of microorganisms. This is a method of clipping the DNA of two or more different species and applying them to the host to deliver new genetic combinations. The launch of such techniques and commodities would be overpriced, but once established, this method could considerably decrease the cost behind the whole process by avoiding the many steps involved. Advanced technologies such as clustered regularly interspaced short palindromic repeats (CRISPR)/Cas9 allow fast and effectual genome editing for producing suitable genes that can enhance the hydrolysis capacity of a wide range of substrates, irrespective of the pretreatment method, devoid of inhibitor tolerance, and so on (Cho et al., 2018). An isopropanol-ethanol-butanol mixture has already been synthesized from mutants produced by genome editing of *Clostridium acetobutylicum* ATCC 824 by CRISPR/Cas9 technology (Wasels et al., 2017).

## 2.7 CONCLUSION

The striving for the substitution of fossil fuels by a renewable source has heightened the importance of converting lignocellulosic biomass to biofuel. Efforts have been made from the early decades of the 20th century until today to synthesize biofuel in the best possible way for commercialization by precluding the existing fuel. Many valorization techniques exist with a long directory of pretreatment techniques, conventional chemical and biological hydrolysis techniques, and also the final fermentation and purification. Among these, hydrolysis plays a pivotal role in the challenge of valorization of lignocellulosic biomass. Chemical homogenous processes like dilute acid hydrolysis, strong acid hydrolysis, and alkali pretreated hydrolysis dictated development for a long time. Since enzymatic hydrolysis started flourishing, the bulk of the hydrolysis has been executed with enzymatic aid.

Enzymatic hydrolysis is carried out by culturing microorganisms to produce enzymes which can hydrolyze the cellulose, hemicellulose, and lignin of the biomass. Enzymatic hydrolysis is more highly expedient than conventional hydrolytic processes in terms of selectivity, yield, and simplicity. However, there are still barriers to improvement, in terms of cost and time. Progressive research studies with incremental innovations are valuable, but there is a long way to go to replace current fossil fuel usage, which is still cheaper. For this, a breakthrough discovery in the field is necessary and it can be achieved by biotechnologists modifying the genes of microorganisms to produce an enzyme with scrupulous activity, by chemists studying the detailed mechanism involved in each process, and by chemical engineers developing new methods for pretreatment and substrate modification. The pace of innovation in research and development departments around the world offers the hope of commercializing biofuel over the existing fossil fuel in the near future.

## ACKNOWLEDGMENT

We gratefully acknowledge the Indian Institute of Technology Palakkad for their financial support.

## REFERENCES

Abdel-Hamid, A. M., Solbiati, J. O., & Cann, I. K. O. (2013). Insights into lignin degradation and its potential industrial applications. *Advances in Applied Microbiology*, *82*, 1–28.

Agrawal, K., Chaturvedi, V., & Verma, P. (2018). Fungal laccase discovered but yet undiscovered. *Bioresources and Bioprocessing*, *5*(4), 1–12. https://doi.org/10.1186/s40643-018-0190-z

Agrawal, K., & Verma, P. (2020). Laccase-Mediated Synthesis of Bio-material Using Agro-residues. In: Sadhukhan, P. & Premi, S. (eds.) *Biotechnological applications in human health*, pp. 87–93. Springer, Singapore.

Alvira, P., Tomás-Pej6, E., Ballesteros, M., & Negro, M. J. (2010). Pretreatment technologies for an efficient bioethanol production process based on enzymatic hydrolysis: A review. *Bioresource Technology*, *101*(13), 4851–4861.

Baciocchi, E., Gerini, M. F., Lanzalunga, O., Lapi, A., Lo Piparo, M. G., & Mancinelli, S. (2001). Isotope-effect profiles in the oxidative N-Demethylation of N, N-Dimethylanilines catalysed by lignin peroxidase and a chemical model. *European Journal of Organic Chemistry*, *2001*(12), 2305–2310.

Bensah, E. C., Mensah, M., Antwi, E., & others. (2011). Status and prospects for household biogas plants in Ghana--lessons, barriers, potential, and way forward. *International Journal of Energy and Environmental*, *2*(5), 887–898.

Bergius, F. (1937). Conversion of wood to carbohydrates. *Industrial & Engineering Chemistry*, *29*(3), 247–253.

Beukes, N., & Pletschke, B. I. (2010). Effect of lime pre-treatment on the synergistic hydrolysis of sugarcane bagasse by hemicellulases. *Bioresource Technology*, *101*(12), 4472–4478.

Beukes, N., & Pletschke, B. I. (2011). Effect of alkaline pre-treatment on enzyme synergy for efficient hemicellulose hydrolysis in sugarcane bagasse. *Bioresource Technology*, *102*(8), 5207–5213.

Bhardwaj, N., & Verma, P. (2021). Xylanases: A Helping Module for the Enzyme Biorefinery Platform. In: Srivastava, N. & Srivastava, M. (eds.) *Bioenergy research: Revisiting latest development*. 7, pp. 161–179, Springer, Singapore.

Bhardwaj, N., Kumar, B., Agrawal, K., & Verma, P. (2021b). Current perspective on production and applications of microbial cellulases: A review. *Bioresources and Bioprocessing*, *8*, 95. https://doi.org/10.1186/s40643-021-00447-6

Bisaria, V. S., & others. (1991). Bioprocessing of agro-residues to glucose and chemicals. *Bioconversion of Waste Materials to Industrial Products*, 187–223.

Charles, E. W., Stephen, R. D., Michael, E. H., John, W. B., Catherine, E. S., & Liisa, V. (2004). *Hydrolysis of cellulose and hemicellulose, polysaccharides: Structural tructural diversity and functional versatility*. CRC Press.

Chaturvedi, V., & Verma, P. (2013). An overview of key pretreatment processes employed for bioconversion of lignocellulosic biomass into biofuels and value added products. *3 Biotech*, *3*(5), 415–431. https://doi.org/10.1007/s13205-013-0167-8

Cho, S., Shin, J., & Cho, B.-K. (2018). Applications of CRISPR/Cas system to bacterial metabolic engineering. *International Journal of Molecular Sciences*, *19*(4), 1089.

Converse, A. O., Matsuno, R., Tanaka, M., & Taniguchi, M. (1988). A model of enzyme adsorption and hydrolysis of microcrystalline cellulose with slow deactivation of the adsorbed enzyme. *Biotechnology and Bioengineering*, *32*(1), 38–45.

Costa, T. H. F., Kadić, A., Chylenski, P., Várnai, A., Bengtsson, O., Lidén, G., Eijsink, V. G. H., & Horn, S. J. (2020). Demonstration-scale enzymatic saccharification of sulfite-pulped spruce with addition of hydrogen peroxide for LPMO activation. *Biofuels, Bioproducts and Biorefining*, *14*(4), 734–745.

Du, R., Huang, R., Su, R., Zhang, M., Wang, M., Yang, J., Qi, W., & He, Z. (2013). Enzymatic hydrolysis of lignocellulose: SEC-MALLS analysis and reaction mechanism. *RSC Advances*, *3*(6), 1871–1877.

Eriksson, T., Börjesson, J., & Tjerneld, F. (2002). Mechanism of surfactant effect in enzymatic hydrolysis of lignocellulose. *Enzyme and Microbial Technology*, *31*(3), 353–364.

Faith, W. L. (1945). Development of the Scholler process in the United States. *Industrial & Engineering Chemistry*, *37*(1), 9–11.

Faulds, C. B., Robertson, J. A., & Waldron, K. W. (2008). Effect of pH on the solubilization of brewers' spent grain by microbial carbohydrases and proteases. *Journal of Agricultural and Food Chemistry*, *56*(16), 7038–7043.

Hall, M., Bansal, P., Lee, J. H., Realff, M. J., & Bommarius, A. S. (2010). Cellulose crystallinity--a key predictor of the enzymatic hydrolysis rate. *The FEBS Journal*, *277*(6), 1571–1582.

Han, L., Feng, J., Zhang, S., Ma, Z., Wang, Y., & Zhang, X. (2012). Alkali pretreated of wheat straw and its enzymatic hydrolysis. *Brazilian Journal of Microbiology*, *43*(1), 53–61.

Harmsen, P. F. H., Huijgen, W., Bermudez, L., & Bakker, R. (2010). Literature review of physical and chemical pretreatment processes for lignocellulosic biomass. *Wageningen UR Food and Biobased Research*.

Hu, J., Mok, Y. K., & Saddler, J. N. (2018). Can we reduce the cellulase enzyme loading required to achieve efficient lignocellulose deconstruction by only using the initially absorbed enzymes. *ACS Sustainable Chemistry & Engineering*, *6*(5), 6233–6239.

Jennings, E. W., & Schell, D. J. (2011). Conditioning of dilute-acid pretreated corn stover hydrolysate liquors by treatment with lime or ammonium hydroxide to improve conversion of sugars to ethanol. *Bioresource Technology*, *102*(2), 1240–1245.

Kaar, W. E., & Holtzapple, M. T. (1998). Benefits from Tween during enzymic hydrolysis of corn stover. *Biotechnology and Bioengineering*, *59*(4), 419–427.

Katzen, R., & Othmer, D. F. (1942). Wood hydrolysis. A continuous process. *Industrial & Engineering Chemistry*, *34*(3), 314–322.

Katzen, R., & Schell, D. J. (2006). Lignocellulosic Feedstock Biorefinery: History and Plant Development for Biomass Hydrolysis. In: *Biorefineries – industrial processes and products: Status quo and future directions*, pp. 129–137. WILEY-VCH.

Khare, S. K., Pandey, A., & Larroche, C. (2015). Current perspectives in enzymatic saccharification of lignocellulosic biomass. *Biochemical Engineering Journal*, *102*, 38–44.

Kim, M. H., Lee, S. B., & Ryu, D. D. (1982). Surface deactivation of cellulase and its prevention. *Enzyme Microbial Technology*, *4*, 99–103.

Kirk, T. K., Tien, M., Kersten, P. J., Mozuch, M. D., & Kalyanaraman, B. (1986). Ligninase of *Phanerochaete chrysosporium*. Mechanism of its degradation of the non-phenolic arylglycerol β-aryl ether substructure of lignin. *Biochemical Journal*, *236*(1), 279–287.

Krazzig, H., & Schurz, J. (2008). *Ullmann's Encyclopedia of Industrial Chemistry* (Sixth Ed.). Wiley-VCH.

Kristensen, J. B., Felby, C., & Jørgensen, H. (2009). Yield-determining factors in high-solids enzymatic hydrolysis of lignocellulose. *Biotechnology for Biofuels*, *2*(1), 1–10.

Kumar, A., & Chandra, R. (2020). Ligninolytic enzymes and its mechanisms for degradation of lignocellulosic waste in environment. *Heliyon*, *6*(2), e03170.

Kumar, B., & Verma, P. (2020). Enzyme mediated multi-product process: A concept of bio-based refinery. *Industrial Crops and Products*, *154*, 112607.

Kumar, B., Bhardwaj, N., Agrawal, K., Chaturvedi, V., & Verma, P. (2020). Current perspective on pretreatment technologies using lignocellulosic biomass: An emerging biorefinery concept. *Fuel Processing Technology*, *199*, 106244. https://doi.org/10.1016/j.fuproc.2019.106244

LaForge, F. B., & Hudson, C. S. (1918). The preparation of several useful substances from corn cobs. *Industrial & Engineering Chemistry*, *10*(11), 925–927.

Loow, Y.-L., Wu, T. Y., Jahim, J. M., Mohammad, A. W., & Teoh, W. H. (2016). Typical conversion of lignocellulosic biomass into reducing sugars using dilute acid hydrolysis and alkaline pretreatment. *Cellulose*, *23*(3), 1491–1520.

Lopes, A. de M., Ferreira Filho, E. X., & Moreira, L. R. S. (2018). An update on enzymatic cocktails for lignocellulose breakdown. *Journal of Applied Microbiology*, *125*(3), 632–645.

Mandels, M., Hontz, L., & Nystrom, J. (1974). Enzymatic hydrolysis of waste cellulose. *Biotechnology and Bioengineering*, *16*(11), 1471–1493.

McKillip, W. J., & Collin, G. (2002). *Ullmann's Encyclopedia of Industrial Chemistry* (Sixth Ed.). Wiley-VCH.

Merino, S. T., & Cherry, J. (2007). Progress and challenges in enzyme development for biomass utilization. *Biofuels*, *108*, 95–120.

Mosier, N., Wyman, C., Dale, B., Elander, R., Lee, Y. Y., Holtzapple, M., & Ladisch, M. (2005). Features of promising technologies for pretreatment of lignocellulosic biomass. *Bioresource Technology*, *96*(6), 673–686.

Murashima, K., Kosugi, A., & Doi, R. H. (2003). Synergistic effects of cellulosomal xylanase and cellulases from *Clostridium cellulovorans* on plant cell wall degradation. *Journal of Bacteriology*, *185*(5), 1518.

Østby, H., Hansen, L. D., Horn, S. J., Eijsink, V. G. H., & Várnai, A. (2020). Enzymatic processing of lignocellulosic biomass: Principles, recent advances and perspectives. *Journal of Industrial Microbiology \& Biotechnology: Official Journal of the Society for Industrial Microbiology and Biotechnology*, *47*(9–10), 623–657.

Persson, I., Tjerneld, F., & Hahn-Hägerdal, B. (1991). Fungal cellulolytic enzyme production: A review. *Process Biochemistry*, *26*(2), 65–74.

Pinto, R., Carvalho, J., Mota, M., & Gama, M. (2006). Large-scale production of cellulose-binding domains. Adsorption studies using CBD-FITC conjugates. *Cellulose*, *13*(5), 557–569.

Ragg, P. L., & Fields, P. R. (1987). The development of a process for the hydrolysis of lignocellulosic waste. *Philosophical Transactions of the Royal Society of London Series A, Mathematical and Physical Sciences*, *321*(1561), 537–547.

Rai, R., Bibra, M., Chadha, B. S., & Sani, R. K. (2019). Enhanced hydrolysis of lignocellulosic biomass with doping of a highly thermostable recombinant laccase. *International Journal of Biological Macromolecules*, *137*, 232–237.

Reid, I. D. (1989). Solid-state fermentations for biological delignification. *Enzyme and Microbial Technology*, *11*(12), 786–803.

Rocha-Meneses, L., Raud, M., Orupõld, K., & Kikas, T. (2017). Second-generation bioethanol production: A review of strategies for waste valorisation. *Agronomy Research*, *15*(3), 830–847.

Selig, M. J., Knoshaug, E. P., Adney, W. S., Himmel, M. E., & Decker, S. R. (2008a). Synergistic enhancement of cellobiohydrolase performance on pretreated corn stover by addition of xylanase and esterase activities. *Bioresource Technology*, *99*(11), 4997–5005.

Selig, M., Weiss, N., & Ji, Y. (2008b). *Enzymatic Saccharification of Lignocellulosic Biomass*. National Renewable Energy Laboratory.

Sherrard, E. C., & Kressman, F. W. (1945). Review of processes in the United States prior to World War II. *Industrial & Engineering Chemistry*, *37*(1), 5–8.

Song, B., Lin, R., Lam, C. H., Wu, H., Tsui, T.-H., & Yu, Y. (2021). Recent advances and challenges of inter-disciplinary biomass valorization by integrating hydrothermal and biological techniques. *Renewable and Sustainable Energy Reviews*, *135*, 110370.

Sun, Y., & Cheng, J. (2002). Hydrolysis of lignocellulosic materials for ethanol production: A review. *Bioresource Technology*, *83*(1), 1–11.

Swatloski, R. P., Spear, S. K., Holbrey, J. D., & Rogers, R. D. (2002). Dissolution of cellose with ionic liquids. *Journal of the American Chemical Society*, *124*(18), 4974–4975.

Tarkow, H., & Feist, W. C. (1969). A mechanism for improving the digestibility of lignocellulosic materials with dilute alkali and liquid ammonia. In Hajny, G.H., Reese, E.T. (eds) Cellulases and their Applications *ACS Publications*. pp 197–218.

Tsai, C.-T., & Meyer, A. S. (2014). Enzymatic cellulose hydrolysis: Enzyme reusability and visualization of β-glucosidase immobilized in calcium alginate. *Molecules*, *19*(12), 19390–19406.

Tu, M., Chandra, R. P., & Saddler, J. N. (2007). Recycling cellulases during the hydrolysis of steam exploded and ethanol pretreated lodgepole pine. *Biotechnology Progress*, *23*(5), 1130–1137.

Van Dyk, J. S., & Pletschke, B. I. (2012). A review of lignocellulose bioconversion using enzymatic hydrolysis and synergistic cooperation between enzymes—factors affecting enzymes, conversion and synergy. *Biotechnology Advances*, *30*(6), 1458–1480.

Wahlström, R. M., & Suurnäkki, A. (2015). Enzymatic hydrolysis of lignocellulosic polysaccharides in the presence of ionic liquids. *Green Chemistry*, *17*(2), 694–714.

Wang, W., Wang, C., Zahoor, Chen, X., Yu, Q., Wang, Z., Zhuang, X., & Yuan, Z. (2020). Effect of a nonionic surfactant on enzymatic hydrolysis of lignocellulose based on lignocellulosic features and enzyme adsorption. *ACS Omega*, *5*(26), 15812–15820.

Wasels, F., Jean-Marie, J., Collas, F., López-Contreras, A. M., & Ferreira, N. L. (2017). A two-plasmid inducible CRISPR/Cas9 genome editing tool for *Clostridium acetobutylicum*. *Journal of Microbiological Methods*, *140*, 5–11.

Wright, J. D., Power, A. J., & Bergeron, P. W. (1985). *Evaluation of concentrated halogen acid hydrolysis processes for alcohol fuel production*. United States.

Wyman, C. E., Balan, V., Dale, B. E., Elander, R. T., Falls, M., Hames, B., Holtzapple, M. T., Ladisch, M. R., Lee, Y. Y., Mosier, N., & others. (2011). Comparative data on effects of leading pretreatments and enzyme loadings and formulations on sugar yields from different switchgrass sources. *Bioresource Technology*, *102*(24), 11052–11062.

Zhang, Y.-H. P., & Lynd, L. R. (2006). A functionally based model for hydrolysis of cellulose by fungal cellulase. *Biotechnology and Bioengineering*, *94*(5), 888–898.

# 3 Overview of the Mechanism of Hydrolytic Enzymes and Their Application in Waste Treatment

*Pradipta Patra, Manali Das, Bharat Manna, Amit Ghosh, and P. K. Sinha*
Indian Institute of Technology Kharagpur, Kharagpur, India

## CONTENTS



DOI: 10.1201/9781003187684-3

## 3.1 INTRODUCTION

A gradual rise in the requirement for petroleum alternatives and a steady increase in the demand for food products has resulted in the generation of huge agricultural and industrial wastes like food wastes, lignocellulosic wastes, kitchen wastes, effluents, and bioplastics. Waste generation is a rapidly growing global problem, particularly in urban regions across many nations. Approximately 2.01 billion tonnes of municipal solid waste are generated annually, out of which approximately 33% are not disposed of in an environmentally-safe manner (*Trends in Solid Waste Management*, n.d.). It is expected that by 2050 municipal waste generation will have increased roughly by 70% to almost 3.4 billion tonnes (*Global Waste Generation – Statistics & Facts | Statista*, n.d.). Additionally, food wastes account for an estimated 1.3 billion tonnes while agricultural and lignocellulosic wastes contribute approximately 2 billion tonnes generated globally per year (Millati et al., 2019; *Technical Platform on the Measurement and Reduction of Food Loss and Waste | Food and Agriculture Organization of the United Nations*, n.d.). Consequently, there has been greater emphasis on the treatment of wastes to reduce environmental hazards as well as valorize them into renewable and commercially important molecules.

Since wastes generally consist of a heterogeneous blend of different biological and synthetic polymers, along with several toxic chemicals, physicochemical-based waste treatment operation methods are commonly used. Several physicochemical techniques such as flocculation, coagulation, activated carbon adsorption, and other sophisticated techniques like nanofiltration, reverse osmosis, ion exchange, and photolysis are available, but these methods require technical skills and high operation costs (with high chemical and energy demand). Compared to physicochemical methods, biological treatment methods using enzymes provide an efficient and environmentally friendly alternative for waste management (Kumar & Verma 2020a).

Enzymes are highly efficient biocatalysts that are extensively used in myriad manufacturing and processing sectors, due to their higher order of catalytic efficiencies over inorganic catalysts under mild conditions. On the basis of different categories of waste (carbohydrates, proteins, lipids), specific enzymes are selected for their degradation and valorization. Widely used classes include glycoside hydrolases (GH), proteases, and lipases for catalytic breakdown. GHs include exoglucanase, endoglucanase, and beta-glucosidase, whereas proteases include exopeptidases and endopeptidases as broader classes. Lipases, on the other hand, are a class of ubiquitous enzymes that can catalyze triglycerides into glycerol and fatty acids, and have great potential in waste recycling. Following enzymatic hydrolysis, the complex biomatter is converted to a simpler form, which can be easily repurposed. Depending on the source and composition of the waste matter, enzymatic hydrolysis produces sugars, such as glucose, xylose, galactose, fructose, glucan, ribose, xylan, mannan, galactan, and arabinan (Anwar Saeed et al., 2018), or oligopeptides and amino acids, or even glycerol and long-chain fatty acids. One of the widely adapted methods for reuse of biowastes is through fermentation of these simple molecules into high-value products such as biofuels, commercially important organic acids (lactic acid, fumaric acid, succinic acid, butyric acid, malic acid, and itaconic acid), and other such valuable molecules like polyhydroxyalkanoates and triacylglycerols. (Diaz et al., 2018; Marzo et al., 2019; Pandey et al., 2000); Kumar & Verma, 2020b. In this chapter, we focus on different classes of enzymes and their respective catalytic mechanisms for waste treatments. Also, the application of enzyme-based waste management is depicted. Moreover, current advances in increasing the catalytic property and stability of the enzymes for enhancing the efficiency of the waste recycling process are highlighted. This work is envisioned to facilitate the search for tailored enzymes that will be robust and can be used in a cost-effective manner.

## 3.2 DIFFERENT CLASSES OF HYDROLYTIC ENZYMES

The major classes of hydrolytic enzymes which are used in the treatment of waste materials are GHs, proteases, and lipases. The enzyme classes are described in detail in the following sections.

### 3.2.1 Cellulolytic and Hemicellulolytic Enzymes

In a waste management platform, saccharolytic enzymes, (*i.e.*, both cellulases and hemicellulases) are effective for depolymerizing carbohydrates to extract pentose and hexose sugars. Cellulose-degrading enzymes can be classified into three broad categories: exo-glucanases (EG) or cellobiohydrolase (CBH), endoglucanases (EG), and $\beta$-glucosidase (BGL) which belong to the EC 3.2.1.X class (Saini et al., 2015; Bhardwaj et al., 2021a). Exo-glucanases or Exo-β-(1,4)-glucanase (EC 3.2.1.74) or CBH can cleave the glucan chain from the non-reducing end (Figure 3.1). Endo-glucanases (EC 3.2.1.4) or Endo-$\beta$-(1,4)-glucanases hydrolyze the $\beta$-(1,4)-glucosidic linkages in the amorphous region of cellulose, producing oligosaccharides of different degrees of polymerization such as cellobiose and cellotriose. However, the rate of hydrolysis is greatly enhanced on cellodextrins as the degree of polymerization increases. Cellobiohydrolases (EC 3.2.1.176) are another class of glucanases that can degrade cellulose by releasing cellobiose units from the non-reducing end. $\beta$-glucosidases (EC 3.2.1.21) can catalytically breakdown short chain $\beta$-1,4-oligoglucosides from cellobiose up to cellohexaose to obtain a fermentable glucose

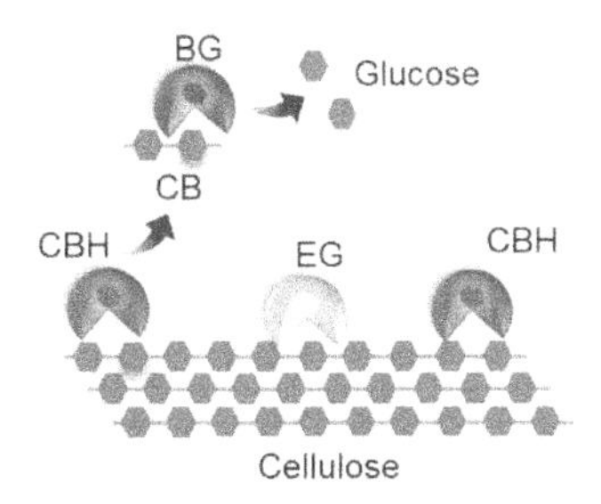


**FIGURE 3.1** Different types of cellulase enzymes. EG = endoglucanase, CBH = cellobiohydrolase, and BG = β-glucosidase. The EG cuts the cellulose chain in the middle, whereas CBH cleaves at the terminal end. The BG takes up the cellobiose to break down it into glucose monomers.

monomer. Hemicellulolytic enzymes such as $\beta$-xylosidase, α-L-arabinofuranosidase, endo-1,4-$\beta$-xylanase, α-glucuronidase, acetyl xylan esterase are found to be effective for hemicellulose degradation (Niehaus et al., 1999). Efficient cellulose hydrolysis can be achieved by the synergistic action of exoglucohydrolases (EC 3.2.1.74), endoglucanases (EC 3.2.1.4), cellobiohydrolases (EC 3.2.1.176), and $\beta$-glucosidases (EC 3.2.1.21).

#### 3.2.1.1 The Carbohydrate-active Enzymes Database

The CAZy database is an invaluable resource in carbohydrate enzymology research (Cantarel et al., 2009). It hosts a large number of manually curated enzymes, capable of deconstructing carbohydrates. A more descriptive overview of the entries in the CAZy database can be found in a sister database called CAZypedia ("Ten Years of CAZypedia: A Living Encyclopedia of Carbohydrate-active Enzymes," 2018). The major enzyme classes found in the CAZy database are glycoside hydrolases (GHs), glycosyl transferases (GTs), polysaccharide lyases (PL), carbohydrate esterases (CEs), auxiliary activities (AAs), and carbohydrate-binding modules (CBMs). GHs are the primary enzymatic machinery for the degradation of polysaccharides found in nature. To date, there are 171 GH families, hosted in the CAZy database. The AA families mostly comprise redox enzymes that can act independently or along with other polysaccharide-degrading enzymes in a cooperative fashion. Currently, there are 17 enzyme families under AAs, out of which nine are known to actively take part in delignification whereas seven families are found to directly degrade the polysaccharides, namely lytic polysaccharide monooxygenases (LPMOs). Enzymes acting on carbohydrates generally display a noncatalytic module that assists in adhesion to the carbohydrate surface. There are 88 classified families of CBMs so far. The CAZy database hosts genomes associated with the carbohydrate-active enzymes from a wide range of kingdoms such as bacteria (8776), viruses (334), archaea (290) and eukaryota (218) can also be found in the database. Moreover, the CAZy database provides a huge resource for the study of carbohydrate-active enzymes, and plays a major role in designing cost-effective saccharification processes.

### 3.2.2 Proteases

Proteases are a class of enzymes with diverse activity and having potential characteristics like high specificity, rapid action, needing mild operational conditions, high biodegradability, and reduced waste generation, which makes them lucrative enzymes for a wide range of biotechnological applications. Globally, the protease market is very promising as it is estimated to grow to 3 billion USD at a compounded annual growth rate by the year 2024. Currently, proteases are being used in diverse applications in industries like chemical, food, and pharmaceutical as well as in wastewater treatment purposes (Naveed et al., 2021). Proteases are usually categorized in different classes on the basis of their source, catalytic property, and the nature of amino acid present in the active site. Naturally, they are found in various sources like plants, animals, fungi, and bacteria. Two major classes of proteases are exopeptidase and endopeptidase, where exopeptidases react at polypeptide chain terminal and endopeptidases react randomly in the inner portions of the polypeptide chains (Barrett, 1994).

#### 3.2.2.1 Exopeptidases

Exopeptidases (EC 3.4.r.x.) cleave distinctive peptide bonds adjacent to the terminal carboxyl (EC 3.4.11-14.x) or amino group (EC 3.4.15-17.x) present in the substrate (Garcia-Carreno & Navarrete Del Toro, 1997). According to their catalyzing properties and site of action, they are classified into two major classes: aminopeptidases and carboxypeptidases (Mótyán et al., 2013).

##### *3.2.2.1.1 Aminopeptidases*

Aminopeptidases catalyze the splitting of amino acids from the amino terminus of smaller peptide or protein substrates resulting in the release of single amino acid, dipeptide and tripeptide residue. Specifically, they recognize and remove the N-terminal methionine of the peptide chain which is not present in natural proteins. Amino peptidases are widely distributed in several species of bacteria and fungi as well as in animal and plant species. Based on their catalytic properties, aminopeptidases are classified into three types: cysteine, serine, and metalloaminopeptidases where metallo-type is the most abundant in bacteria (Gonzales & Robert-Baudouy, 1996). Interestingly, some of the aminopeptidases can catabolize exogenously supplied peptides and are crucial for the last stages of protein turnover.

##### *3.2.2.1.2 Carboxypeptidases*

The carboxypeptidases are a class of exopeptidase that hydrolyze the C-terminal amino acid from the peptide chain to release dipeptide and single amino acid. Usually, they are not considered as endopeptidase as after catalysis the amino acid is left at target protein. Thus, they have an advantageous property in the removal of tags incorporated at the C-terminal of the target protein. Carboxypeptidases are classified on the basis of preference for amino acids, as aromatic or branched are called carboxypeptidase A, whereas those which cleave positively charged amino acids, like arginine and lysine, are called carboxypeptidase B (Mótyán et al., 2013).

#### 3.2.2.2 Endopeptidases

Endopeptidases (EC 3.4.21-24.x) are the class of enzymes that help in catalytic removal of peptide bonds at a distant region of the protein substrate. Endopeptidases are classified as serine, aspartic, cysteine, metalloprotease, and threonine based on specific functional groups located on active sites which are discussed below.

##### *3.2.2.2.1 Serine Proteases*

Serine proteases are a class of protein-digesting enzymes that are diversely distributed in nature. Their occurrence is not only in cellular organisms but also in viral genomes. Thus, they are one-third of all known proteolytic enzymes. Mostly, they are endopeptidases that cleave the bond at the inner portion of the chain, whereas some of them act as exopeptidases too, which split the amino acids from the terminal part of the peptide sequence. This enzyme has Ser residue located at the nucleophilic active site. During the catalytic activity, an aryl-enzyme intermediate is produced as the Ser residue attacks the carbonyl end of the substrate's peptide bond (Hedstrom, 2002).

##### *3.2.2.2.2 Aspartic Proteases*

Aspartic proteases are less common than the endopeptidase that is found in animals, plants, and microbes. They are structurally bilobed with a central catalytic region made up of aspartates. These enzymes show their optimum activity in acidic pH, having isoelectric points in the pH range of 3 to 4.5, and mostly work on hydrophobic amino acids present near to the dipeptide bonds (Souza et al., 2017; Theron & Divol, 2014). Functional residues in these enzymes, which predominantly activate the water molecule, initiate the nucleophilic attacks (Cuesta et al., 2020). Besides, Adenosine triphosphate (ATP)-dependent proteases are an exception to their general mechanism of action as they require energy for their activation. Microbes are one of the significant sources of aspartic protease enzymes which are usually categorized into two classes, namely, pepsin-like enzymes obtained from *Penicillium, Aspergillus, Neurosprora*, and *Rhizopus*, whereas the other class is renin-like enzymes which are produced by mucor and *Endothia*, among others (de Souza et al., 2015). These enzymes have numerous applications in beverage and food industries like the degradation of protein turbidity complex in alcoholic liqueurs and fruit juices, modifying gluten in bread, and other waste degradation purposes too.

##### *3.2.2.2.3 Cysteine/Thiol Proteases*

Thiol proteases are also known as cysteine proteases as their active site is made up of cysteine. They are naturally found in prokaryotic as well as eukaryotic organisms. They work optimally in pH 6–8 at 50–70°C. Their activity can be inhibited by oxidizing agents and catalytic activity is sensitive to sulfhydryl compounds (Koch et al., 2009).

##### *3.2.2.2.4 Metalloproteases*

Metalloproteases are endopeptidases majorly found in bacteria and fungi where zinc is present in its catalytic site. Water is also needed for hydrolysis by these enzymes. However, various metal ions like calcium, zinc, and cobalt are significantly involved in their reactivation. Zinc is essential for enzymatic activity and calcium is necessary for the stability of the enzyme structure. Metalloproteases have optimum pH from 5–9 and are sensitive to chelating agents like EDTA (Ellaiah et al., 2002).

### 3.2.3 Lipolytic Enzymes

Lipases (triacylglycerol acyl hydrolase, E.C. 3.1.1.3) are a ubiquitous class of key enzymes, with a swiftly growing application in both biotechnology as well as waste management, owing to their multifaceted properties (Gupta et al., 2004; Javed et al., 2018). Lipases are responsible for the esterification of water-insoluble substrates to generate mono- and diacylglycerides, fatty acids, and glycerol. These enzymes belong to the $\alpha/\beta$ hydrolase fold super-family (Kapoor & Gupta, 2012), containing a triad of Ser-Asp(Glu)-His at their active site that is responsible for a network of hydrogen (Faouzi et al., 2015; Schrag et al., 1991). These enzymes are substrate-specific and can catalyze heterogeneous reactions in both hydrophilic and hydrophobic conditions. Owing to their wide catalytic abilities, these are extensively used in industries like tanning, agrochemical, and waste treatment (Ajit Kumar et al., 2012; Sarac & Ugur, 2016).

Lipases can be classified into three types based on their positional specificity (regiospecificity): non-specific lipases, 1,3-specific lipases, and fatty-acid-specific lipases. The first class is the non-specific lipases, which break down triglycerides into free fatty acids and glycerol. Monoglycerides and diglycerides are formed as intermediates during this process. This lipase class can remove fatty acid from any carbon position in the substrate and has the affinity to hydrolyze mono- and diglycerides faster than triglycerides (Utama et al., 2019). The second class, 1,3-specific lipases, releases fatty acids from carbon positions 1 and 3 of the triglycerides. These cannot hydrolyze the ester bonds at position 2 and allow faster release of diglycerides than monoglycerides from triglycerides (Utama et al., 2019). The final class is the fatty-acid specific lipases, which specifically catalyzes the hydrolysis of esters of triglycerides that have long-chain fatty acid-containing double bonds in the cis position between C-9 and C-10 (Utama et al., 2019) (Table 3.1).

Lipases, in general, do not require cofactors and can remain active even in organic solvents (Ashok Kumar et al., 2016; Utama et al., 2019). Since lipases have a low reaction time, high production of fatty acids in nonaqueous media, resistance to low pH, as well as the ability to degrade different types of lipid-based substrates, they are highly desirable as biocatalysts (Ashok Kumar et al., 2016). These enzymes can be of microbial, plant, or animal origin, although microbe-derived lipases are generally preferred in industry due to features like higher catalytic activities, ease in genetic manipulation, and their ability to use cheaper carbon substrates. (Vishnoi et al., 2020). As these enzymes are tolerant to organic solvents, they have scope for infinite application in industries such as fragrance, food, cosmetics, pharmaceuticals, and biofuel.

## 3.3 MECHANISM OF ACTION OF HYDROLASE ENZYMES

### 3.3.1 Glycosidic Hydrolases (GHs)

Present industrial-scale saccharification is mostly dependent on the noncomplexed cellulase catalyzed reactions as found in *Trichoderma reesei*. The overall process is hypothesized to have four distinct steps: adsorption, complexation, hydrolysis, and desorption (Jeoh et al., 2017). In the adsorption step, cellulases diffuse to the solid cellulose surface by the help of its carbohydrate-binding module (CBM) (Figure 3.2). In the next step, the enzyme-substrate complex is formed by binding to a single cellulose molecule in the active site of the catalytic domain (CD). This productively bound enzyme catalyzes the hydrolysis of the glycosidic bonds in the glucan chain. Finally, dissociation leads to the transfer of the enzyme from the solid surface to the liquid phase.

Basically, there are two mechanisms of action that most of the GHs follow for the glycosidic bond cleavage (Raich et al., 2016). These are known as inverting hydrolysis and retaining hydrolysis (Figure 3.3). In the inverting mechanism, the whole process takes place in a single step where the anomeric carbon of glucose in a polysaccharide faces a nucleophilic attack by a water molecule resulting in a proton

**TABLE 3.1**
**The Broad Classes of Hydrolytic Enzymes and Their Applications in Different Industries**

| Enzyme | Mechanism of Action | Microbial Source | Function | References |
|---|---|---|---|---|
| Glucosidase | Catalyze the breakdown of glycosidic bonds of proteins | *Bacillus licheniformis* | Bioflocculant degradation | Z. Chen et al. (2017) |
| | | *Geobacillus sp.* | Starch liquefaction | Novik et al. (2019) |
| | | *Aspergillus niger and Aspergillus oryzae* | Sugarcane bagasse degradation | El-Deen et al. (2014) |
| | | *Bacillus subtilis* | Fruit biocompost degradation | Nawawi et al. (2017) |
| Protease | Catalyze the breakdown of peptide bonds of proteins | *Bacillus subtilis* | Degradation of casein, keratin of feathers, crustacean waste | Suh & Lee (2001); J. K. Yang et al. (2000) |
| | | *Chryseobacterium* sp | Degradation of feathers | Riffel et al. (2003) |
| | | *Bacillus pumilus* | Degradation of feathers | El-Refai et al. (2005) |
| | | *Aspergillus sp.* | Keratin degradation | Vermelho et al. (2012) |
| Lipase | Proton transfer between aspartate, histidine, and serine residues of the lipase followed by hydroxyl residue of the serine attacks the carbonyl group of the substrate. During diacylation, nucleophile attacks the enzyme regenerating the enzyme and releasing the product. | *Bacillus subtilis* | Waste water treatment, grease and trough oil degradation | Haniya et al. (2017); Saraswat et al. (2017) |
| | | *Bacillus pumilus* | Palm oil degradation in waste water | Saranya et al. (2019) |
| | | *Sphingobacterium* | Polylactic acid degradation | Satti et al. (2019) |

transfer from the water to the catalytic base, followed by a proton transfer from the catalytic acid to the glycosidic oxygen to break the linkage. In this way, the stereochemistry of the anomeric carbon is inverted. The catalytic acid and base are the amino acid residues of the enzyme, containing carboxylate groups (e.g. aspartic acid or glutamic acid). Retaining hydrolysis takes place in two steps:glycosylation and deglycosylation. In the glycosylation step, anomeric carbon is attacked by nucleophilic amino acid residue and at the same time, a proton transfer takes place from the catalytic acid residue to the glycosidic oxygen. As a result, the glycosidic bond is cleaved and an intermediate glycosyl-enzyme complex is formed. In the deglycosylation step, a water molecule attacks the anomeric carbon to break the

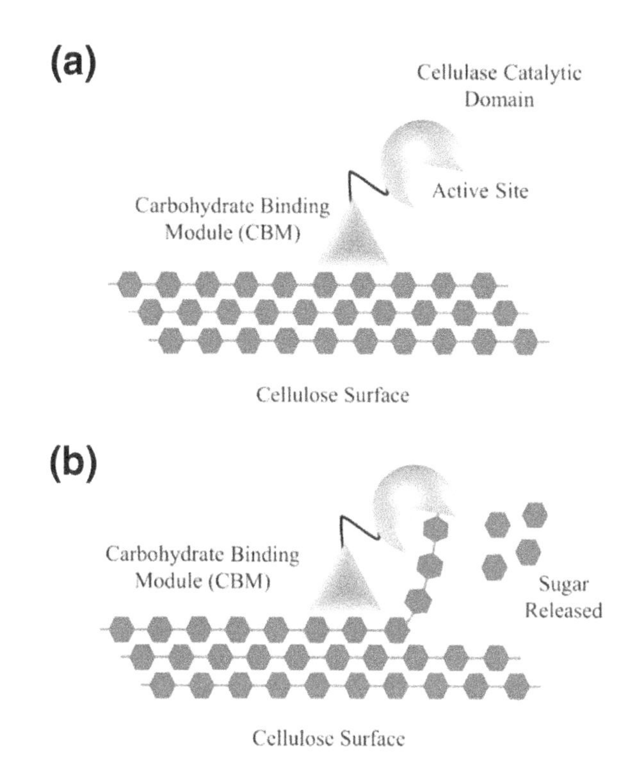


**FIGURE 3.2** Schematic diagram of the cellulase hydrolysis process: (a) Cellulase binds to the cellulose surface with the help of carbohydrate binding module (CBM) via specific or nonspecific binding. Nonspecific binding results in nonproductive enzyme-glucan complex whereas specific binding gives rise to productive-enzyme complex. (b) However, only the productive enzyme complex is active and leads to glycosidic bond cleavage releasing fermentable sugars.

complex. Simultaneously a proton from the water molecule is transferred to the catalytic base which restores the initial catalytic state of the enzyme active site. However, many GHs have been discovered that do not follow this mechanism of action (Petersen, 2010).

### 3.3.2 Proteases

Proteases are hydrolase enzymes that break protein molecules into shorter peptides. Endopeptidases catalyze the hydrolysis of amide bonds in the middle of polypeptide chains while exopeptidases cleave the terminal amino acid from polypeptide chains. The general mode of proteases is the hydrolysis of the peptide bond with the addition of water molecules. Based on the mechanism of action, proteases are categorized into four main classes: serine, cysteine, aspartyl, and metalloproteases. Among these, the mode of action of serine proteases is well studied. Chymotrypsin was the first serine protease to be investigated in detail. The overall process consists of: (a) an acylation step, and (b) a deacylation step (Figure 3.4). More specifically, the His57, Asp102,

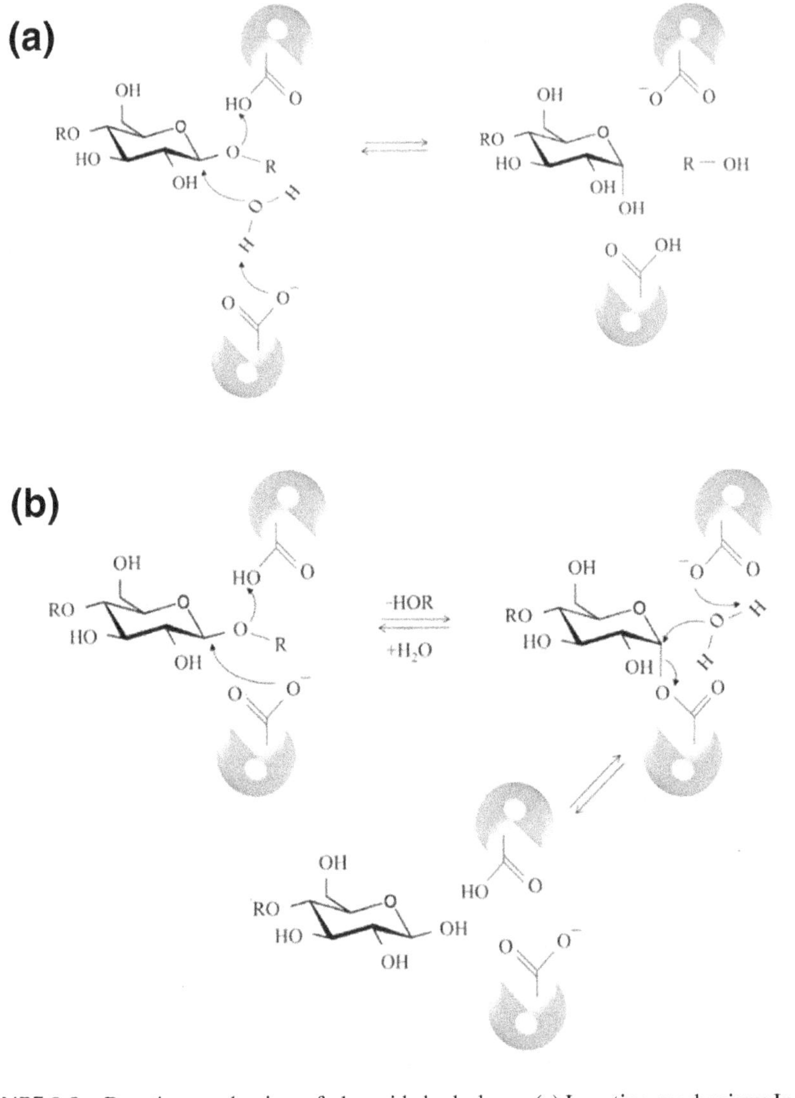

**FIGURE 3.3** Reaction mechanism of glycoside hydrolases. (a) Inverting mechanism: In this single displacement catalysis a catalytic basic residue of the enzyme withdraws a proton from a water molecule and the water hydroxyl group attacks the C1 carbon of the target sugar while at the same time another catalytic acid residue of the enzyme transfers a proton to the gylcosidic oxygen resulting in the cleavage of the bond. In the whole process, the stereochemistry of the anomeric carbon is inverted. R represents the long glucan chain. Only the enzyme catalytic domain is shown for better clarity. (b) Retaining mechanism: This is double displacement catalysis with two distinct steps, i.e., glycosylation and deglycosylation. In the glycosylation step, a basic residue of the enzyme attacks the anomeric carbon, and at the same time another acidic residue transfers a proton to the glycosidic linked oxygen, resulting in the breakage of glycosidic bond and formation of an enzyme-substrate complex. In the deglycosylation step, a water molecule attacks the C1 position of the sugar and transfers a proton to the new catalytic base residue of the glycosylation step. In this way, the catalytic state of the enzyme is retained.

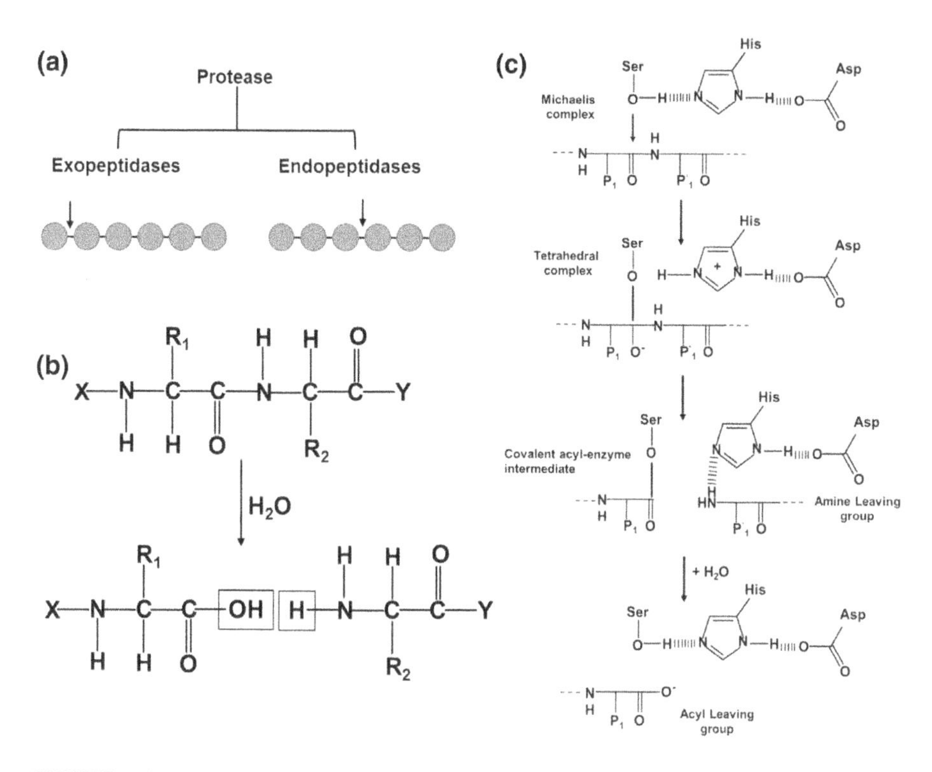


**FIGURE 3.4** (a) Classification of proteases, in terms of the site of peptide bond cleavage. They are divided into exo- and endo-peptidases according to their action at or away from the termini, respectively. (b) The general mechanism of protease enzymes action. The proteases cleave proteins by a hydrolysis reaction by the addition of a water molecule to the peptide bond. (c) Reaction mechanism of serine protease.

and Ser195 form a catalytic triad and charge the active site in the chymotrypsin (Neitzel, 2010). The His57 acts as a proton acceptor (base), whereas the Ser195 donates a proton from its OH group. Initially, the His57 forms a hydrogen bond with the hydroxyl group of Ser195. The negatively charged Asp102 further stabilizes the positively charged His57. During the acylation step, His57 extracts the proton of the hydroxyl group in Ser195, whereas Ser195 reacts with the carbon atom of the peptide bond in the incoming substrate. A proton is donated from the His57 to the N of the scissile bond, leading to the peptide bond cleavage. The Ser195 is still bonded to the peptide by forming an acyl-enzyme intermediate. During the deacylation step, a water molecule comes and donates a proton to the His57 which in turn is transferred to Ser195 to form the hydrogen bond. At the same time, the hydroxyl group of the water is added to the carbonyl carbon of the peptide, releasing the product and leaving the enzyme to its initial state.

Cysteine proteases have a Cys–His–Asn(Gln) residue combination at their active site, out of which the His residue works as the proton donor to increase the nucleophilicity of Cys. The Cys residue initiates the nucleophilic attack on the carboxyl carbon present in the peptide bond, resulting in a tetrahedral thioester intermediate and the release of an amino terminus on the substrate fragment (Coulombe et al., 1996). The intermediate tetrahedral structure is then stabilized by the hydrogen bonds formed between a highly conserved Gln and the substrate oxyanion. Finally, the thioester bond is hydrolyzed, resulting in a carboxylic acid terminus on the substrate fragment.

This mechanism is generally observed in both serine and cysteine proteases, whereas the aspartyl proteases and metalloproteases employ a water molecule as the nucleophile. However, all protease classes follow an overall similar mechanism for the peptide bond cleavage.

### 3.3.3 Lipases

Lipases catalyze the hydrolysis of the ester bonds in lipid chains (Svendsen, 2000), and are responsible for biochemical reactions like esterification, interesterification, and transesterification in a nonaqueous medium. These enzymes can hydrolyze triglycerides into mono- and diglycerides, free fatty acids, and glycerol.

Lipases (triacylglycerols ester hydrolases EC3.1.1.3) are mainly involved in hydrolyzing the triacylglycerols (TAG). These take part in reactions such as esterification and transesterification. The triacylglycerol lipase consists of an Asp-His-Ser catalytic triad at its active site (Jegannathan et al., 2008). The His is hydrogen-bonded to the carboxylate group of the Asp. At the same time, the nitrogen atom of the His forms another hydrogen bond with the Ser by pulling off the proton from the alcohol group. Thus, His helps the Ser alcohol to become a better nucleophile forming an oxyanion. Next, the Ser attacks the carbonyl carbon of the incoming substrate leading to the formation of a tetrahedral intermediate 1 (Figure 3.5).

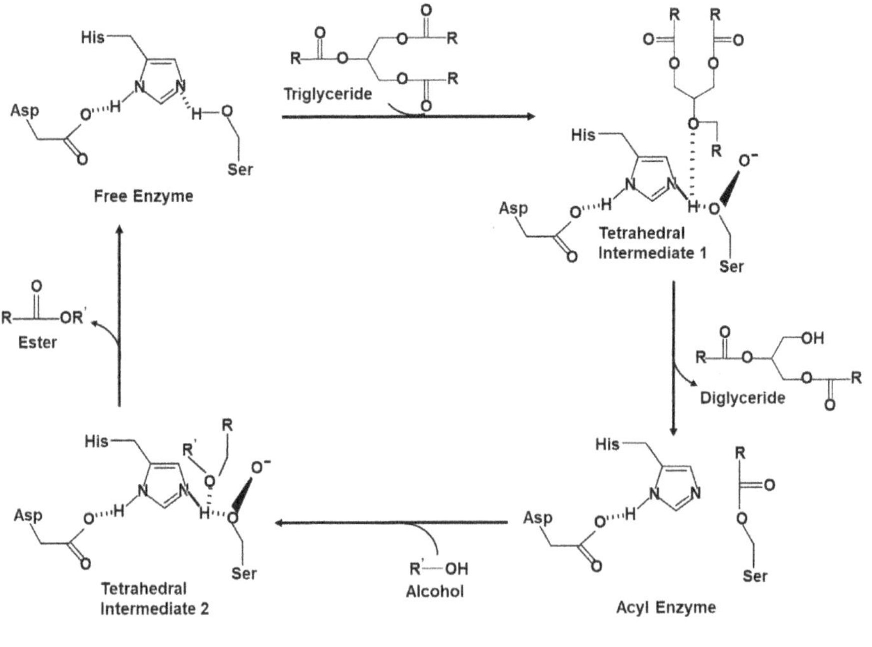

**FIGURE 3.5** Reaction mechanism of lipase in transesterification.

During the process, the electron on the Ser oxyanion is transferred to the carbonyl carbon leading to the acyl-enzyme ester complex, whereas the His's proton is taken up by the releasing diglyceride (Al-Zuhair, 2005). Next, a hydrogen molecule is removed from the incoming alcohol molecule by the His, forming the tetrahedral intermediate 2 followed by the release of free fatty acid. At the end of the process, the active site of the enzyme is returned to the free enzyme state for the next cycle.

## 3.4 ROLE OF HYDROLYTIC ENZYMES IN WASTE MANAGEMENT

The possibility of applying enzymes for waste treatment has been ventured into over recent last decades. Previously, wastes used to be treated through the hydrolytic enzymes secreted by either bacteria or fungi added to the biomatter. Recently, the use of commercially available enzymes has provided a growing opportunity in the area of waste management (Bhardwaj et al., 2021b). In this section, we detail the current scenario of enzymatic hydrolysis of feedstocks rich in different biologically important macromolecules.

### 3.4.1 Treatment of Carbohydrate Wastes

Carbohydrates, predominantly in the form of starch, cellulose, and hemicellulose, constitute the major portion of the chemical composition of wastes derived from food disposal and agricultural lands. In fact, food wastes contain about 30–60% starch depending on the type of food (Salimi et al., 2019). Over 30% of edible materials are generally disposed of globally, provoking huge expenditure costs to the world economy as well as serious environmental issues (Kumar & Longhurst, 2018). In addition to food wastes, agricultural wastes are also one of the most prolifically generated wastes that need to be treated efficiently to minimize environmental hazards. Lignocellulosic wastes contain 50–90% of their biomass as carbohydrate polymers, depending on the source, and have been recognized as a potential source of carbon for valorization (Bernal et al., 2017). Valorization of carbohydrate wastes for either sugar production or other valuable molecules, like ethanol, has been considered as a strategic method to increase production from low-cost feedstocks (Kumar & Verma, 2021a; Kumar & Verma, 2021b).

For releasing the sugar from the waste feedstocks, both thermochemical and enzyme-mediated methods are available (Wahlström & Suurnäkki, 2015). Biological valorization methods like microbial enzymatic hydrolysis have emerged as an advantageous process toward the hydrolysis of sludge (Teo & Wong, 2014). Hydrolytic enzymes can disrupt the extracellular polymeric substances and dissolve the insoluble materials in the sludge. Lysozymes, and amylase are two of the commonly used enzymes for hydrolysis and biodegradability of carbohydrate wastes in an environment-friendly manner (He et al., 2014; S. Yu et al., 2013b). Amylases and other glucoamylases can hydrolyze carbohydrates to simpler ones, thus aiding in dissolving the complex carbohydrates in sludge. In a study, the $\alpha$-amylase from *Bacillus subtilis*, $\beta$-amylase from sweet potato, and glucoamylase from *Aspergillus niger* were used to hydrolyze starch derived from maize, rice, and potato. It was found that hydrolysis is dependent on the surface area of the starchy granules and not on the substrate mass (J. C. Kim et al., 2008). In another study, a crude mixture of the xylanosidases *Penicillium sp.* AHT-1 and *Rhizomucor pusillus* HHT-1 was used to treat either raw or pretreated (milling or steam explosion) barley bran, sunflower seed peels, and shells of different nuts. Enzymatic hydrolysis of pretreated biomass resulted in the release of ~13 g of D-xylose from 100 g feedstock (Cho et al., 2002). A multi-enzyme solution containing glucoamylase and protease was developed by Melikoglu et al. to hydrolyze waste bread, which was then used in solid-state fermentation by *Aspergillus awamori* (Melikoglu et al., 2015). Also, a nutrient-complete feedstock was created by enzymatic hydrolysis of mixed food waste for microalgal cultivation (Lau et al., 2014) as well as succinic acid production (Sun et al., 2014).

Over the last decade, in addition to naturally available amylases and other glucosidases, commercialized concoctions of enzymes have found a niche in the treatment of food and lignocellulosic wastes. Han et al. hydrolysed food wastes using commercial glucoamylase, which made the production of biohydrogen feasible by facilitating the release of simpler substrates for microbial consumption (Han et al., 2016). Similarly, Novozymes® enzymes NS50013 and NS50010 were used to treat waste paper in laboratory-scale trials. Different types of paper wastes such as cellulose pulp, filter paper, recycled paper, and cardboard were applied for direct hydrolysis, and the highest glucose yield was recorded as 48.3 ± 2.7% per 10% cellulase loading with a 0.25% $H_3PO_4$ and 2% NaOH (Brummer et al., 2014).

Although enzymatic hydrolysis has excellent potential to address the management of biowastes, the high cost of enzymes and slow reaction rates in presence of inhibitors or toxic molecules can hinder the commercial application of this technique on an optimal scale (Subhedar & Gogate, 2013). To address these issues, enzyme engineering is a promising area of research to improve enzyme properties and functions, which is discussed in detail in section 3.5 of this chapter.

### 3.4.2 Treatment of Protein Wastes

In addition to carbohydrate wastes, proteins are also a major constituent of the wastes generated from food and agro-based industries, tanneries, wastes from fisheries, and animal husbandry. Most of the protein wastes generated from these industries include skin trimmings, fur, feathers, fins, and other keratinous exoskeletons, as well as visceral parts that are often deemed unfit for human consumption. Generally, these wastes are either thrown away or are used as cheap fertilizer, or in silage (Fao, 2016). However, over the last two decades, these protein-rich biowastes have been considered as a rich source for the production of different commercially important by-products like oils, hydrolysates, bioactive peptides, gelatin, and collagen (Ishak & Sarbon, 2017; Ovissipour et al., 2009).

Proteases released by different bacteria or fungi have been considered to degrade the animal wastes to allow both efficient treatment processes as well as valorization into important compounds (Whiteley et al., 2002). Proteases were used to increase deflocculating in sludges as well as reduce both solid concentration and pathogens like *Salmonella* and other coliform bacteria (Parmar et al., 2001; Rugabber & Talley, 2006). To date, a wide range of alkaline proteases have found their applications for wastewater treatment. Keratinolytic proteases were reported to degrade keratinous waste materials like hair and feathers from poultry industries. After hydrolysis, these degraded keratinous exoskeletons are converted into feedstuffs, glues, fertilizers, films, and as the source of rare amino acids, namely serine, proline, and cysteine (Cao et al., 2009). Protease-producing *B. subtilis* were also used to treat crustacean wastes (J. K. Yang et al., 2000). Additionally, chitinases were used to obtain chitin from shrimp and crab shell wastes (Thadathil & Velappan, 2014). Shrimp waste was treated with a bacterial protease to obtain caroteno protein, a carotenoid protein complex (Sowmya et al., 2014). Salt-tolerant proteases from *Pseudomonas aeruginosa* were reported to be used on tannery saline wastewater treatment (Sivaprakasam et al., 2011).

In addition to direct treatment of protein wastes using microbes, commercially available protease-mixes have also been used to valorize these wastes. A combination of commercial proteases have been used to hydrolyse chicken, pork, and beef by-products as well as salmon viscera from Norwegian food industries to create a rich ingredient for fermentation media. Yeast growth experiments have been conducted using the media developed from wastes, and it was found to be a promising alternative for expensive nitrogen sources often used in yeast fermentation experiments (Lapeña et al., 2018). Fillet by-products from seabream and seabass fishes were hydrolysed to obtain both fish oils as well as fish protein hydrolysates. These hydrolysates have been shown to have an excellent amino acid profile and deemed suitable as food supplements/additives (Valcarcel et al., 2020). Bovine cruor has been hydrolysed using commercial proteases to generate bioactive peptides having antibacterial and antioxidant properties (Abou-Diab et al., 2020).

Although the high cost of protein production often inhibits the widespread use of proteases for the valorization of wastes, improving both the stability and longevity of the enzymes can aid in making this process widely used. Particularly, the availability of engineered proteins is creating a niche in the bioremediation and hydrolysis of wastes, details of which have been provided in section 3.5 of this chapter. Since enzyme-induced valorization is less invasive and more environment-friendly compared to either physical or chemical methods, the development of a range of proteases having the ability to function in a range of temperature, pH and hazardous conditions is of utmost importance.

### 3.4.3 Treatment of Lipid Wastes

Lipids constitute approximately 10–40% of food wastes (Salimi et al., 2019). Application of biological degradation of fats using lipolytic enzymes has been rapidly considered for accelerating the degradation process with less time (Table 3.2). Of course, the source of the feedstock will affect the time and efficiency of degradation. While treating effluents and other wastewaters, it is essential to use low-cost enzymatic concoctions, since using a high-cost counterpart will render the process costlier and commercially unfeasible. Lipases are comprehensively used in the treatment of wastewater in both aerobic and anaerobic conditions. One common aerobic treatment device is known as the activated sludge process where the thin layers of fat from the aerated tank surface are constantly removed to maintain the biomass

**TABLE 3.2**
**Lipases from Microbial and Commercial Sources for Degradation of Different Oil-Based Wastes**

| Source of Lipid-Rich Waste | Enzymatic Treatment Method | Reference |
|---|---|---|
| Coconut oil waste | *Staphylococcus pasteuri* lipase | Kanmani et al. (2015) |
| Waste cooking oil | Immobilized *Candida* lipase | Y. Chen et al. (2009) |
| Waste sunflower and soybean oil | *Thermomyces lanuginosus* lipase | Dizge et al. (2009) |
| Waste cooking oil | *Geotrichum sp.* lipase | Yan et al. (2011) |
| Waste cooking oil | *Candida antarctica* lipase | Chesterfield et al. (2012) |
| Waste cooking oil | *Pseudomonas cepacia* lipase | C. Y. Yu et al. (2013a) |
| Waste cooking oil | *C. antarctica* type B lipase and *T. lanuginosus* lipase | Vescovi et al. (2016) |
| Waste cooking oil | *Pseudomonas aeruginosa* FW_SH-1 lipase | Ali et al. (2017) |
| Waste cooking oil | *T. lanuginosus* lipase | Sarno & Iuliano (2019) |
| Industrial liquid wastes | Immobilized lipase B from *C. antarctica* (CALB) and Novozyme 435 | Foukis et al. (2017) |
| Dairy wastewater | Crude enzyme extract and lipase from rawmilk | Bhange & Suke (2018) |
| Slaughterhouse and other poultry wastes | *Penicillium restrictum* lipase | Valladão et al. (2007) |

(Samer, 2015). Lipases from different microbial sources like *Acinetobacter, C. rugosa, Pseudomonas, Bacillus*, are usually used to digest the fat-rich impurities.

Thus, they are often used for treatment of industrial wastewater having dairy waste, food waste, and waste from oil mills by anaerobic conditions (Ken Ugo et al., 2017; Porwal et al., 2015). The effluent water produced from the tannery, automobile industries, food processing, restaurants, and fast food outlets are treated using various lipase enzymes from microbial sources (Maier & Lindner, 2007). During this process of waste treatment of fat using enzymes, triglycerides are hydrolyzed up to 90% filth and the process can be further enhanced by immobilizing lipase synthesizing bacteria (Chandra et al., 2020). *Candida rugosa* and *P. aeroginosa* lipases have been used to treat domestic wastewaters as well as clearing sewer systems and sinkholes (Dharmsthiti & Kuhasuntisuk, 1998; Jaeger & Reetz, 1998; S. Kumar et al., 2011). Lipases have also been used in the degradation of polymers (Sivalingam et al., 2003; Takamoto et al., 2001), and synthetic ester emulsions in water from oil-wells (*US5725771A - Process for Enzyme Pretreatment of Drill Cuttings - Google Patents*, n.d.), among others. Currently, genetic engineering of microbes for effluent treatment is being applied, too (Singh et al., 2011).

In addition to degradation, extracellular lipases released from microbes have been used to valorize oil wastes into fine chemicals. Two lipases were purified from *Yarrowia lipolytica* isolates that were used to hydrolyze oily wastes and agro-industrial hydrocarbons into citric acid (Mafakher et al., 2010). Fungal lipases from *Penicillium* sp. were evaluated for degradation of greasy feedstocks obtained from dairy wastewater under anaerobic conditions (Rosa et al., 2009). Lipase isolated from *A. awamori* was shown to remove ~92% oils and fat from mill effluent (Basheer et al., 2011). Another study showed that *Geotrichum candidum* lipase could be used for the degradation of olive mill wastewater (Asses et al., 2009). It should also be mentioned that upon hydrolysis, the lipids in food wastes produce long-chain fatty acids (LCFA) and glycerol, which often inhibits the fermentation rate of the microorganisms (Angelidaki & Ahring, 1992). These molecules have been reported to reduce ATP availability in the fermenting microbes (Perle et al., 1995). The inhibitory effects due to LCFA is directly proportional to the number of double bonds present and the availability of cis-isomers; something which is abundantly present in natural lipids (Rinzema et al., 1994). Hanaki et al. demonstrated that LCFA level affected the hydrogen production by acetogenic bacteria, who are also involved in the $\beta$-oxidation of LCFA (Hanaki et al., 1981). Inhibition of acetotrophic methanogens and acetogens due to LCFA resulted in a pronounced lag phase in the batch experiments. It was further observed that an increase in lipid concentration caused a reduction in hydrogen conversion rate by hydrogenotrophic methanogens (Rinzema et al., 1994). Consequently, researchers have studied the effect of treating the lipid-rich waste into sodium oleate as the sole carbon source. This approach increased the LCFA biodegradation capacity as well as the tolerance of acetoclastic bacteria to oleic acid (Cammarota & Freire, 2006). Another approach for increasing the lipid digestibility by microbes is the degradation of fats by lipolytic enzymes under thermophilic conditions (Becker et al., 1999). Under high temperatures, favorable changes in most of the physical properties of hydrophobic compounds take place, thus making the thermophilic conditions advantageous. Under these conditions, both fat solubility and diffusion coefficients are increased significantly, thus allowing these molecules to be more easily accessible to the microbes for digestion.

Several research reports and patents have been made available to date applying enzymes from different microbes for the treatment of fat and oil-rich biological effluents. A lipase from *C. regosa* was used by a Japanese company to treat fat residues in the equipment of several effluent treatment plants in the US. Also, Neozyme International Inc. manufactures and sells patented biocatalytic combinations that can efficiently and quickly break down organic oleaginous contaminants.

In addition to degrading lipid wastes, lipases have also found themselves utilized as a potential degradation of bioplastics. To curb environmental complications, biodegradable plastics are widely used for green and clean processes. Moreover, biodestructible plastics are also used interchangeably without considering their transformation (Song et al., 2009). The extent and rate of degradation is the main variation between biodegradable and biodestructible plastic where the latter needs further processing, unlike biodegradable plastic. Efficient degradation of plastic is dependent upon the potential of the lipase to break polycaprolactone (PUR) which is an aliphatic polyester (Abdel-Motaal et al., 2014). *Pseudomonas sp.* is the most common bacterial species to act on PUR (Danso et al., 2019). Some of the examples of lipase-producing bacterial species are *P. protegens, P. floresecens, P. chlororaphis* which degrades polyurethanes (Wilkes & Aristilde, 2017). Naturally, PueB lipase from *P. chlororaphis* was the first enzyme to react on polycaprolactone. Moreover, a lipase has been identified in *P. pelagia* using in silico genome mining which can act on polyesters (Restrepo-Flórez et al., 2014). Furthermore, polyesterases have been developed via protein engineering to create rapid and effective polyester-degrading lipases (Gricajeva et al., 2021).

Lipases have great potential in the transformation of organic lipid-rich effluents into valuable molecules. Besides, the simpler molecules derived after hydrolysis can be degraded by the microbial community present in the subsequent biological treatment. Further research on improving the performance and stability of these enzymes can allow a greener and more sustainable management of wastes.

## 3.5 RECENT ADVANCES IN ENZYMATIC HYDROLYSIS OF WASTES

As has been described in the preceding sections, the current enzymatic hydrolysis methods rely on naturally evolved site-specific cleavage-inducing enzymes (proteases, lipases, etc.). Over the years, engineering of these enzymes towards increased solubility and stability in feedstocks containing a high percentage of inhibitors and other toxic substances has been effectively done. Particularly, proteases have been targeted to tailor them to specific needs, which brings forth the need for creating the next generation of engineered proteases with high specificity toward individual hydrolysis reactions. It has been observed that the key idea behind the engineering of proteases was the apparent plasticity of their active sites that allowed the evolving of these enzymes with varying degrees of substrate selectivity and specificity, thus making them a promising framework for engineering them toward useful and unique activities.

In industrial bioprocesses, the use of native biocatalysts is often limited by their loss of activity under aggressive industrial conditions. Particularly, the high temperature used in pretreatment can affect enzyme activity (Lonhienne et al., 2000). Thus, improved thermostability and catalytic performance of proteases are highly desired characteristics in industry. In addition to using thermotolerant enzymes, protein engineering can allow the development of cost-efficient strategies for large-scale applications of enzymes (Hasunuma et al., 2013). Three major strategies in protein engineering for improving catalytic features have been attempted so far: (a) rational design; (b) *de novo* design; and (c) directed evolution. Of these, rational protein design is mostly preferred due to smaller variant libraries and hastened processing time. Applying rational modification of serine alkaline protease in *B. pumilis* through N99Y mutation resulted in increased thermostability (B. Jaouadi et al., 2010). Additionally, mutations at L31 and T33 of *B. pumilus* serine alkaline protease increased biocatalytic activity and substrate specificity and were suggested as promising biocatalysts for leather curing (N. Z. Jaouadi et al., 2014). A similar approach has also been tried in *P. aeruginosa* where its serine peptidase has been mutated (V336I and A29G) to improve its catalytic potentiality and thermostability (Ashraf et al., 2019). These mutations resulted in a 1.4-fold increase in catalytic activity compared to the native enzyme, while the $T_m$ of both mutants was increased by 5°C.

On the other hand, *de novo* protein design and engineering involve the creation of synthetic proteins that are not available in nature with atomic-level accuracy. Owing to the development in computational methodology, it has been possible to explore full protein sequence space and design possible motifs guided by physical principles governing protein folding. *De novo* protein engineering relies heavily on extensive computing setup as this is computationally expensive. So far, *de novo* protein engineering has been used for developing antibodies (Kries et al., 2013; Richter et al., 2011; H. Yang et al., 2019), but this method can be extended to designing proteases for waste treatment.

The third most commonly used enzyme-engineering method in biotechnology is directed evolution, which allows not only catalytic engineering, but also elucidates the relation between sequence, structures, and functions of proteins (Arnold et al., 2001; Porter et al., 2015). In this approach, a library of mutants is constructed and the variants which exhibit desired traits are selected via a screening method. This process is repeated until a variant with the target attributes is achieved. This approach has notable benefits over rational design because it does not require meticulous prior knowledge of the target protein while mutating its regions. In a recent study, directed evolution was applied to enhance the resistance of the PT121 metalloproteases in organic solvents. This resulted in the generation of mutants (H224F, H224Y, and T46Y) that rendered the ability to tolerate high levels of acetone and acetonitrile (Zhu et al., 2020). Also, compared to the native protein, the half-life of the mutant enzymes was improved by 1.2–3.5-folds after directed evolution. It was found that mutual variants T46Y/H224Y and T46Y/H224F possessed tremendous caseinolytic performance and catalytic stability.

The above studies have shown that engineered proteases have enormous potential to address the problem of waste management, due to their attributes like stability and specificity. Industries that are based on proteases can look forward to new engineered or fusion proteins that can have numerous activities united in one protein.

## 3.6 CURRENT CHALLENGES IN THE ENZYMATIC HYDROLYSIS OF WASTES

The key factors affecting efficient enzymatic hydrolysis of wastes are: effective depolymerization of complex macromolecular chains; high cellulose crystallinity index in lignocellulosic wastes; the effectiveness of delignification during the pretreatment; separation of nonfermentable particulate matter and other toxic metals; total enzyme accessible surface area of the macromolecular chains; average particle size; and the porosity of the material, among others. Investigations are being carried out to probe the synergistic effects of different enzyme combinations to find out the optimal combination for higher yield with improved economics of this process. During hydrolysis of lignocellulosic wastes, enzymatic catalysis is very much affected by the complex network formation by cellulose, hemicellulose, and lignin, making it recalcitrant in nature and leading to the adsorption due to non-specific linkages in condensed lignin. Besides, several degradation by-products of the lignocellulosic biomass may inhibit or significantly reduce the enzyme performance (Rasmussen et al., 2014). For example, inhibitory by-products like HMF, formic acid, levulinic acid, and phenolic compounds are produced from glucose, while hemicellulose derivatives like xylose, furan-2-carbaldehyde (furfural), and/or multiple C-1 and C-4 compounds are formed during the pretreatment process. In addition, monomeric sugars obtained from the biomass can interact and lead to the synthesis of pseudo-lignin and humins. Also, several aldehydes and ketones and many different organic acids (Vanillic acid,

Syringic acid, Ferulic acid, Tannic acid, Gallic acid, Cinnamic acid, p-Coumaric acid, 4-Hydroxybenzoic acid, etc.) and aromatic lignin derivatives from the degradation of lignocellulosic biomass can inhibit the activity of cellulase and hemicellulase in the saccharification step (Du et al., 2010; Y. Kim et al., 2011; Tejirian & Xu, 2011; Ximenes et al., 2011). Currently, researchers are searching for novel, robust, and efficient bacterial and fungal enzymes as they have high commercial value. Protein engineering is the best possible way of producing enzymes with desirable properties. A large number of protease enzymes have been engineered.

## 3.7 CONCLUSION

Remediation of wastes have found widespread use as they are a cheap reservoir of fermentable carbon. However, microbial degradation often faces the challenge of colonization on their surface, thus reducing the surface area for efficient hydrolysis. Consequently, the use of commercial enzymes has found widespread application to fast-forward the waste degradation and valorization, without affecting the microbial consortia. This chapter has highlighted the role of enzymatic pretreatment as a prerequisite step for microbial degradation of biowaste. Moreover, the mechanism of action of microbial degradation necessitates further studies to detect an effective enzymatic system that can efficiently break down biological macromolecules like carbohydrates, proteins, and lipids. Finally, ecofriendly, effective, and cost-effective waste management and valorization methods have been highlighted through the development of new technologies.

## ACKNOWLEDGMENTS

The authors are thankful to the Department of Science and Technology (Grant No. CRG/2020/002080), the Department of Biotechnology (Grant No. BT/RLF/Re-entry/06/2013) and the Scheme for Promotion of Academic and Research Collaboration (SPARC), the MHRD, and the Govt. of India (Grant No. SPARC/2018-2019/P265/SL). Pradipta Patra appreciates the support from the Department of Science and Technology (DST) and INSPIRE, India for the award of fellowships. Manali Das thanks the Council of Scientific and Industrial Research (CSIR) for their support.

## REFERENCES

Abdel-Motaal, F. F., El-Sayed, M. A., El-Zayat, S. A., & Ito, S. I. (2014). Biodegradation of poly (ε-caprolactone) (PCL) film and foam plastic by *Pseudozyma japonica* sp. nov., a novel cutinolytic ustilaginomycetous yeast species. *3 Biotech*, *4*(5). https://doi.org/10.1007/s13205-013-0182-9

Abou-Diab, M., Thibodeau, J., Deracinois, B., Flahaut, C., Fliss, I., Dhulster, P., Bazinet, L., & Nedjar, N. (2020). Bovine hemoglobin enzymatic hydrolysis by a new eco-efficient process-part ii: Production of bioactive peptides. *Membranes*, *10*(10). https://doi.org/10.3390/membranes10100268

Al-Zuhair, S. (2005). Production of biodiesel by lipase-catalyzed transesterification of vegetable oils: A kinetics study. *Biotechnology Progress*, *21*(5). https://doi.org/10.1021/bp050195k

Ali, C. H., Qureshi, A. S., Mbadinga, S. M., Liu, J. F., Yang, S. Z., & Mu, B. Z. (2017). Biodiesel production from waste cooking oil using onsite produced purified lipase from *Pseudomonas aeruginosa* FW_SH-1: Central composite design approach. *Renewable Energy, 109*. https://doi.org/10.1016/j.renene.2017.03.018

Angelidaki, I., & Ahring, B. K. (1992). Effects of free long-chain fatty acids on thermophilic anaerobic digestion. *Applied Microbiology and Biotechnology, 37*(6). https://doi.org/10.1007/BF00174850

Anwar Saeed, M., Ma, H., Yue, S., Wang, Q., & Tu, M. (2018). Concise review on ethanol production from food waste: Development and sustainability. *Environmental Science and Pollution Research, 25*(29). https://doi.org/10.1007/s11356-018-2972-4

Arnold, F. H., Wintrode, P. L., Miyazaki, K., & Gershenson, A. (2001). How enzymes adapt: Lessons from directed evolution. *Trends in Biochemical Sciences, 26*(2). https://doi.org/10.1016/S0968-0004(00)01755-2

Ashraf, N. M., Krishnagopal, A., Hussain, A., Kastner, D., Sayed, A. M. M., Mok, Y. K., Swaminathan, K., & Zeeshan, N. (2019). Engineering of serine protease for improved thermostability and catalytic activity using rational design. *International Journal of Biological Macromolecules, 126*. https://doi.org/10.1016/j.ijbiomac.2018.12.218

Asses, N., Ayed, L., Bouallagui, H., Ben Rejeb, I., Gargouri, M., & Hamdi, M. (2009). Use of Geotrichum candidum for olive mill wastewater treatment in submerged and static culture. *Bioresource Technology, 100*(7). https://doi.org/10.1016/j.biortech.2008.10.048

Barrett, A. J. (1994). Classification of peptidases. *Methods in Enzymology, 244*. https://doi.org/10.1016/0076-6879(94)44003-4

Basheer, S. M., Chellappan, S., Beena, P. S., Sukumaran, R. K., Elyas, K. K., & Chandrasekaran, M. (2011). Lipase from marine *Aspergillus awamori* BTMFW032: Production, partial purification and application in oil effluent treatment. *New Biotechnology, 28*(6). https://doi.org/10.1016/j.nbt.2011.04.007

Becker, P., Köster, D., Popov, M. N., Markossian, S., Antranikian, G., & Märkl, H. (1999). The biodegradation of olive oil and the treatment of lipid-rich wool scouring wastewater under aerobic thermophilic conditions. *Water Research, 33*(3). https://doi.org/10.1016/S0043-1354(98)00253-X

Bernal, M. P., Sommer, S. G., Chadwick, D., Qing, C., Guoxue, L., & Michel, F. C. (2017). Current approaches and future trends in compost quality criteria for agronomic, environmental, and human health benefits. In *Advances in Agronomy*, 144. https://doi.org/10.1016/bs.agron.2017.03.002

Bhange, V. P., & Suke, S. G. (2018). Effect of lipase from different source on high fat content wastewater of dairy industry. *Indian Journal of Biotechnology, 17*(2), 24–50.

Bhardwaj, N., Kumar, B., Agrawal, K., & Verma, P. (2021a). Current perspective on production and applications of microbial cellulases: A review. *Bioresources and Bioprocessing, 8*, 1–34.

Bhardwaj, N., Agrawal, K., Kumar, B., & Verma, P. (2021b). Role of Enzymes in Deconstruction of Waste Biomass for Sustainable Generation of Value-added Products. In: Thatoi, H., Mohapatra, S., & Das, S. K. (eds.) *Bioprospecting of Enzymes in Industry, Healthcare and Sustainable Environment* (pp. 219–250). Springer, Singapore.

Brummer, V., Jurena, T., Hlavacek, V., Omelkova, J., Bebar, L., Gabriel, P., & Stehlik, P. (2014). Enzymatic hydrolysis of pretreated waste paper - Source of raw material for production of liquid biofuels. *Bioresource Technology, 152*. https://doi.org/10.1016/j.biortech.2013.11.030

Cammarota, M. C., & Freire, D. M. G. (2006). A review on hydrolytic enzymes in the treatment of wastewater with high oil and grease content. *Bioresource Technology, 97*(17). https://doi.org/10.1016/j.biortech.2006.02.030

Cantarel, B. I., Coutinho, P. M., Rancurel, C., Bernard, T., Lombard, V., & Henrissat, B. (2009). The Carbohydrate-Active EnZymes database (CAZy): An expert resource for glycogenomics. *Nucleic Acids Research, 37*(SUPPL. 1). https://doi.org/10.1093/nar/gkn663

Cao, Z. J., Zhang, Q., Wei, D. K., Chen, L., Wang, J., Zhang, X. Q., & Zhou, M. H. (2009). Characterization of a novel Stenotrophomonas isolate with high keratinase activity and purification of the enzyme. *Journal of Industrial Microbiology and Biotechnology*, *36*(2). https://doi.org/10.1007/s10295-008-0469-8

Chandra, P., Singh, R., & Arora, P. K. (2020). Microbial lipases and their industrial applications: A comprehensive review. *Microbial Cell Factories*, *19*(1). https://doi.org/10.1186/s12934-020-01428-8

Chen, Y., Xiao, B., Chang, J., Fu, Y., Lv, P., & Wang, X. (2009). Synthesis of biodiesel from waste cooking oil using immobilized lipase in fixed bed reactor. *Energy Conversion and Management*, *50*(3). https://doi.org/10.1016/j.enconman.2008.10.011

Chen, Z., Meng, T., Li, Z., Liu, P., Wang, Y., He, N., & Liang, D. (2017). Characterization of a beta-glucosidase from *Bacillus licheniformis* and its effect on bioflocculant degradation. *AMB Express*, *7*(1). https://doi.org/10.1186/s13568-017-0501-3

Chesterfield, D. M., Rogers, P. L., Al-Zaini, E. O., & Adesina, A. A. (2012). Production of biodiesel via ethanolysis of waste cooking oil using immobilised lipase. *Chemical Engineering Journal*, *207–208*. https://doi.org/10.1016/j.cej.2012.07.039

Cho, C. H., Hatsu, M., & Takamizawa, K. (2002). The production of D-xylose by enzymatic hydrolysis of agricultural wastes. *Water Science and Technology*, *45*(12). https://doi.org/10.2166/wst.2002.0414

Coulombe, R., Grochulski, P., Sivaraman, J., Ménard, R., Mort, J. S., & Cygler, M. (1996). Structure of human procathepsin L reveals the molecular basis of inhibition by the prosegment. *EMBO Journal*, *15*(20). https://doi.org/10.1002/j.1460-2075.1996.tb00934.x

Cuesta, S. A., Mora, J. R., Zambrano, C. H., Torres, F. J., & Rincón, L. (2020). Comparative study of the nucleophilic attack step in the proteases catalytic activity: A theoretical study. *Molecular Physics*, *118*(14). https://doi.org/10.1080/00268976.2019.1705412

Danso, D., Chow, J., & Streita, W. R. (2019). Plastics: Environmental and biotechnological perspectives on microbial degradation. *Applied and Environmental Microbiology*, *85*(19). https://doi.org/10.1128/AEM.01095-19

de Souza, P. M., de Assis Bittencourt, M. L., Caprara, C. C., de Freitas, M., de Almeida, R. P. C., Silveira, D., Fonseca, Y. M., Filho, E. X. F., Pessoa Junior, A., & Magalhães, P. O. (2015). A biotechnology perspective of fungal proteases. *Brazilian Journal of Microbiology*, *46*(2). https://doi.org/10.1590/S1517-838246220140359

Dharmsthiti, S., & Kuhasuntisuk, B. (1998). Lipase from *Pseudomonas aeruginosa* LP602: Biochemical properties and application for wastewater treatment. *Journal of Industrial Microbiology and Biotechnology*, *21*(1–2). https://doi.org/10.1038/sj.jim.2900563

Diaz, A. B., Blandino, A., & Caro, I. (2018). Value added products from fermentation of sugars derived from agro-food residues. *Trends in Food Science and Technology*, *71*. https://doi.org/10.1016/j.tifs.2017.10.016

Dizge, N., Aydiner, C., Imer, D. Y., Bayramoglu, M., Tanriseven, A., & Keskinler, B. (2009). Biodiesel production from sunflower, soybean, and waste cooking oils by transesterification using lipase immobilized onto a novel microporous polymer. *Bioresource Technology*, *100*(6). https://doi.org/10.1016/j.biortech.2008.10.008

Du, B., Sharma, L. N., Becker, C., Chen, S. F., Mowery, R. A., van Walsum, G. P., & Chambliss, C. K. (2010). Effect of varying feedstock-pretreatment chemistry combinations on the formation and accumulation of potentially inhibitory degradation products in biomass hydrolysates. *Biotechnology and Bioengineering*, *107*(3), 430–440. https://doi.org/10.1002/bit.22829

El-Deen, A. M. N., Shata, H. M. A. H., & Farid, M. A. F. (2014). Improvement of β-glucosidase production by co-culture of *Aspergillus niger* and *A. oryzae* under solid state fermentation through feeding process. *Annals of Microbiology*, *64*(2). https://doi.org/10.1007/s13213-013-0696-8

El-Refai, H. A., AbdelNaby, M. A., Gaballa, A., El-Araby, M. H., & Abdel Fattah, A. F. (2005). Improvement of the newly isolated *Bacillus pumilus* FH9 keratinolytic activity. *Process Biochemistry*, *40*(7). https://doi.org/10.1016/j.procbio.2004.09.006

Ellaiah, P., Srinivasulu, B., & Adinarayana, K. (2002). A review on microbial alkaline proteases. *Journal of Scientific and Industrial Research*, *61*(9).

Fao. (2016). *The State of World Fisheries and Aquaculture 2016*. www.fao.org/publications

Faouzi, L., Fatimazahra, E. B., Moulay, S., Adel, S., Wifak, B., Soumya, E., Iraqui, M., & Saad, K. I. (2015). Higher tolerance of a novel lipase from *Aspergillus flavus* to the presence of free fatty acids at lipid/water interface. *African Journal of Biochemistry Research*, *9*(1). https://doi.org/10.5897/ajbr2014.0804

Foukis, A., Gkini, O. A., Stergiou, P. Y., Sakkas, V. A., Dima, A., Boura, K., Koutinas, A., & Papamichael, E. M. (2017). Sustainable production of a new generation biofuel by lipase-catalyzed esterification of fatty acids from liquid industrial waste biomass. *Bioresource Technology*, *238*. https://doi.org/10.1016/j.biortech.2017.04.028

Garcia-Carreno, F. L., & Navarrete Del Toro, M. A. (1997). Classification of proteases without tears. *Biochemical Education*, *25*(3). https://doi.org/10.1016/S0307-4412(97)00005-8

*Global Waste Generation - Statistics & Facts | Statista*. (n.d.). Retrieved July 19, 2021, from https://www.statista.com/topics/4983/waste-generation-worldwide/#dossierSummary__chapter1

Gonzales, T., & Robert-Baudouy, J. (1996). Bacterial aminopeptidases: Properties and functions. *FEMS Microbiology Reviews*, *18*(4). https://doi.org/10.1016/0168-6445(96)00020-4

Gricajeva, A., Nadda, A. K., & Gudiukaite, R. (2021). Insights into polyester plastic biodegradation by carboxyl ester hydrolases. *Journal of Chemical Technology and Biotechnology*. https://doi.org/10.1002/jctb.6745

Gupta, R., Gupta, N., & Rathi, P. (2004). Bacterial lipases: An overview of production, purification and biochemical properties. *Applied Microbiology and Biotechnology*, *64*(6). https://doi.org/10.1007/s00253-004-1568-8

Han, W., Yan, Y., Shi, Y., Gu, J., Tang, J., & Zhao, H. (2016). Biohydrogen production from enzymatic hydrolysis of food waste in batch and continuous systems. *Scientific Reports*, *6*. https://doi.org/10.1038/srep38395

Hanaki, K., Matsuo, T., & Nagase, M. (1981). Mechanism of inhibition caused by long-chain fatty acids in anaerobic digestion process. *Biotechnology and Bioengineering*, *23*(7). https://doi.org/10.1002/bit.260230717

Haniya, M., Naaz, A., Sakhawat, A., Amir, S., Zahid, H., & Syed, S. A. (2017). Optimized production of lipase from *Bacillus subtilis* PCSIRNL-39. *African Journal of Biotechnology*, *16*(19). https://doi.org/10.5897/ajb2017.15924

Hasunuma, T., Okazaki, F., Okai, N., Hara, K. Y., Ishii, J., & Kondo, A. (2013). A review of enzymes and microbes for lignocellulosic biorefinery and the possibility of their application to consolidated bioprocessing technology. *Bioresource Technology*, *135*. https://doi.org/10.1016/j.biortech.2012.10.047

He, J. G., Xin, X. D., Qiu, W., Zhang, J., Wen, Z. D., & Tang, J. (2014). Performance of the lysozyme for promoting the waste activated sludge biodegradability. *Bioresource Technology*, *170*. https://doi.org/10.1016/j.biortech.2014.07.095

Hedstrom, L. (2002). Serine protease mechanism and specificity. *Chemical Reviews*, *102*(12). https://doi.org/10.1021/cr000033x

Ishak, N. H., & Sarbon, N. M. (2017). Optimization of the enzymatic hydrolysis conditions of waste from shortfin scad (*Decapterus macrosoma*) for the production of angiotensin I-converting enzyme (ACE) inhibitory peptide using response surface methodology. *International Food Research Journal*, *24*(4).

Jaeger, K. E., & Reetz, M. T. (1998). Microbial lipases form versatile tools for biotechnology. *Trends in Biotechnology*, *16*(9). https://doi.org/10.1016/S0167-7799(98)01195-0

Jaouadi, B., Aghajari, N., Haser, R., & Bejar, S. (2010). Enhancement of the thermostability and the catalytic efficiency of *Bacillus pumilus* CBS protease by site-directed mutagenesis. *Biochimie*, *92*(4). https://doi.org/10.1016/j.biochi.2010.01.008

Jaouadi, N. Z., Jaouadi, B., Hlima, H. Ben, Rekik, H., Belhoul, M., Hmidi, M., Ben Aicha, H. S., Hila, C. G., Toumi, A., Aghajari, N., & Bejar, S. (2014). Probing the crucial role of leu31 and thr33 of the *Bacillus pumilus* cbs alkaline protease in substrate recognition and enzymatic depilation of animal hide. *PLoS ONE, 9*(9). https://doi.org/10.1371/journal.pone.0108367

Javed, S., Azeem, F., Hussain, S., Rasul, I., Siddique, M. H., Riaz, M., Afzal, M., Kouser, A., & Nadeem, H. (2018). Bacterial lipases: A review on purification and characterization. *Progress in Biophysics and Molecular Biology, 132.* https://doi.org/10.1016/j.pbiomolbio.2017.07.014

Jegannathan, K. R., Abang, S., Poncelet, D., Chan, E. S., & Ravindra, P. (2008). Production of biodiesel using immobilized lipase - A critical review. *Critical Reviews in Biotechnology, 28*(4). https://doi.org/10.1080/07388550802428392

Jeoh, T., Cardona, M. J., Karuna, N., Mudinoor, A. R., & Nill, J. (2017). Mechanistic kinetic models of enzymatic cellulose hydrolysis—A review. *Biotechnology and Bioengineering, 114*(7), 1369–1385. https://doi.org/10.1002/bit.26277

Kanmani, P., Kumaresan, K., & Aravind, J. (2015). Utilization of coconut oil mill waste as a substrate for optimized lipase production, oil biodegradation and enzyme purification studies in *Staphylococcus pasteuri. Electronic Journal of Biotechnology, 18*(1). https://doi.org/10.1016/j.ejbt.2014.11.003

Kapoor, M., & Gupta, M. N. (2012). Lipase promiscuity and its biochemical applications. *Process Biochemistry, 47*(4). https://doi.org/10.1016/j.procbio.2012.01.011

Ken Ugo, A., Vivian Amara, A., Igwe, C. N., & Kenechuwku, U. (2017). Microbial lipases: A prospect for biotechnological industrial catalysis for green products: A review. *Fermentation Technology, 06*(02). https://doi.org/10.4172/2167-7972.1000144

Kim, J. C., Kong, B. W., Kim, M. J., & Lee, S. H. (2008). Amylolytic hydrolysis of native starch granules affected by granule surface area. *Journal of Food Science, 73*(9). https://doi.org/10.1111/j.1750-3841.2008.00944.x

Kim, Y., Ximenes, E., Mosier, N. S., & Ladisch, M. R. (2011). Soluble inhibitors/deactivators of cellulase enzymes from lignocellulosic biomass. *Enzyme and Microbial Technology, 48*(4–5), 408–415. https://doi.org/10.1016/j.enzmictec.2011.01.007

Koch, S., Volkmar, C. M., Kolb-Bachofen, V., Korth, H. G., Kirsch, M., Horn, A. H. C., Sticht, H., Pallua, N., & Suschek, C. V. (2009). A new redox-dependent mechanism of MMP-1 activity control comprising reduced low-molecular-weight thiols and oxidizing radicals. *Journal of Molecular Medicine, 87*(3). https://doi.org/10.1007/s00109-008-0420-5

Kries, H., Blomberg, R., & Hilvert, D. (2013). De novo enzymes by computational design. *Current Opinion in Chemical Biology, 17*(2). https://doi.org/10.1016/j.cbpa.2013.02.012

Kumar, Ajit, Parihar, S. S., & Batra, N. (2012). Enrichment, isolation and optimization of lipase-producing *Staphylococcus* sp. from oil mill waste (Oil cake). *Journal of Experimental Sciences, 3*(8), 26–30.

Kumar, Ashok, Dhar, K., Kanwar, S. S., & Arora, P. K. (2016). Lipase catalysis in organic solvents: Advantages and applications. *Biological Procedures Online, 18*(1). https://doi.org/10.1186/s12575-016-0033-2

Kumar, B., & Verma, P. (2020a). Enzyme mediated multi-product process: A concept of bio-based refinery. *Industrial Crops and Products, 154,* 112607. https://doi.org/10.1016/j.indcrop.2020.112607

Kumar, B., & Verma, P. (2020b). Application of Hydrolytic Enzymes in Biorefinery and Its Future Prospects. In: Srivastava, N., Srivastava, M., Mishra, P. K., & Gupta, V. K. (eds.) *Microbial Strategies for Techno-economic Biofuel Production* (pp. 59–83). Springer, Singapore.

Kumar, B., & Verma, P. (2021a). Techno-Economic Assessment of Biomass-Based Integrated Biorefinery for Energy and Value-Added Product. In: Verma, P. (ed.) *Biorefineries: A Step Towards Renewable and Clean Energy* (pp. 581–616). Springer, Singapore.

Kumar, B., & Verma, P. (2021b). Biomass-based biorefineries: An important architype towards a circular economy. *Fuel*, *288*, 119622, Elsevier.

Kumar, S., Katiyar, N., Ingle, P., & Negi, S. (2011). Use of evolutionary operation (EVOP) factorial design technique to develop a bioprocess using grease waste as a substrate for lipase production. *Bioresource Technology*, *102*(7). https://doi.org/10.1016/j.biortech.2010.12.114

Kumar, V., & Longhurst, P. (2018). Recycling of food waste into chemical building blocks. *Current Opinion in Green and Sustainable Chemistry*, *13*. https://doi.org/10.1016/j.cogsc.2018.05.012

Lapeña, D., Vuoristo, K. S., Kosa, G., Horn, S. J., & Eijsink, V. G. H. (2018). Comparative assessment of enzymatic hydrolysis for valorization of different protein-rich industrial by-products. *Journal of Agricultural and Food Chemistry*, *66*(37). https://doi.org/10.1021/acs.jafc.8b02444

Lau, K. Y., Pleissner, D., & Lin, C. S. K. (2014). Recycling of food waste as nutrients in Chlorella vulgaris cultivation. *Bioresource Technology*, *170*. https://doi.org/10.1016/j.biortech.2014.07.096

Lonhienne, T., Gerday, C., & Feller, G. (2000). Psychrophilic enzymes: Revisiting the thermodynamic parameters of activation may explain local flexibility. *Biochimica et Biophysica Acta - Protein Structure and Molecular Enzymology*, *1543*(1). https://doi.org/10.1016/S0167-4838(00)00210-7

Mafakher, L., Mirbagheri, M., Darvishi, F., Nahvi, I., Zarkesh-Esfahani, H., & Emtiazi, G. (2010). Isolation of lipase and citric acid producing yeasts from agro-industrial wastewater. *New Biotechnology*, *27*(4). https://doi.org/10.1016/j.nbt.2010.04.006

Maier, N. M., & Lindner, W. (2007). Chiral recognition applications of molecularly imprinted polymers: A critical review. *Analytical and Bioanalytical Chemistry*, *389*(2). https://doi.org/10.1007/s00216-007-1427-4

Marzo, C., Díaz, A. B., Caro, I., & Blandino, A. (2019). Valorization of agro-industrial wastes to produce hydrolytic enzymes by fungal solid-state fermentation. *Waste Management and Research*, *37*(2). https://doi.org/10.1177/0734242X18798699

Melikoglu, M., Lin, C. S. K., & Webb, C. (2015). Solid state fermentation of waste bread pieces by *Aspergillus awamori*: Analysing the effects of airflow rate on enzyme production in packed bed bioreactors. *Food and Bioproducts Processing*, *95*. https://doi.org/10.1016/j.fbp.2015.03.011

Millati, R., Cahyono, R. B., Ariyanto, T., Azzahrani, I. N., Putri, R. U., & Taherzadeh, M. J. (2019). Agricultural, industrial, municipal, and forest wastes: An overview. *Sustainable Resource Recovery and Zero Waste Approaches*. https://doi.org/10.1016/B978-0-444-64200-4.00001-3

Mótyán, J., Tóth, F., & Tőzsér, J. (2013). Research applications of proteolytic enzymes in molecular biology. *Biomolecules*, *3*(4). https://doi.org/10.3390/biom3040923

Naveed, M., Nadeem, F., Mehmood, T., Bilal, M., Anwar, Z., & Amjad, F. (2021). Protease—A versatile and ecofriendly biocatalyst with multi-industrial applications: An updated review. *Catalysis Letters*, *151*(2). https://doi.org/10.1007/s10562-020-03316-7

Nawawi, M. H., Mohamad, R., Tahir, P. M., & Saad, W. Z. (2017). Extracellular xylanopectinolytic enzymes by *Bacillus subtilis* ADI1 from EFB's compost. *International Scholarly Research Notices*, *2017*. https://doi.org/10.1155/2017/7831954

Neitzel, J. J. (2010). Enzyme catalysis: The serine proteases. *Nature Education*, *3*(9), 21.

Niehaus, F., Bertoldo, C., Kähler, M., & Antranikian, G. (1999). Extremophiles as a source of novel enzymes for industrial application. *Applied Microbiology and Biotechnology*, *51*(6), 711–729. https://doi.org/10.1007/s002530051456

Novik, G., Savich, V., & Meerovskaya, O. (2019). Geobacillus Bacteria: Potential Commercial Applications in Industry, Bioremediation, and Bioenergy Production. In *Growing and Handling of Bacterial Cultures*. https://doi.org/10.5772/intechopen.76053

Ovissipour, M., Abedian, A., Motamedzadegan, A., Rasco, B., Safari, R., & Shahiri, H. (2009). The effect of enzymatic hydrolysis time and temperature on the properties of protein hydrolysates from *Persian sturgeon* (*Acipenser persicus*) viscera. *Food Chemistry, 115*(1). https://doi.org/10.1016/j.foodchem.2008.12.013

Pandey, A., Soccol, C. R., Nigam, P., Soccol, V. T., Vandenberghe, L. P. S., & Mohan, R. (2000). Biotechnological potential of agro-industrial residues. II: Cassava bagasse. *Bioresource Technology, 74*(1). https://doi.org/10.1016/S0960-8524(99)00143-1

Parmar, N., Singh, A., & Ward, O. P. (2001). Characterization of the combined effects of enzyme, pH and temperature treatments for removal of pathogens from sewage sludge. *World Journal of Microbiology and Biotechnology, 17*(2). https://doi.org/10.1023/A:1016606020993

Perle, M., Kimchie, S., & Shelef, G. (1995). Some biochemical aspects of the anaerobic degradation of dairy wastewater. *Water Research, 29*(6). https://doi.org/10.1016/0043-1354(94)00248-6

Petersen, L. (2010). *Catalytic strategies of glycoside hydrolases*. https://doi.org/10.31274/ETD-180810-675

Porter, J. L., Boon, P. L. S., Murray, T. P., Huber, T., Collyer, C. A., & Ollis, D. L. (2015). Directed evolution of new and improved enzyme functions using an evolutionary intermediate and multidirectional search. *ACS Chemical Biology, 10*(2). https://doi.org/10.1021/cb500809f

Porwal, H. J., Mane, A. V., & Velhal, S. G. (2015). Biodegradation of dairy effluent by using microbial isolates obtained from activated sludge. *Water Resources and Industry, 9*. https://doi.org/10.1016/j.wri.2014.11.002

Raich, L., Nin-Hill, A., Ardèvol, A., & Rovira, C. (2016). Enzymatic cleavage of glycosidic bonds: Strategies on how to set up and control a QM/MM metadynamics simulation. *Methods in Enzymology, 577*. https://doi.org/10.1016/bs.mie.2016.05.015

Rasmussen, H., Sørensen, H. R., & Meyer, A. S. (2014). Formation of degradation compounds from lignocellulosic biomass in the biorefinery: Sugar reaction mechanisms. *Carbohydrate Research, 385*, 45–57. https://doi.org/10.1016/j.carres.2013.08.029

Restrepo-Flórez, J. M., Bassi, A., & Thompson, M. R. (2014). Microbial degradation and deterioration of polyethylene - A review. *International Biodeterioration and Biodegradation*, 88. https://doi.org/10.1016/j.ibiod.2013.12.014

Richter, F., Leaver-Fay, A., Khare, S. D., Bjelic, S., & Baker, D. (2011). De novo enzyme design using Rosetta3. *PLoS ONE, 6*(5). https://doi.org/10.1371/journal.pone.0019230

Riffel, A., Lucas, F., Heeb, P., & Brandelli, A. (2003). Characterization of a new keratinolytic bacterium that completely degrades native feather keratin. *Archives of Microbiology, 179*(4). https://doi.org/10.1007/s00203-003-0525-8

Rinzema, A., Boone, M., van Knippenberg, K., & Lettinga, G. (1994). Bactericidal effect of long chain fatty acids in anaerobic digestion. *Water Environment Research, 66*(1). https://doi.org/10.2175/wer.66.1.7

Rosa, D. R., Duarte, I. C. S., Katia Saavedra, N., Varesche, M. B., Zaiat, M., Cammarota, M. C., & Freire, D. M. G. (2009). Performance and molecular evaluation of an anaerobic system with suspended biomass for treating wastewater with high fat content after enzymatic hydrolysis. *Bioresource Technology, 100*(24). https://doi.org/10.1016/j.biortech.2009.06.089

Rugabber, T. P., & Talley, J. W. (2006). Enhancing bioremediation with enzymatic processes: A review. *Practice Periodical of Hazardous, Toxic, and Radioactive Waste Management, 10*(2). https://doi.org/10.1061/(ASCE)1090-025X(2006)10:2(73)

Saini, J. K., Saini, R., & Tewari, L. (2015). Lignocellulosic agriculture wastes as biomass feedstocks for second-generation bioethanol production: Concepts and recent developments. *3 Biotech, 5*(4), 337–353. https://doi.org/10.1007/s13205-014-0246-5

Salimi, E., Saragas, K., Taheri, M. E., Novakovic, J., Barampouti, E. M., Mai, S., Moustakas, K., Malamis, D., & Loizidou, M. (2019). The role of enzyme loading on starch and cellulose hydrolysis of food waste. *Waste and Biomass Valorization*, *10*(12). https://doi.org/10.1007/s12649-019-00826-3

Samer, M. (2015). Biological and chemical wastewater treatment processes. *Wastewater Treatment Engineering*. https://doi.org/10.5772/61250

Sarac, N., & Ugur, A. (2016). A green alternative for oily wastewater treatment: Lipase from *Acinetobacter haemolyticus* NS02-30. *Desalination and Water Treatment*, *57*(42). https://doi.org/10.1080/19443994.2015.1106346

Saranya, P., Selvi, P. K., & Sekaran, G. (2019). Integrated thermophilic enzyme-immobilized reactor and high-rate biological reactors for treatment of palm oil-containing wastewater without sludge production. *Bioprocess and Biosystems Engineering*. https://doi.org/10.1007/s00449-019-02104-x

Saraswat, R., Verma, V., Sistla, S., & Bhushan, I. (2017). Evaluation of alkali and thermotolerant lipase from an indigenous isolated *Bacillus* strain for detergent formulation. *Electronic Journal of Biotechnology*, *30*. https://doi.org/10.1016/j.ejbt.2017.08.007

Sarno, M., & Iuliano, M. (2019). Biodiesel production from waste cooking oil. *Green Processing and Synthesis*, *8*(1), 828–836. https://doi.org/10.1515/GPS-2019-0053

Satti, S. M., Abbasi, A. M., Salahuddin, Rana, Q. ul A., Marsh, T. L., Auras, R., Hasan, F., Badshah, M., Farman, M., & Shah, A. A. (2019). Statistical optimization of lipase production from *Sphingobacterium* sp. strain S2 and evaluation of enzymatic depolymerization of Poly(lactic acid) at mesophilic temperature. *Polymer Degradation and Stability*, *160*. https://doi.org/10.1016/j.polymdegradstab.2018.11.030

Schrag, J. D., Li, Y., Wu, S., & Cygler, M. (1991). Ser-His-Glu triad forms the catalytic site of the lipase from *Geotrichum candidum*. *Nature*, *351*(6329). https://doi.org/10.1038/351761a0

Singh, J. S., Abhilash, P. C., Singh, H. B., Singh, R. P., & Singh, D. P. (2011). Genetically engineered bacteria: An emerging tool for environmental remediation and future research perspectives. *Gene*, *480*(1–2). https://doi.org/10.1016/j.gene.2011.03.001

Sivalingam, G., Chattopadhyay, S., & Madras, G. (2003). Solvent effects on the lipase catalyzed biodegradation of poly (ε-caprolactone) in solution. *Polymer Degradation and Stability*, *79*(3). https://doi.org/10.1016/S0141-3910(02)00357-9

Sivaprakasam, S., Dhandapani, B., & Mahadevan, S. (2011). Optimization studies on production of a salt-tolerant protease from *Pseudomonas aeruginosa* strain BC1 and its application on tannery saline wastewater treatment. *Brazilian Journal of Microbiology*, *42*(4). https://doi.org/10.1590/S1517-83822011000400038

Song, J. H., Murphy, R. J., Narayan, R., & Davies, G. B. H. (2009). Biodegradable and compostable alternatives to conventional plastics. *Philosophical Transactions of the Royal Society B: Biological Sciences*, *364*(1526). https://doi.org/10.1098/rstb.2008.0289

Souza, P. M., Werneck, G., Aliakbarian, B., Siqueira, F., Ferreira Filho, E. X., Perego, P., Converti, A., Magalhães, P. O., & Junior, A. P. (2017). Production, purification and characterization of an aspartic protease from Aspergillus foetidus. *Food and Chemical Toxicology*, *109*. https://doi.org/10.1016/j.fct.2017.03.055

Sowmya, R., Ravikumar, T. M., Vivek, R., Rathinaraj, K., & Sachindra, N. M. (2014). Optimization of enzymatic hydrolysis of shrimp waste for recovery of antioxidant activity rich protein isolate. *Journal of Food Science and Technology*, *51*(11). https://doi.org/10.1007/s13197-012-0815-8

Subhedar, P. B., & Gogate, P. R. (2013). Intensification of enzymatic hydrolysis of lignocellulose using ultrasound for efficient bioethanol production: A review. *Industrial and Engineering Chemistry Research*, *52*(34). https://doi.org/10.1021/ie401286z

Suh, H. J., & Lee, H. K. (2001). Characterization of a keratinolytic serine protease from Bacillus subtilis KS-1. *Journal of Protein Chemistry*, *20*(2). https://doi.org/10.1023/A:1011075707553

Sun, Z., Li, M., Qi, Q., Gao, C., & Lin, C. S. K. (2014). Mixed food waste as renewable feedstock in succinic acid fermentation. *Applied Biochemistry and Biotechnology*, *174*(5). https://doi.org/10.1007/s12010-014-1169-7

Svendsen, A. (2000). Lipase protein engineering. *Biochimica et Biophysica Acta - Protein Structure and Molecular Enzymology*, *1543*(2). https://doi.org/10.1016/S0167-4838(00)00239-9

Takamoto, T., Shirasaka, H., Uyama, H., & Kobayashi, S. (2001). Lipase-catatyzed hydrolytic degradation of polyurethane in organic solvent. *Chemistry Letters*, *6*, 492–493. https://doi.org/10.1246/CL.2001.492

*Technical Platform on the Measurement and Reduction of Food Loss and Waste | Food and Agriculture Organization of the United Nations*. (n.d.). Retrieved July 19, 2021, from http://www.fao.org/platform-food-loss-waste/en/

Tejirian, A., & Xu, F. (2011). Inhibition of enzymatic cellulolysis by phenolic compounds. *Enzyme and Microbial Technology*, *48*(3), 239–247. https://doi.org/10.1016/j.enzmictec.2010.11.004

Ten years of CAZypedia: A living encyclopedia of carbohydrate-active enzymes. (2018). *Glycobiology*, *28*(1), 3–8. https://doi.org/10.1093/glycob/cwx089

Teo, C. W., & Wong, P. C. Y. (2014). Enzyme augmentation of an anaerobic membrane bioreactor treating sewage containing organic particulates. *Water Research*, *48*(1). https://doi.org/10.1016/j.watres.2013.09.041

Thadathil, N., & Velappan, S. P. (2014). Recent developments in chitosanase research and its biotechnological applications: A review. *Food Chemistry 150*. https://doi.org/10.1016/j.foodchem.2013.10.083

Theron, L. W., & Divol, B. (2014). Microbial aspartic proteases: Current and potential applications in industry. *Applied Microbiology and Biotechnology*, *98*(21). https://doi.org/10.1007/s00253-014-6035-6

*Trends in Solid Waste Management*. (n.d.). Retrieved July 19, 2021, from https://datatopics.worldbank.org/what-a-waste/trends_in_solid_waste_management.html

*US5725771A - Process for enzyme pretreatment of drill cuttings - Google Patents*. (n.d.). Retrieved July 19, 2021, from https://patents.google.com/patent/US5725771

Utama, Q. D., Sitanggang, A. B., Adawiyah, D. R., & Hariyadi, P. (2019). Lipase-catalyzed interesterification for the synthesis of medium-long-medium (MLM) structured lipids - A review. *Food Technology and Biotechnology*, *57*(3). https://doi.org/10.17113/ftb.57.03.19.6025

Valcarcel, J., Sanz, N., & Vázquez, J. A. (2020). Optimization of the enzymatic protein hydrolysis of by-products from seabream (*Sparus aurata*) and seabass (dicentrarchus labrax), chemical and functional characterization. *Foods*, *9*(10). https://doi.org/10.3390/foods9101503

Valladão, A. B. G., Freire, D. M. G., & Cammarota, M. C. (2007). Enzymatic pre-hydrolysis applied to the anaerobic treatment of effluents from poultry slaughterhouses. *International Biodeterioration and Biodegradation*, *60*(4). https://doi.org/10.1016/j.ibiod.2007.03.005

Vermelho, A. B., Supuran, C. T., & Guisan, J. M. (2012). Microbial enzyme: Applications in industry and in bioremediation. *Enzyme Research*, *2012*. https://doi.org/10.1155/2012/980681

Vescovi, V., Rojas, M. J., Baraldo, A., Botta, D. C., Santana, F. A. M., Costa, J. P., Machado, M. S., Honda, V. K., de Lima Camargo Giordano, R., & Tardioli, P. W. (2016). Lipase-catalyzed production of biodiesel by hydrolysis of waste cooking oil followed by esterification of free fatty acids. *JAOCS, Journal of the American Oil Chemists' Society*, *93*(12). https://doi.org/10.1007/s11746-016-2901-y

Vishnoi, N., Dixit, S., & Mishra, J. (2020). *Microbial Lipases and Their Versatile Applications.* https://doi.org/10.1007/978-981-15-1710-5_8

Wahlström, R. M., & Suurnäkki, A. (2015). Enzymatic hydrolysis of lignocellulosic polysaccharides in the presence of ionic liquids. *Green Chemistry, 17*(2), 694–714. https://doi.org/10.1039/c4gc01649a

Whiteley, C. G., Heron, P., Pletschke, B., Rose, P. D., Tshivhunge, S., Van Jaarsveld, F. P., & Whittington-Jones, K. (2002). The enzymology of sludge solubilisation utilising sulphate reducing systems: Properties of proteases and phosphatases. *Enzyme and Microbial Technology, 31*(4). https://doi.org/10.1016/S0141-0229(02)00100-X

Wilkes, R. A., & Aristilde, L. (2017). Degradation and metabolism of synthetic plastics and associated products by *Pseudomonas* sp.: Capabilities and challenges. *Journal of Applied Microbiology, 123*(3). https://doi.org/10.1111/jam.13472

Ximenes, E., Kim, Y., Mosier, N., Dien, B., & Ladisch, M. (2011). Deactivation of cellulases by phenols. *Enzyme and Microbial Technology, 48*(1), 54–60. https://doi.org/10.1016/j.enzmictec.2010.09.006

Yan, J., Yan, Y., Liu, S., Hu, J., & Wang, G. (2011). Preparation of cross-linked lipase-coated micro-crystals for biodiesel production from waste cooking oil. *Bioresource Technology, 102*(7). https://doi.org/10.1016/j.biortech.2011.01.006

Yang, H., Li, Y. C., Zhao, M. Z., Wu, F. L., Wang, X., Xiao, W. Di, Wang, Y. H., Zhang, J. L., Wang, F. Q., Xu, F., Zeng, W. F., Overall, C. M., He, S. M., Chi, H., & Xu, P. (2019). Precision de novo peptide sequencing using mirror proteases of ac-lysarginase and trypsin for large-scale proteomics. *Molecular and Cellular Proteomics, 18*(4). https://doi.org/10.1074/mcp.TIR118.000918

Yang, J. K., Shih, I. L., Tzeng, Y. M., & Wang, S. L. (2000). Production and purification of protease from a Bacillus subtilis that can deproteinize crustacean wastes. *Enzyme and Microbial Technology, 26*(5–6). https://doi.org/10.1016/S0141-0229(99)00164-7

Yu, C. Y., Huang, L. Y., Kuan, I. C., & Lee, S. L. (2013a). Optimized production of biodiesel from waste cooking oil by lipase immobilized on magnetic nanoparticles. *International Journal of Molecular Sciences, 14*(12). https://doi.org/10.3390/ijms141224074

Yu, S., Zhang, G., Li, J., Zhao, Z., & Kang, X. (2013b). Effect of endogenous hydrolytic enzymes pretreatment on the anaerobic digestion of sludge. *Bioresource Technology, 146.* https://doi.org/10.1016/j.biortech.2013.07.087

Zhu, F., He, B., Gu, F., Deng, H., Chen, C., Wang, W., & Chen, N. (2020). Improvement in organic solvent resistance and activity of metalloprotease by directed evolution. *Journal of Biotechnology, 309.* https://doi.org/10.1016/j.jbiotec.2019.12.014

# 4 Cellulase in Waste Valorization

*Manswama Boro, Ashwani Kumar Verma, Dixita Chettri, and Anil Kumar Verma*
Sikkim University, Gangtok, India

## CONTENTS



## 4.1 INTRODUCTION

Cellulose is one of the most abundant natural polymers on earth. Billions of tons of biodegradable lignocellulosic waste are accumulated worldwide every year. Along with this, cellulosic waste is abundant throughout the world. Cellulose is the major component of lignocellulosic waste biomass with about 50% of the total composition, and can be found in complexes with other components such as hemicellulose and lignin (Shen et al. 2013). Cellulosic wastes are a part of the biodegradable waste from a whole range of sources such as: agricultural wastes such as straw, seeds, stalks, leaves, roots, and husks; industrial wastes such as pomace, residues from the food industry, breweries, wood industry, jute industry and textile industry; municipal wastes such as household biodegradable wastes, office paper, and packaging board (Kumar et al. 2020a) (Figure 4.1).

These wastes are primarily disposed of by landfilling or by composting. Landfilling such a large amount of waste from so many sources leads to various problems,

DOI: 10.1201/9781003187684-4

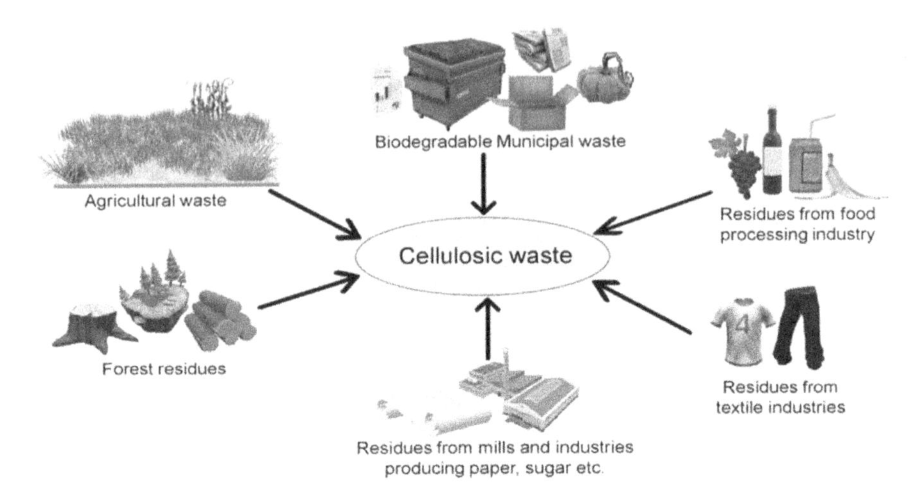


**FIGURE 4.1** Different sources of cellulosic waste.

including environmental pollution. Burning of this waste also causes serious environmental issues, including the release of greenhouse gases. Hydrolysis and processing such waste to produce valuable products are good solutions to this problem. Since waste material is a cheap source, it is also a cost-effective option for industrial production of many valuable products (Chaturvedi & Verma 2013).

Cellulosic waste can be degraded using the cellulase enzyme complex, which hydrolyzes the cellulose molecule and produce glucose as the final product. The glucose thus produced can be further used for the production of valuable products such as biofuel, chemicals and organic acids such as acetic acid. Biofuel includes bioethanol, biogas, hydrogen, biobutanol, and biodiesel. Bioethanol is an important fuel used as an alternative to depleting fossil fuels (Kumar et al. 2020b). Biodiesel is also a good alternative to fossil fuels due to its environmental benefits (Demirbas 2009a). Biogas is another biofuel that is produced by decomposing organic waste and sludge and can be used as cooking gas (Abbasi et al. 2011). Biohydrogen is also a type of biofuel that is used as an energy source in motor vehicles (Demirbas 2009b). The production of biofuels and chemicals as an alternative to traditional nonrenewable fossil fuels by using lignocellulosic waste material is considered as one of the best strategies for a better future (Kumar & Verma 2021).

## 4.2 CELLULOSE

Cellulose is a natural polymer associated with human civilization since ancient times. From the construction of monuments, ships, ropes, linen and cotton to musical instruments and paper for writing manuscripts, everything is linked to the history of the civilization of human beings and the growing importance of cellulose associated with it.

Cellulose, a glucose polysaccharide, one of the most abundant organic biological compounds on earth, was first discovered in the world of science by Anselme Payen in 1838 during his study of plant tissue by treatment with acid ammonia. After several decades of study, the molecular structure of cellulose homopolymer was determined to be composed of β-D-glucopyranose units linked by (1→4) glycosidic bonds (Marchessault & Sundararajan 1983). There are six polymorphic, interconvertible structures of cellulose (I, II, III1, III11, IV1, and IV11), of which cellulose I (the native cellulose) and cellulose II are the most thoroughly studied (O'Sullivan 1997). Cellulose is one of the major structural components in the cell wall of plant cells and is also produced by bacteria, fungi, and algae (Brigham 2018).

Cellulose has various industrial applications, such as in the paper, textile, food and beverage, and pharmaceutical industries, and more recently in biofuel production, which aims to replace rapidly depleting fossil fuels (Lavanya et al. 2011). For the conversion of cellulosic waste to valuable products, the cellulose must first be hydrolyzed to simple sugars and then fermented to produce ethanol and other by-products. Due to the complex structures of lignocellulose and the densely packed crystalline structure of cellulose, it needs to be hydrolyzed for proper use of cellulose to produce value-added products. It is important to release cellulose from lignocellulose complexes for proper cellulose hydrolysis (Michelin et al. 2015). The hydrolysis of cellulose is conducted both chemically and enzymatically. The chemical hydrolysis method is not always preferred as this method uses strong acids for hydrolysis that result in the release of toxic by-products. Therefore, cellulose hydrolyzing enzymes (cellulase) are a better option for this purpose (Bhardwaj et al. 2021). These enzymes are more specific to cellulose and there is a negligible reaction to impurities in the waste materials (Mandels & Sternberg 1976).

## 4.3 CELLULASE

Cellulases are complex enzymes efficiently synthesized by various microorganisms, including bacteria, fungi, and protozoa, and by other living organisms such as animals and plants (Lee & Koo 2001; Kumar et al. 2018). Cellulase production is done either aerobically or anaerobically. Bacteria such as *Pseudomonas fluorescens, E. coli, Bacillus subtilis, Serratia marcescens, Thermomonospora* sp., and *Cellulomonas* sp. produce a large amount of cellulase under optimal environmental conditions (Saranraj et al. 2012; Sethi et al. 2013). Fungi are the best producers of cellulase. *Aspergillus* and *Trichoderma* are some of the fungal genera that produce a significant amount of cellulase (Singh et al. 2019).

Cellulases are types of hydrolase enzymes that catalyze the hydrolysis of cellulose polysaccharides to their monomeric glucose units that can be further processed for production of different products (Figure 4.2). Cellulases hydrolyze the β-1,4-glycosidic bond in cellulose homopolymer chains (Sukumaran et al. 2005). These enzymes are inducible as their production depends on the presence of cellulosic material in the growth medium (Lee & Koo 2001).

Based on guidelines issued by the International Union of Biochemistry and Molecular Biology (IUBMB), substrate specificity determines the name of the enzyme. The cellulase enzymes have been grouped with various hemicellulases and different polysaccharides as O-glycoside hydrolase (EC 3.2.1.x). However, the classification based on substrate specificity is not meaningful because the full range of

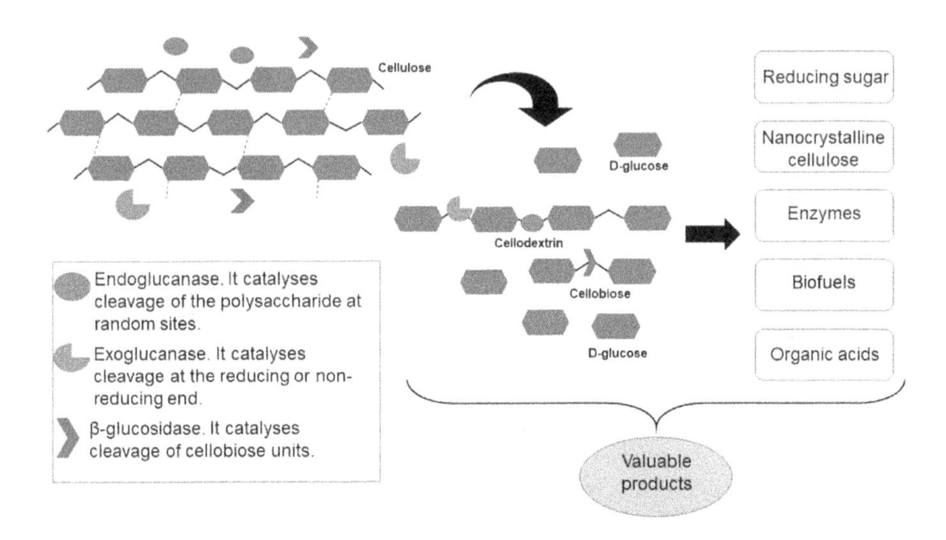


**FIGURE 4.2** Hydrolysis of cellulose catalyzed by cellulase enzymes.

substrates is not always determined, so another basis for classification has become accepted as an alternative. Glycoside hydrolases' (GH) classification based on homology of the amino acid sequence has been proposed (Henrissat 1991; Henrissat & Bairoch 1996). In a new type of nomenclature proposed by Henrissat et al. the first three letters indicate the substrate of interest, the glycoside hydrolase family is indicated by the number, and the order in which the enzymes are indicated is indicated by the last capital letters. For example, *Trichoderma reesei* EGI, CBHI, and CBHII are designated as Cel6B, Cel7A, and Cel6A, respectively (Henrissat et al. 1998).

An integrated database has been created by Henrissat and Coutinho to cover the rapidly growing number of glycoside hydrolases (http://www.cazy.org/) (Coutinho & Henrissat 1999). Cellulase enzymes are discovered in several GH families such as 5,6,7,8,9,12,44,45,48, 51, 61, and 74. GH family 9 includes cellulase enzymes produced by fungi, plants, animals (protozoa and termites), and bacteria (aerobic and anaerobic). GH family 7 includes fungal hydrolases, while GH family 8 includes bacterial hydrolases. The cellulase enzyme produced by the same microorganism may be located in different GH families. For example, endoglucanase and exoglucanase isolated from the bacterium *Clostridium thermocellum* cellulosome have been found in GH families 5, 8, 9, 44, and 48 (Shoham et al. 1999). GH family members have been classified into two major enzyme subfamilies: E1 and E2. Those categorized under E1 present a close association of an immunoglobulin-like domain along with the catalytic domain. E2 subfamily members are associated with a carbohydrate-binding module (CBM) which is classified in the 3c family (Béguin 1990). Molecular observation of various cellulolytic enzymes has provided in-depth knowledge of their activity. Such studies showed that within each GH family, different enzyme members share similarity in protein folding pattern, similar catalytic amino acid residues, and the same mechanism of action, that is, inversion of configuration after a single substitution or β-configuration retention at the anomeric carbon position after a double substitution reaction (Béguin & Aubert 1994). In addition to the catalytic domain of the cellulase enzyme system, there is another important domain, the carbohydrate-binding domain (CBD), which is not specifically involved in catalysis but plays a key role in substrate binding. CBDs mediate the hydrolysis of cellulose by bringing the insoluble cellulose substrate close to the catalytic domain (Guillén et al. 2010). Such CBDs are normally present at the carboxyl and amino terminus of the polypeptide chain. The carbohydrate-binding domains are separated from the catalytic domain utilizing glycosylated linker segments rich in amino acids such as proline, threonine, and serine. CBDs are better recognized than carbohydrate-binding modules (CBMs). The presence of CBMs significantly increases the rate of cellulose hydrolysis (Hilden & Johansson 2004).

### 4.3.1 Types of Cellulase

Cellulase in its natural form is mainly composed of three types of enzymes present in a complex mixture: endoglucanase/EG/endo-β-1,4-glucanase (EC 3.2.14), exoglucanase/CBH/cellobiohydrolase (EC 3.2.1.91), and β-glucosidase/BG/cellobiase (EC 3.2.1.21). These enzymes are individually separable and may be present in different compositions among themselves (Sukumaran et al. 2005). These three types of cellulases catalyze the hydrolysis of cellulose to produce primary products: glucose, cello-oligosaccharides, and cellobiose (Li et al. 2009).

EG facilitates hydrolysis or cleavage of the amorphous regions of the cellulose polysaccharide at readily accessible random sites, resulting in the formation of new shorter polysaccharide chains with one reducing and one non-reducing end. These enzymes reduce the overall length of the polymer and increase the concentration of reducing sugars. CBH cleaves at the reducing or nonreducing end of the polysaccharide chain, releasing cellobiose (a disaccharide composed of two glucose molecules) or cyclodextrin units. This cleavage exposes a new end group in the polysaccharide chain that is available for CBH hydrolysis. BG further cleaves the cellobiose units at the resulting non-reducing end to glucose units (Cavaco-Paulo et al. 1996). These three enzymes act simultaneously and synergistically to mediate the degradation of cellulose into its degradation products. These enzymes vary in specificity depending on the macroscopic structure of the substrate. They hydrolyze 1,4-β-glycosidic bonds in the substrate cellulose.

### 4.3.2 Catalytic Mechanism

The catalytic mechanism of the enzyme cellulase, which catalyzes the hydrolysis of cellulose, has attracted the interest of many researchers because the mechanism is beyond the optimization of years of evolution. Many researchers have proposed a lysozyme type of mechanism that works for different cellulases based on various chemical and kinetic studies on enzyme modification. The lysosome-type mechanism was concluded on the basis of the amino acid sequence homology of cellulase enzyme to the active sites of lysozyme (Coughlan 1991). The lysozyme enzyme functions by a double acid hydrolysis mechanism in which an ionized aspartic acid residue and nonionized glutamic acid residue act as proton acceptor and donor, respectively.

Koshland proposed two stereospecific mechanisms by which glycoside hydrolase catalyzes glycosidic bond cleavage in 1953. These two reaction mechanisms for the enzyme cellulase include retaining and inverting (Koshland 1953). In the first hypothetical retaining mechanism, the first residue functions as an acid catalyst (AC) that serves to protonate glycosidic oxygen, and the second residue, the base (B-), provides nucleophilic support for the leaving group. A water molecule hydrolyzes the resulting glycosyl enzyme and, similar to the lysozyme enzyme, the second nucleophilic substitution reaction at the site of anomeric carbon atom yields a product that has the same stereospecificity compared to the processed substrates. In the second hypothetical inversion mechanism, an acid residue performs the protonation of the glycosidic oxygen molecule, followed by a water molecule activated by a base residue, which attacks the leaving molecule. Similar to the products obtained from β-amylase hydrolysis, the formation of an end product with opposite stereospecificity to substrate results from a single nucleophilic substitution reaction (Kumar et al. 2016).

In simple terms, we can say two mechanisms of catalysis by cellulase for the hydrolysis of cellulose have been presented considering the aromatic center of cellulose. The catalysis depends on acidic and basic catalytic residues present on the enzyme for the hydrolysis of the glycosidic bond. The first mechanism is the inversion mechanism, in which the acid and base catalysis is carried out by the cellulase enzyme working on the catalysis of the exit of the leaving unit and the attack on the aromatic center, resulting in an inversion of the stoichiometry of the aromatic center. Another mechanism is that of maintaining the stoichiometry of the aromatic center through a two-step process by acid/base catalysts. In the first step, the facilitation of the exit of the leaving unit occurs followed by stabilization of the enzyme by the second residue. Subsequently, in the second step, the attack on the aromatic center is facilitated by water molecules (Vocadlo & Davies 2008).

The steps of enzymatic hydrolysis of cellulose by cellulase enzyme depend upon the structural complexity of cellulose fibers, the interaction between cellulose and cellulase, the capacity of enzymes to function, and the ability of enzymes to inhibit product formation (Walker & Wilson 1991). The distance between the respective carboxyl groups represents the key structural difference in various glycosidic enzymes mediating retaining or inverting mechanisms to facilitate the glycosidic bond cleavage in the substrate molecule. For example, in the retaining mechanisms, the nucleophiles and the acid-base catalysts are separated by about 5–5.5 Å, whereas in molecules performing the inverting mechanism, the difference between the acid and base residues is about 9–10 Å (Morana et al. 2011). Comprehensive knowledge of the mechanisms and structural differences enables researchers to implement studies to evaluate the glycosidase systems. With in-depth knowledge, site-specific mutations can be introduced at the nucleophile site, enabling the conversion of retainers to inverters and vice versa (Vocadlo & Davies 2008).

## 4.4 ROLE OF CELLULASE IN WASTE VALORIZATION

Cellulosic wastes constitute a large percentage of the waste generated in the environment, both naturally and by human actions. Cellulosic and lignocellulosic

**TABLE 4.1**
**Cellulose Content in the Various Waste Cellulosic Substrate**

| Cellulosic Waste Substrate | Cellulose Composition (gm/100 gm) | Reference |
|---|---|---|
| Rice straw | 38.3 | Lu and Hsieh (2012) |
| Rice husk | 35 | Johar et al. (2012) |
| Wheat straw | 34–40 | Liu et al. (2005) |
| Barley straw | 56.2 | Fortunati et al. (2016) |
| Barley husk | 45.7 | Fortunati et al. (2016) |
| Corn cob | 45.3 | Ai et al. (2021) |
| Corn straw | 31.7 | Liu et al. (2018) |
| Soyabean hulls | 30–40 | Merci et al. (2019) |
| Grape pomace | 19.30 ± 0.67 | Coelho et al. (2018) |
| Banana peel | 55.48 ± 1.57 | Harini et al. (2018) |
| Banana bract | 64.67 ± 2.36 | Harini et al. (2018) |
| Banana plant | 57.5 | Reddy and Yang (2015) |
| Potato peel | 6.31 | Zhang et al. (2016) |
| Citrus peel | 12.7–13.6 | Ververis et al. (2007) |
| Tomato pomace | 10.2–13.1 | Jiang and Hsieh (2015) |
| Pineapple peel | 17 | Madureira et al. (2018) |
| Pineapple leaves | 70–82 | Asim et al. (2015) |
| Sugarcane bagasse | 39 | Leão et al. (2017) |
| Apple pomace | 32.48 ± 0.33 | Melikoğlu et al. (2019) |
| Coconut husk | 24.7 | Cabral et al. (2016) |
| Kelp waste | 42 | Liu et al. (2017) |
| Cassava waste | 33–44 | Veiga et al. (2016) |
| Olive pomace | 18.33 | Haddadin et al. (2009) |
| Jackfruit peel | 20.08 | Trilokesh and Uppuluri (2019) |
| Cucumber peel | 18.22 | Prasanna and Mitra (2020) |
| Newspaper | 45 | Byadgi and Kalburgi (2016) |

waste is abundant worldwide; it is naturally renewable and can be used as a cost-effective source for the production of valuable products such as biofuel, food additives, organic acids, and many others (Maitan-Alfenas et al. 2015). Cellulosic waste includes residues from agricultural fields, residues from food processing industries, paper industries, forest residues, and renewable municipal solid wastes. Some examples of different sources of cellulosic waste and their cellulose content have been tabulated in Table 4.1. These cellulosic residues can be treated with cellulase to produce a wide range of products that add value to otherwise discarded waste material.

For efficient hydrolysis of the lignocellulosic waste material for the production of valuable products, a mixture of enzymes is required along with cellulase as such waste materials consist of lignin, hemicellulose, and many other organic molecules in addition to cellulose (Agrawal et al. 2018). Due to the complex structure of lignocellulosic waste, waste materials are treated before enzymatic hydrolysis. This

**TABLE 4.2**
**Advantages and Limitations of Using Cellulosic Waste as a Substrate**

| Advantages | Limitations |
|---|---|
| Cellulosic waste is abundant worldwide and a cheap source of glucose. | The complex crystalline structure of cellulose is tough to hydrolyse. |
| The composition of cellulosic waste makes it a good source for energy production in the form of biofuels. | Cellulosic waste contains other polymers like lignin and hemicellulose in lignocellulose complexes that act as a barrier, hindering cellulose accessibility and degradation. |
| Biofuels like bioethanol, biobutanol produced from cellulosic waste can be utilized as an alternative to fossil fuels. | Cellulase enzyme complex to be employed must be able to perform maximum cellulose hydrolysis with minimum enzyme use. Poor degradation efficiency is one of the major bottlenecks associated with cellulosic waste biodegradation. |
| Valorization of cellulosic waste reduces environmental damage caused by burying, piling up and dumping of such waste. | Proper waste pretreatment methods must be adapted in order to valorize cellulosic waste, which is sometimes a chaotic task. |

step is known as the pretreatment step and it helps expose the cellulose polymer for efficient hydrolysis. Cellulase enzymes then break down the long-chain homopolymer of cellulose and release glucose sugar unit, which can be used as a substrate for fermentation for the production of various products. There are several advantages and limitations of using cellulosic waste material as a source for the production of value-added products upon hydrolysis that have been tabulated in Table 4.2.

### 4.4.1 Cellulase in the Valorization of Agricultural Wastes

Agriculture is one of the most important practices worldwide and is responsible for a large amount of waste generation. Tons of agricultural wastes are generated worldwide every year. Agricultural waste not only includes agricultural residues, but also forest residues. Agricultural wastes contain husks, stems, leaves, seeds, straw, roots, bagasse, pulp, peels, and so forth, which consist of cellulosic and hemicellulosic materials. These waste materials can be hydrolyzed with the help of enzymes like cellulase to produce valuable products and there are various reports of studies regarding this. In an experiment, agriculture-based industrial wastes such as rice straw, wheat straw, sugarcane bagasse, corncob, and jute stalks were hydrolyzed to produce glucose which can be used as a raw material for the production of various products. Hydrolysis was conducted using a cocktail of cellulase enzymes produced by *Talaromyces verruculosus* IIPC 324 mutant UV-8 (Jain et al. 2020).

### 4.4.2 Cellulase in Valorization of Cellulosic Waste from Industries

Industries such as the agro-based food processing industry, edible oil industry, wood, and paper industries, textile industries, among others produce a large amount of cellulosic waste. The growing world population is increasing the demand for food and processed goods from industries and hence there is an increase in the production of waste materials (Ng et al. 2020). Agro-based industries like those that produce rice, flour, sugar, fruit juice, vegetable oil, and so on produce high quantities of cellulosic wastes that are renewable and can be hydrolyzed. Textile waste, including both preconsumer and postconsumer waste, is abundant, and since the cotton fibers that make up textile materials contain about 99% cellulose, such waste is a good source for renewable energy (Li et al. 2019).

Some examples of the use of industrial waste for the production of valuable products have been provided here. Cassava is widely known for its use in the production of bioethanol, which leads to the formation of a large number of cassava residues. Cassava residues can be further hydrolyzed to produce valuable products such as biosurfactants, biogas, organic acids, and more biofuel (Zhang et al. 2016). *Penicillium oxalicum* was used for the production of cellulase with high β-glucosidase potential, which could degrade cassava residues up to about 95%, and the product could be used in various processes to produce valuable products (Su et al. 2017). Apple pomace was used for the production of ethanol and acetic acid using cellulase, β-glucosidase, and pectinase (Parmar & Rupasinghe 2013). In one report, pulp and sludge from a paper mill were hydrolyzed using the commercial cellulase mixture Cellic® CTec2 and the hydrolysate was further treated with *S. cerevisiae* and *Cutaneotrichosporon oleaginosum* for bioethanol and biolipid production (Zambare & Christopher 2020).

### 4.4.3 Cellulase in the Valorization of Renewable Municipal Waste

The accumulation of biodegradable and solid municipal waste is a major problem in the environment. Landfilling with untreated municipal waste poses a great threat to the environment and human health. However, the usage of municipal solid waste (MSW) for the production of various valuable products is a much better step as large amounts of MSW is generated every day. The production of biofuels such as ethanol, biogas, and hydrogen using enzymes such as cellulase to hydrolyze cellulosic material in municipal waste has received much attention recently (Matsakas et al. 2017).

The increasing human population has been increasing the production of municipal waste worldwide. Biodegradable municipal waste includes large quantities of cellulosic materials such as paper, newspaper, food materials, household waste such as fruit and vegetable peel, and packing cardboard. About 30%–40% of municipal waste consists of paper waste (Byadgi and Kalburgi 2016). In one study, thermophilic cellulase was produced using the fungus *Myceliophthora thermophila* from municipal waste as a cellulose source. This cellulase thus formed was further used for ethanol production using municipal waste, which showed a threefold increase in ethanol yield compared to ethanol production without the addition of the previously produced cellulase enzyme (Matsakas & Christakopoulos 2015).

## 4.5 VALUABLE PRODUCTS FROM HYDROLYZED CELLULOSIC WASTE

Cellulosic and lignocellulosic wastes, when hydrolyzed, can produce some high-value products. Since they are cheap and abundant worldwide, such waste materials are cost-effective sources of sugar for industrial purposes. There are several products that can be made from the glucose obtained by hydrolysis of cellulosic waste. Some of them are detailed below.

### 4.5.1 Cellulase Enzyme

Cellulosic waste can be used for large-scale cellulase production. Cellulase has a high position in the world market of enzymes. It is an important enzyme required in many industries such as textile, paper, bioethanol, and biobutanol production. Therefore, it is necessary to produce cost-effective and efficient cellulase for proper usage of cellulosic waste in industries (Verma et al. 2021). Biological studies have shown that there are many species of bacteria, fungi, and actinomycetes that are good cellulase producers and can be used industrially. Waste can be used as a substrate to produce industrially important cellulase by using fermentation techniques employing cellulase-producing microorganisms. Cellulase production with such substrate is cost-effective as it reduces the cost of cellulosic raw material and reduces pollution caused by various measures of dumping unused agricultural wastes (Bhardwaj et al. 2021).

Cellulase and hemicellulose were produced with a natural *Aspergillus niger* P-19 variant using rice straw as a lignocellulosic source for use in ethanol production in a low-cost process (Kaur et al. 2020). *Bacillus licheniformis* 2D55 isolated from chicken manure compost (a thermophilic bacterial species) could produce cellulase enzymes such as CMCase, FPase, xylanase, and β-glucosidase by degrading a cocktail of agro-wastes such as sugarcane bagasse, rice husk, and rice straw (Kazeem et al. 2017). When *Trichoderma reesei* was used on *Litchi chinensis* husk and *Luffa cylindrica* wastes, they proved to be good sources of cellulase production, with *L. cylindrica* showing higher cellulase production ability in comparison (Verma et al. 2018). *Trichoderma harzanium* was used for cellulase production using waste surgical cotton and cardboard used for packaging as cellulose sources by a solid phase submersion method (Ramamoorthy et al. 2019). In a study, polyester and cotton textile wastes were used for cellulase production using *Aspergillus niger* CKB. Additionally, the cellulase produced was used to treat textile waste, resulting in the recovery of usable glucose and polyester (Hu et al. 2018). A study on the use of office paper waste for cellulase production was reported. In this experiment, *Bacillus velezensis* ASN1 was used for cellulase production by the fermentation technique (Nair et al. 2018). The use of packaging cardboard waste treated with 0.1% $H_2SO_4$ using *B. subtilis* S1 isolated from soil samples for cellulase production was reported in an experiment (Al Azkawi et al. 2018).

### 4.5.2 Biofuel

The continuous depletion of fossil fuel resources, rising demand for energy, and increasing environmental pollution have led to the need to search for alternative energy sources worldwide. In this context, biofuel is considered the fuel of the future. It is a form of renewable energy and can be used as a replacement for vigorously depleting

**TABLE 4.3**
**Some Cellulosic Substrates with High Potential for Biofuel Production**

| Cellulosic Substrate | Biofuel Product | Biofuel Yield | References |
|---|---|---|---|
| Newspaper | Bioethanol | 6.9% v/v | Byadgi and Kalburgi (2016) |
| Potato peel waste | | 0.46 gm/gm | Arapoglou et al. (2010) |
| | | 0.2 gm/gm | Hijosa-Valsero et al. (2018) |
| Cassava bagasse | | 0.53gm/gm | Aruwajoye et al. (2020) |
| Sugarcane waste | | 6.72% | Braide et al. (2016) |
| Maize plant waste | | 4.6% | Braide et al. (2016) |
| Rice straw | Bioethanol | 189gm/kg | Jin et al. (2020) |
| | | 87.4 gm/kg | Elsayed et al. (2018) |
| | Biogas | 372.4 L/kg | Elsayed et al. (2018) |
| | Biobutanol | 0.17 gm/gm | Tsai et al. (2020) |
| Soybean residue | Biogas | 560.47 ml | Onthong and Juntarachat (2017) |
| Papaya peel | | 404.24 ml | Onthong and Juntarachat (2017) |
| Sugarcane bagasse | | 263.04 ml | Onthong and Juntarachat (2017) |
| Rice straw | | 4.26 ml | Onthong and Juntarachat (2017) |
| Greater galangal | | 45.83 ml | Onthong and Juntarachat (2017) |
| Sugarcane bagasse | Biobutanol | 0.16 gm/gm | Tsai et al. (2020) |
| | Biohydrogen | 0.874 mol $H_2$/mol sugar | Reddy et al. (2017) |
| Coconut husk | Biohydrogen | 0.279 mol $H_2$/mol sugar | Muharja et al. (2018) |

and polluting fossil fuels. As the demand for biofuels has increased with time and need, the use of abundant cellulosic waste for biofuel production has attracted a lot of attention. Biofuels include bioethanol, biobutanol, biogas, and biohydrogen. Some examples of different cellulosic sources as potential substrates for biofuel production are listed in Table 4.3.

Bioethanol is the most popular biofuel and currently one of the most widely used biofuels for vehicles. Worldwide demand has been increasing rapidly. Currently, bioethanol is mainly produced using common agricultural waste substrates such as rice straw, wheat straw, corn straw, and sugarcane bagasse. Bioethanol produced from agricultural and agro-industrial wastes belongs to the second-generation bioethanol and undergoes multiple-step intermediate processes, including substrate pretreatment, cellulosic hydrolysis, and fermentation. Various microbial species such as those belonging to the bacterial genus *Clostridium, Bacillus*, and fungal genus *Trichoderma* (the most studied cellulase-producing fungal genus) are extensively studied for cellulase production to conduct the enzymatic hydrolysis of cellulose to ultimately produce bioethanol as an end product (Sarkar et al. 2012). Rice straw pretreated with alkali was used for bioethanol production by hydrolysis and fermentation with a mixture of cellulase enzyme and the fungus *Mucor circinelloides* (Takano & Hoshino 2018). Bioethanol production by hydrolysis of rice straw using a mixture of enzymes containing cellulase, xylanase, and other lignocellulose-degrading enzymes along with a suitable thermostable microbial strain all working in a simultaneous saccharification and fermentation process was reviewed by M. Hansa et al. (Hansa et al. 2019). Ethanol was produced by a single-pot bioprocessing method from municipal waste using cellulase with *Trichoderma reesei* and other enzymes with different microorganisms (Althuri & Mohan 2019).

The fuel properties of biobutanol are similar to those of gasoline and therefore biobutanol can be used as an alternative to gasoline as a fuel for running vehicles and is already used as a biofuel in many countries. Biobutanol can also be used in a blend with gasoline. Several agro-based cellulosic wastes such as corncob, sugarcane bagasse, rice straw, wheat straw, cassava waste, and barley straw have been found to be good substrates for biobutanol production (Huzir et al. 2018). Biobutanol has some advantages over bioethanol because it is insoluble in water and can produce more energy, although the production cost of biobutanol is higher than that of bioethanol. Due to the higher production cost, the commercial popularity of biobutanol production is low. Acetone and ethanol are common by-products of biobutanol production that increase the economy of production.

Biogas is an age-old, well-known waste vaporizing process that provides renewable energy. Biogas production to generate heat and power has been known by humans for hundreds of years. Biogas is produced by the anaerobic digestion of organic material by microorganisms. Cellulosic wastes, which are abundant worldwide, can be used as the primary substrate for large-scale biogas production. Biogas is composed of about 50–75% methane, 25–50% $CO_2$, and small proportions of other gases and water (Plugge 2017). Household kitchen wastes such as vegetable peels; agricultural wastes and residues such as straw, husks; and agro-industrial wastes such as pomace and bagasse are all considered good substrates for biogas production.

Biohydrogen is another renewable energy source that can be produced from cellulosic wastes. It is a biofuel that can be used as an alternative to fossil fuels and can be used as a fuel for vehicles and a source of green energy that is both environmentally friendly and sustainable. Various sources of cellulosic waste such as agricultural waste, industrial waste, and municipal waste can be effectively used as a substrate for biohydrogen production. Although biohydrogen has proved to be a great biofuel with high potential, a lot of research development work is still required for its effective large-scale production (Kamaraj et al. 2020).

### 4.5.3 Organic Acids

Various agricultural and food processing wastes can be hydrolyzed with cellulase and used for the production of valuable organic acids like acetic acid, citric acid, and tartaric acid. Acetic acid can be prepared by hydrolysis of mango-processing waste, pineapple-processing waste, citrus peels, and molasses. Citric acid is produced using pineapple peels, apple pomace, grape pomace, rice bran, wheat bran, and many other wastes from agro-food industries using microorganisms such as *Aspergillus* species. The production of tartaric acid from grape pomace, oxalic acid from apple pomace, and fumaric acid from cassava bagasse have also been reported by various researchers (Das & Singh 2004).

## 4.6 CONCLUSION

Cellulosic waste is an abundant and renewable sugar source that can be hydrolyzed by cellulase enzymes to form glucose units that can be further fermented by employing microorganisms or can be processed by several other techniques to ultimately produce value-added products like biofuels. With the depletion and increasing cost of fossil fuel, it is wise to start using biofuel as an alternative, provided proper technologies are introduced by researchers for the production of biofuels at a much cheaper production rate. The usage of cellulosic waste is not only a good gesture toward the environment, it is also a cost-effective source of cellulosic material for many industrial processes. So, it is important to conduct more research on the degradation of cellulosic waste using cellulase enzymes in a much effective way with maximum product generation and minimum waste by-product formation, for utilization of these wastes instead of discarding them at dumping sites. Though such waste usage practices have been taken up by many countries, currently they are less used in developing countries.

## ACKNOWLEDGMENTS

The authors want to thank the Department of Microbiology, Sikkim University for providing the computational infrastructure and central library facilities for procuring references.

## REFERENCES

Abbasi T, Tauseef S M, Abbasi S A. 2011. "Biogas and Biogas Energy: An Introduction." *Biogas Energy* 2: 1–10. Springer Science & Business Media.

Agrawal R, Semwal S, Kumar R, Mathur A, Gupta R P, Tuli D K, Satlewal A. 2018. "Synergistic Enzyme Cocktail to Enhance Hydrolysis of Steam Exploded Wheat Straw at Pilot Scale." *Frontiers in Energy Research* 6: 122.

Ai J, Yang S, Sun Y, Liu M, et al. 2021. "Corncob Cellulose-Derived Hierarchical Porous Carbon for High Performance Supercapacitors." *Journal of Power Sources* 484: 229221.

Michelin M, Ruiz H A, Silva D P da, Ruzene D S, et al. 2015. "Cellulose from Lignocellulosic Waste." In Ramawat K G, Mérillon J-M (eds.) *Polysaccharides - Bioactivity and Biotechnology*. Berlin: Springer International Publishing, 1–33.

Althuri A, Mohan S V. 2019. "Single Pot Bioprocessing for Ethanol Production from Biogenic Municipal Solid Waste." *Bioresource Technology* 283: 159–67.

Arapoglou D, Varzakas T, Vlyssides A, Israilides C. 2010. "Ethanol Production from Potato Peel Waste (PPW)." *Waste Management* 30(10): 1898–902.

Aruwajoye G S, Faloye F D, Kana E G. 2020. "Integrated Economic and Environmental Assessment of Biogas and Bioethanol Production from Cassava Cellulosic Waste." *Waste and Biomass Valorization* 11: 2409–20.

Asim M, Abdan K, Jawaid M, Nasir M. 2015. "A Review on Pineapple Leaves Fibre and Its Composites." *International Journal of Polymer Science* 2015: 1–16.

Al Azkawi A S, Sivakumar N, Al Bahry S. 2018. "Bioprocessing of Cardboard Waste for Cellulase Production." *Biomass Conversion and Biorefinery volume* 8: 597–606.

Béguin, P. 1990. "Molecular Biology of Cellulose Degradation." *Annual Review of Microbiology* 44(1): 219–48.

Béguin P, Aubert J P. 1994. "The Biological Degradation of Cellulose." *FEMS Microbiology Reviews* 13(1): 25–58.

Bhardwaj N, Kumar B, Agrawal, K. Verma, P. 2021 "Current perspective on Production and Applications of Microbial Cellulases: A Review." *Bioresources and Bioprocessing*, 8(1): 1–34.

Braide W, Kanu I A, Oranusi U S, Adeleye S A. 2016. "Production of Bioethanol from Agricultural Waste." *Journal of Fundamental and Applied Sciences* 8(2): 372–86.

Brigham, Christopher. 2018. "Biopolymers: Biodegradable Alternatives to Traditional Plastics." *Green Chemistry*: 753–70. https://doi.org/10.1016/B978-0-12-809270-5.00027-3.

Byadgi S A, Kalburgi P B. 2016. "Production of Bioethanol from Waste Newspaper." *Procedia Environmental Sciences* 35: 555–62.

Cabral M M S, Abud A K S, Silva C E F, Almeida R M R G. 2016. "Bioethanol Production from Coconut Husk Fiber." *Ciência Rural* 46(10): 1872–77.

Cavaco-Paulo A, Almeida L, Bishop D. 1996. "Effects of Agitation and Endoglucanase Pretreatment on the Hydrolysis of Cotton Fabrics by a Total Cellulase." *Textile Research Journal* 64: 287–94.

Chaturvedi, V, Verma, P. 2013. "An Overview of Key Pretreatment Processes Employed for Bioconversion of Lignocellulosic Biomass into Biofuels and Value Added Products." *Biotech* 3(5): 415–31.

Coelho C C S, Michelin M, Cerqueira M A, Gonçalves C, et al. 2018. "Cellulose Nanocrystals from Grape Pomace: Production, Properties and Cytotoxicity Assessment." *Carbohydrate Polymers* 192: 327–36.

Coughlan M P 1991. "Mechanisms of Cellulose Degradation by Fungi and Bacteria." *Animal Feed Science and Technology* 32(1–3): 77–100.

Coutinho P M, Henrissat B 1999. "The Modular Structure of Cellulases and Other Carbohydrate-Active Enzymes: An Integrated Database Approach." In *Genetics, Biochemistry and Ecology of Cellulose Degradation*. Tokyo, Japan: Uni Publishers Co., 15–23.

Demirbas A 2009a. "Biofuels Securing the Planet's Future Energy Needs." *Energy Conversion and Management* 50(9): 2239–49.

Demirbas A 2009b. *Biohydrogen*. Springer, London. https://doi.org/10.1007/978-1-84882-511-6_6.

Elsayed M, Abomohra A E F, Ai P, Wang D, et al. 2018. "Biorefining of Rice Straw by Sequential Fermentation and Anaerobic Digestion for Bioethanol and/or Biomethane Production: Comparison of Structural Properties and Energy Output." *Bioresource Technology* 268: 183–89.

Fortunati E, Benincasa P G, Balestra M, Luzi F, et al. 2016. "Revalorization of Barley Straw and Husk as Precursors for Cellulose Nanocrystals Extraction and Their Effect on PVA_CH Nanocomposites." *Industrial Crops and Products* 92: 201–17.

Guillén D, Sánchez S, Rodríguez-Sanoja R. 2010. "Carbohydrate-Binding Domains: Multiplicity of Biological Roles." *Applied Microbiology and Biotechnology* 85(5): 1241–49.

Das H, Singh S K. 2004. "Useful Byproducts from Cellulosic Wastes of Agriculture and Food Industry—A Critical Appraisal." *Critical Reviews in Food Science and Nutrition* 44(2): 77–89.

Haddadin M S Y, Haddadin J, Arabiyat O I, Hattar B. 2009. "Biological Conversion of Olive Pomace into Compost by Using *Trichoderma Harzianum* and *Phanerochaete Chrysosporium*." *Bioresource Technology* 100(20): 4773–82.

Hansa M, Kumar S, Chandel A K, Polikarpov I. 2019. "A Review on Bioprocessing of Paddy Straw to Ethanol Using Simultaneous Saccharification and Fermentation." *Process Biochemistry* 85: 125–34.

Harini K, Ramya K, Sukumar M. 2018. "Extraction of Nano Cellulose Fibers from the Banana Peel and Bract for Production of Acetyl and Lauroyl Cellulose." *Carbohydrate Polymers* 201: 329–39.

Henrissat B, Teeri T T, Warren R A. 1998. "A Scheme for Designating Enzymes That Hydrolyse the Polysaccharides in the Cell Walls of Plants." *FEBS Letters* 425(2): 352–54.

Henrissat B 1991. "A Classification of Glycosyl Hydrolases Based on Amino Acid Sequence Similarities." *Biochemical Journal* 280(2): 309–16.

Henrissat B, Bairoch A. 1996. "Updating the Sequence-Based Classification of Glycosyl Hydrolases." *Biochemical Journal* 316(2): 695–96.

Hijosa-Valsero M, Paniagua-García A I, Díez-Antolínez R. 2018. "Industrial Potato Peel as a Feedstock for Biobutanol Production." *New Biotechnology* 46: 54–60.

Hilden L, Johansson G. 2004. "Recent Developments on Cellulases and Carbohydrate-Binding Modules with Cellulose Affinity." *Biotechnology Letters* 26(22): 1683–93.

Hu Y, Du C, Pensupa N, Lin C S K. 2018. "Optimisation of Fungal Cellulase Production from Textile Waste Using Experimental Design." *Process Safety and Environmental Protection* 118: 133–42. https://doi.org/10.1016/j.psep.2018.06.009.

Huzir N M, Aziz M M A, Ismail S B, Abdullah B, Mahmood N A N, et al. 2018. "Agro-Industrial Waste to Biobutanol Production: Eco-Friendly Biofuels for next Generation." *Renewable and Sustainable Energy Reviews* 94: 476–85.

Jain L, Kurmi A K, Kumar A, Narani A, Bhaskara T, Agrawal D. 2020. "Exploring the Flexibility of Cellulase Cocktail Obtained from Mutant UV-8 of *Talaromyces Verruculosus* IIPC 324 in Depolymerising Multiple Agro-Industrial Lignocellulosic Feedstocks." *International Journal of Biological Macromolecules* 154: 538–44. https://doi.org/10.1016/j.ijbiomac.2020.03.133.

Jiang F, Hsieh Y L. 2015. "Cellulose Nanocrystal Isolation from Tomato Peels and Assembled Nanofibers." *Carbohydrate Polymers* 122: 60–68.

Jin X, Song J, Liu G Q. 2020. "Bioethanol Production from Rice Straw through an Enzymatic Route Mediated by Enzymes Developed In-House from *Aspergillus fumigatus*." *Energy* 190: 116395.

Johar N, Ahmad I, Dufresne A. 2012. "Extraction, Preparation and Characterization of Cellulose Fibres and Nanocrystals from Rice Husk." *Industrial Crops and Products* 37(1): 93–99.

Kamaraj M, Ramachandran K K, Aravind J. 2020. "Biohydrogen Production from Waste Materials: Benefits and Challenges." *International Journal of Environmental Science and Technology* 17(1): 559–76.

Kaur J, Chugh P, Soni R, Soni S K. 2020. "A Low-Cost Approach for the Generation of Enhanced Sugars and Ethanol from Rice Straw Using in-House Produced Cellulase-Hemicellulase Consortium from A. Niger P-19." *Bioresource Technology Reports* 11: 100469.

Kazeem M O, Md Shah U K, Baharuddin A S, AbdulRahman N A. 2017. "Prospecting Agro-Waste Cocktail: Supplementation for Cellulase Production by a Newly Isolated Thermophilic *B. licheniformis* 2D55." *Applied Biochemistry and Biotechnology* 182: 1318–40.

Koshland Jr D E. 1953. "Stereochemistry and the Mechanism of Enzymatic Reactions." *Biological Reviews* 28(4): 416–36.

Kumar A, Asthana M, Jain K G, Singh V. 2016. "5. Microbial Production of Enzymes: An Overview." In Kumar Gupta, V, Zeilinger, S, Ferreira Filho, E, Carmen Durán-Dominguez-de-Bazu, M, Purchase, D. (eds.) *Microbial Applications: Recent Advancements and Future Developments*. Berlin, Boston: De Gruyter, 107–38.

Kumar B, Bhardwaj N, Agrawal K, Chaturvedi V, Verma P. 2020a. "Current Perspective on Pretreatment Technologies Using Lignocellulosic Biomass: An Emerging Biorefinery Concept." *Fuel Processing Technology* 199: 106244.

Kumar B, Bhardwaj N, Agrawal K, Verma P. 2020b. "Bioethanol Production: Generation-Based Comparative Status Measurements." In Srivastava N, Srivastava N, Mishra P K, Kumar Gupta V (eds.) *Biofuel Production Technologies: Critical Analysis for Sustainability*. Singapore: Springer, 155–201.

Kumar B, Bhardwaj N, Alam A, Agrawal K, Prasad H, Verma P. 2018. "Production, Purification and Characterization of an Acid/Alkali and Thermo Tolerant Cellulase from *Schizophyllum commune* NAIMCC-F-03379 and Its Application in Hydrolysis of Lignocellulosic Wastes." *AMB Express* 8(1): 1–16.

Kumar B, Verma P. 2021 "Biomass-based Biorefineries: An Important Architype Towards a Circular Economy." *Fuel* 288: 119622. Elsevier.

Lavanya D, Kulkarni P K, Dixit M, Raavi P K, Krishna L N V. 2011. "Sources of Cellulose and Their Applications—A Review." *International Journal of Drug Formulation and Research* 2(6): 19–38.

Leão R M, Miléo P C, João Maia M L L, Luz S M. 2017. "Environmental and Technical Feasibility of Cellulose Nanocrystal Manufacturing from Sugarcane Bagasse." *Carbohydrate Polymers* 175: 518–29.

Lee S M, Koo Y M. 2001. "Pilot-Scale Production of Cellulase Using Trichoderma Reesei Rut C-30 Fed-Batch Mode." *Journal of Microbiology and Biotechnology* 11(2): 229–33.

Li X, Hu Y, Du C, Lin C S K. 2019. "Recovery of Glucose and Polyester from Textile Waste by Enzymatic Hydrolysis." *Waste and Biomass Valorization* 10(12): 3763–72.

Li X H, Yang H J, Roy B, Wang D, et al. 2009. "The Most Stirring Technology in Future: Cellulase Enzyme and Biomass Utilization." *African Journal of Biotechnology* 8(11).

Liu R, Yu H, Huang Y. 2005. "Structure and Morphology of Cellulose in Wheat Straw." *Cellulose* 12: 25–34.

Liu Z, Li L, Liu C, Xu A. 2018. "Pretreatment of Corn Straw Using the Alkaline Solution of Ionic Liquids." *Bioresource Technology* 260: 417–20. https://doi.org/10.1016/j.biortech.2018.03.117.

Liu Z, Li X, Xie W, Deng H. 2017. "Extraction, Isolation and Characterization of Nanocrystalline Cellulose from Industrial Kelp (*Laminaria japonica*) Waste." *Carbohydrate Polymers* 173: 353–59.

Lu P, Hsieh Y L. 2012. "Preparation and Characterization of Cellulose Nanocrystals from Rice Straw." *Carbohydrate Polymers* 87(1): 564–73.

Madureira A R, Atatoprak T, Çabuk D, Sousa F, et al. 2018. "Extraction and Characterisation of Cellulose Nanocrystals from Pineapple Peel." *International Journal of Food Studies* 7(1): 24–33.

Maitan-Alfenas G P, Visser E M, Guimarães V M. 2015. "Enzymatic Hydrolysis of Lignocellulosic Biomass: Converting Food Waste in Valuable Products." *Current Opinion in Food Science* 1: 44–49.

Mandels M, Sternberg D. 1976. "Recent Advances in Cellulase Technology." *Hakko Kogaku Zasshi (Japan)* 54(4): 267–86.

Marchessault R H, Sundararajan P R. 1983. "Cellulose." In *The Polysaccharides*. Academic Press, 11–95. https://doi.org/10.1016/B978-0-12-065602-8.50007-8.

Matsakas L, Gao Q, Jansson S, Rovaa U, Christakopoulos P. 2017. "Green Conversion of Municipal Solid Wastes into Fuels and Chemicals." *Electronic Journal of Biotechnology* 26: 69–83.

Matsakas L, Christakopoulos P. 2015. "Ethanol Production from Enzymatically Treated Dried Food Waste Using Enzymes Produced On-Site." *Sustainability* 7(2): 1446–58. https://doi.org/10.3390/su7021446.

Melikoğlu A Y, Bilek S E, Cesur S. 2019. "Optimum Alkaline Treatment Parameters for the Extraction of Cellulose and Production of Cellulose Nanocrystals from Apple Pomace." *Carbohydrate Polymers* 215: 330–37.

Merci A, Marim R G, Urbano A, Mali S. 2019. "Films Based on Cassava Starch Reinforced with Soybean Hulls or Microcrystalline Cellulose from Soybean Hulls." *Food Packaging and Shelf Life* 20: 100321.

Morana A, Maurelli L, Ionata E, La Cara F, Rossi M. 2011. "Cellulases from Fungi and Bacteria and Their Biotechnological Applications." In Adam E. Golan (ed.) *Cellulase: Types and Action, Mechanisms and Uses*. New York (US): Nova Science Publisher, Inc, 1e79.

Muharja M, Junianti F, Ranggina D, Nurtono T. 2018. "An Integrated Green Process: Subcritical Water, Enzymatic Hydrolysis, and Fermentation, for Biohydrogen Production from Coconut Husk." *Bioresource Technology* 249: 268–75.

Nair A S, Al-Battashi H, Al-Akzawi A, Annamalai N, et al. 2018. "Waste Office Paper: A Potential Feedstock for Cellulase Production by a Novel Strain *Bacillus velezensis* ASN1." *Waste Management* 79: 491–500. https://doi.org/10.1016/j.wasman.2018.08.014.

Ng H S, Kee P E, Yim H S, Chen P T, Wei Y H, Lan J C W. 2020. "Recent Advances on the Sustainable Approaches for Conversion and Reutilization of Food Wastes to Valuable Bioproducts." *Bioresource Technology* 302: 122889. https://doi.org/10.1016/j.biortech.2020.122889.

O'Sullivan A C. 1997. "Cellulose: The Structure Slowly Unravels." *Cellulose* 4(3): 173–207.

Onthong U, Juntarachat N. 2017. "Evaluation of Biogas Production Potential from Raw and Processed Agricultural Wastes." *Energy Procedia* 138: 205–10.

Parmar I, Rupasinghe H P V. 2013. "Bio-Conversion of Apple Pomace into Ethanol and Acetic Acid: Enzymatic Hydrolysis and Fermentation." *Bioresource Technology* 130: 613–20. https://doi.org/10.1016/j.biortech.2012.12.084.

Plugge C M. 2017. "Biogas." *Microbial Biotechnology* 10(5): 1128–30. https://doi.org/10.1111/1751-7915.12854.

Prasanna N S, Mitra J. 2020. "Isolation and Characterization of Cellulose Nanocrystals from Cucumis Sativus Peels." *Carbohydrate Polymers* 247: 116706.

Ramamoorthy N K, Sambavi T R, Renganathan S. 2019. "A Study on Cellulase Production from a Mixture of Lignocellulosic Wastes." *Process Biochemistry* 83: 148–58. https://doi.org/10.1016/j.procbio.2019.05.006.

Reddy K, Nasr M, Kumari S, Kumar S, et al. 2017. "Biohydrogen Production from Sugarcane Bagasse Hydrolysate: Effects of PH, S/X, Fe2+, and Magnetite Nanoparticles." *Environmental Science and Pollution Research* 24(9): 8790–804.

Reddy N, Yang Y. 2015. "Fibers from Banana Pseudo-Stems." *Innovative Biofibers from Renewable Resources*: 25–27. https://doi.org/10.1007/978-3-662-45136-6_7

Saranraj P, Stella D, Reetha D. 2012. "Microbial Cellulases and Its Applications: A Review." *International Journal of Biochemistry & Biotech Science* 1: 1–12.

Sarkar N, Ghosh S K, Bannerjee S, Aikat K. 2012. "Bioethanol Production from Agricultural Wastes: An Overview." *Renewable Energy* 37(1): 19–27.

Sethi S, Datta A, Gupta B L, Gupta S. 2013. "Optimization of Cellulase Production from Bacteria Isolated from Soil." *ISRN Biotechnology* 2013: 1–7.

Shen D, Xiao R, Gu S, Zhang H. 2013. "The Overview of Thermal Decomposition of Cellulose in Lignocellulosic Biomass." In Theo van de Ven and John Kadla (eds.), *Cellulose-Biomass Conversion*, 193–226. https://doi.org/10.5772/51883

Shoham Y, Lamed R, Bayer E A. 1999. "The Cellulosome Concept as an Efficient Microbial Strategy for the Degradation of Insoluble Polysaccharides." *Trends in Microbiology* 7(7): 275–81.

Singh R S, Singh T, Pandey A. 2019. "Microbial Enzymes—An Overview." *Advances in Enzyme Technology*: 1–40. https://doi.org/10.1016/B978-0-444-64114-4.00001-7.

Su L H, Zhao S, Jiang S X, Liao Z, Duan C J, Feng J X. 2017. "Cellulase with High β-Glucosidase Activity by *Penicillium oxalicum* under Solid State Fermentation and Its Use in Hydrolysis of Cassava Residue." *World Journal of Microbiology and Biotechnology* 33(Article no. 37). https://doi.org/10.1007/s11274-016-2200-7.

Sukumaran R K, Singhania R R, Pandey A. 2005. "Microbial Cellulases-Production, Applications and Challenges." *Journal of Scientific and Industrial Research* 64(11): 832–44. http://hdl.handle.net/123456789/5375.

Takano M, Hoshino K. 2018. "Bioethanol Production from Rice Straw by Simultaneous Saccharification and Fermentation with Statistical Optimized Cellulase Cocktail and Fermenting Fungus." *Bioresources and Bioprocessing* 5(1): 1–12.

Trilokesh C, Uppuluri K B. 2019. "Isolation and Characterization of Cellulose Nanocrystals from Jackfruit Peel." *Scientific Reports* 9(1): 1–8.

Tsai T Y, Lo Y C, Dong C D, Nagarajan D. 2020. "Biobutanol Production from Lignocellulosic Biomass Using Immobilized *Clostridium acetobutylicum*." *Applied Energy* 277: 115531.

Veiga J P S, Valle T L, Feltran J C, Bizzo W A. 2016. "Characterization and Productivity of Cassava Waste and Its Use as an Energy Source." *Renewable Energy* 93: 691–99.

Verma N, Kumar V, Bansal M C. 2018. "Utility of *Luffa cylindrica* and *Litchi chinensis* Peel, an Agricultural Waste Biomass in Cellulase Production by *Trichoderma reesei* under Solid State Cultivation." *Biocatalysis and Agricultural Biotechnology* 16: 483–92. https://doi.org/10.1016/j.bcab.2018.09.021.

Verma N, Kumar V, Bansal M C. 2021. "Valorization of Waste Biomass in Fermentative Production of Cellulases: A Review." *Waste and Biomass Valorization* 12: 613–40.

Ververis C, Georghiou K, Danielidis D, et al. 2007. "Cellulose, Hemicelluloses, Lignin and Ash Content of Some Organic Materials and Their Suitability for Use as Paper Pulp Supplements." *Bioresource Technology* 98(2): 296–301.

Vocadlo D J, Davies G J. 2008. "Mechanistic Insights into Glycosidase Chemistry." *Current Opinion in Chemical Biology* 12(6): 539–55.

Walker L P, Wilson D B. 1991. "Enzymatic Hydrolysis of Cellulose: An Overview." *Bioresource Technology* 36(1): 3–14.

Zambare V P, Christopher L P. 2020. "Integrated Biorefinery Approach to Utilization of Pulp and Paper Mill Sludge for Value-Added Products." *Journal of Cleaner Production* 274: 122791.

Zhang M, Xie L, Yin Z, Khanal S K, Zhou Q. 2016. "Biorefinery Approach for Cassava-Based Industrial Wastes: Current Status and Opportunities." *Bioresource Technology* 215: 50–62.

# 5 Cellulosic Bioethanol Production from Liquid Wastes Using Enzymatic Valorization

*Surinder Singh, Diksha, Mohd. Aseel Rizwan, and S. K. Kansal*
Panjab University, Chandigarh, India

*S. Suresh*
Maulana Azad National Institute of Technology, Bhopal, India

*Mamta Bhagat*
Deenbandhu Chottu Ram University of Science and Technology, India

*Sarika Verma*
CSIR-Advanced Materials and Processes Research Institute, Hoshangabad Road, Bhopal, India

*S. Arisutha*
Energy Centre, Maulana Azad National Institute of Technology & Eco Science and Technology, India

## CONTENTS



DOI: 10.1201/9781003187684-5

## 5.1 INTRODUCTION

The quest for clean energy to fulfill global energy demand is prompted by the rise in living standards.Conversely, the reliance on fossil fuels as the primary source of energy has resulted in global issues such as pollution and rising temperatures (Hoekman, 2009). As a result of this, the nation, industry, and transport sector have discovered ecologically sound, sustainable, and clean renewable energy. Liquid biofuels have been emphasized among alternative fuels since they account for roughly 40% of global energy demand (Azhar et al., 2017).

Agricultural crops such as sugarcane and rapeseed oil are now used to make biofuels like bioethanol and biodiesel. These biofuels produced from food materials are collectively referred to as first-generation biofuels (Kumar et al., 2020a). Albeit first-generation biofuels could possibly supplant petroleum derivatives as the fundamental source of energy supply, their creation is beset by specific issues like the annihilation of tropical forest areas. In lieu, second-generation bioethanol, which uses non-palatable sources such as lignocellulose biomass to deliver ethanol, has been demonstrated to be more acceptable as the source of sustainable energy. Bioethanol and biodiesel are two lignocellulosic-based liquid mobility fuels on the market that might eventually replace petroleum and diesel. This chapter deals with bioethanol (Tan et al., 2008; Demirbas, 2009).

Biofuels made from inexhaustible raw material have attracted impressive scientific research consideration since they may be utilized to provide energy and elective substitute fuels. Bioethanol is one of the most intriguing biofuels because it is climate-friendly and is the primary fuel utilized as a petroleum alternative for street transport vehicles (Kumar et al., 2020b). Ethanol has the same molecular formula whether or not it is created from carbohydrate-based raw material like maize (as it is fundamentally in the USA), sugar cane (as it is in Brazil), or from cellulosic raw materials, (for example, wood chips and crop deposits). The USA and Brazil use ethanol or ethyl liquor ($C_2H_5OH$) which is a clear colored, transparent liquid that is biodegradable, low in toxicity, and causes minimal natural contamination if spilled. Ethanol consumption creates carbon dioxide and water. Ethanol is a supercharged fuel and has supplanted lead as an octane enhancer in petroleum. By mixing ethanol with gas, we can likewise oxygenate the fuel blend so it burns more completely and minimizes polluting emanations (Bušić et al., 2018; EFB, 2021). Figure 5.1 represents the global

**GLOBAL ETHANOL PRODUCTION BY COUNTRY OR REGION (MILL. GALLONS)**

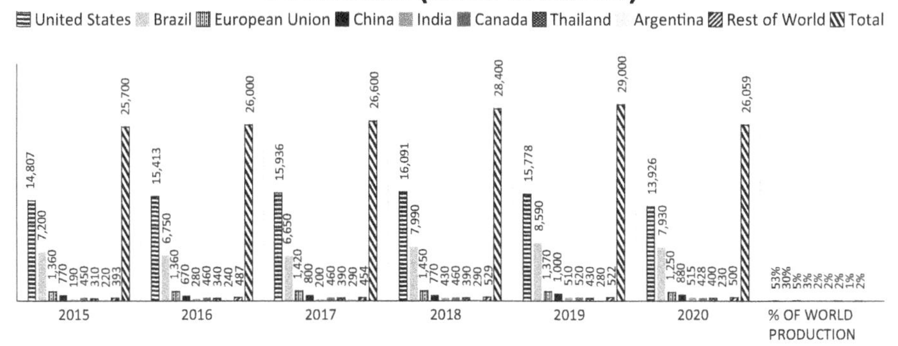


**FIGURE 5.1** Global ethanol production by country or region. [Source: Data from Renewable Fuel Association (RFA, 2021)].

production of ethanol in the past five years showing that the USA is the world's largest producer of ethanol, having produced over 15.7 billion gallons in 2019. Together, the USA and Brazil produce 83% of the world's ethanol (M & D, 2021), and from 2020 to 2027, the worldwide ethanol market is expected to grow at a compound annual growth rate (CAGR) of 4.8% (GVR, 2020).

Corn-to-ethanol, sugarcane-to-ethanol, simple and integrated lignocellulosic biomass-to-ethanol are some of the techniques used to make bioethanol. After numerous pretreatment stages to isolate hemicellulose and lignin from cellulose, and hydrolysis of cellulose to get fermentable sugars, the raw ingredients reach the fermentation stage (fermentation to convert sugars into ethanol,) followed by distillation to separate and purify, where bioethanol is generated. These are the several steps involved in bioethanol production (Anyanwu et al., 2018; Dimian et al., 2014).

The kind of feedstock utilized has an impact on the ethanol manufacturing process. Lignocellulosic biomasses have a complicated structure that may be subdivided into fermentable sugars either enzymatically or chemically using sulfuric or some other acids.

Acid hydrolysis is a nongreen technique that is hampered by the neutralization stages and the creation of inhibitory pollutants. Enzymatic hydrolysis, on the other hand, is a sustainable process that may be beneficial owing to the selectivity of the enzymes (Haldar et al., 2016; Kuila et al., 2017; Kumar and Verma, 2020). Technical feasibility of biomass to biofuels conversion and kinetic modeling of ethanol production has been reported in the literature by Suresh et al., 2018a and 2018b.

It is considered as an essential phase in the conversion of biomass cellulose to glucose, which is performed out by cellulase enzymes within moderate operation conditions (pH 4.8–5.0 and temperature 45–50°C). These conditions use less energy and do not result in the production of toxins or rusting of equipment. Enzymatic hydrolysis, in general, produces large yields of sugars (80–95%) and is ecologically beneficial. Optimal circumstances, such as temperature, time, pH, enzyme loading, and substrate concentration, determine the effectiveness of enzymatic hydrolysis (Bhardwaj et al., 2021; Zeghlouli et al., 2021).

## 5.2 ABOUT LIQUID WASTES AND DIFFERENT ENZYMES

Liquid wastes are described as liquids such as fats, grease or oils, wastewater, sewage sludges, and hazardous household liquids. Fluid wastes consist of domestic or sewage wastewater produced by industrial activities (Goswami et al., 2021). Diverse classes of fluid wastes are described here in detail.

### 5.2.1 Sewage Sludge

Sewage sludge is a semisolid, mud-like slurry that develops after receiving human and other waste from industries and households. Sewage sludges incorporate a comprehensive collection of organic and inorganic compounds. The organic matter in the sewage water includes pesticides, detergents, grease, oil, fat solvents, colourings, and phenols. Generally, two types of sewage sludge are seen: primary and secondary. Figure 5.2 shows a photographic view of sewage sludge. Primary sludge is the mixture of floating grease and solid wastes, whereas secondary sludge, also known as activated sludge, holds the suspended solids and primarily microbial cells (Mondala et al., 2009). Municipal water management has evaluated the amount of sewage water generated in cities or towns, with a population above 50,000 is $3.8254 \times 10^4$ million liters/day (Kamyotra and Bhardwaj, 2011). The segregation of lipids from the sludge results in the productive resource of low-cost feedstock for the production of bioethanol or biodiesel. It is the feasible substitute for sludge management to employ liquid waste as a renewable resource for biofuel production.

### 5.2.2 Animal Dung

Animal manure is the most extensively distributed agro-waste, which is a classical lignocellulosic material. Animals like cows, buffalos, pigs, and so on, discharge manure which comprises undigested and digested organic material, and microorganisms.

**FIGURE 5.2** Sewage sludge for biofuel production.

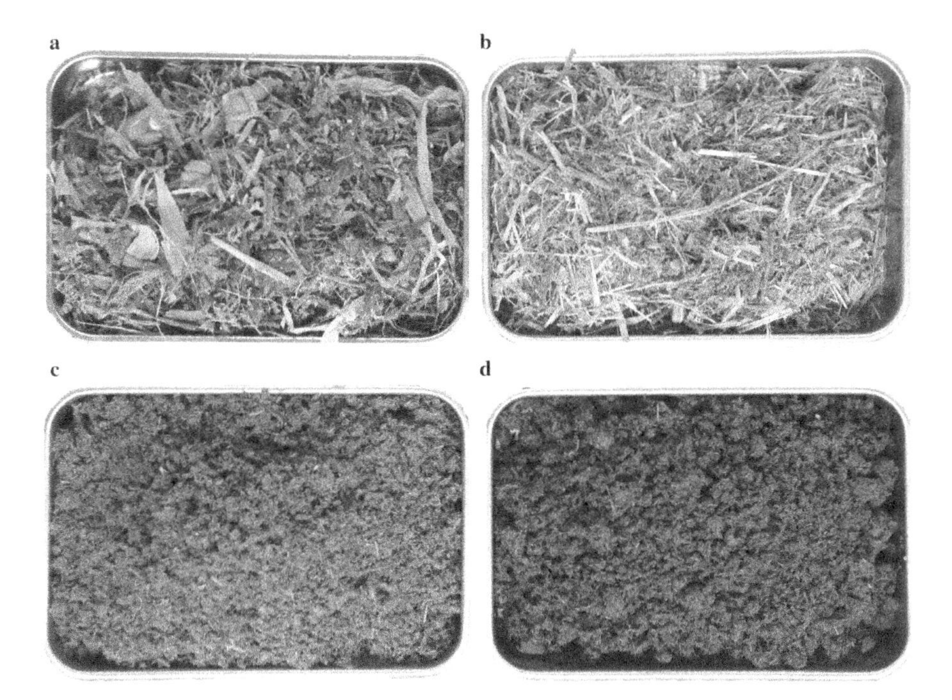


**FIGURE 5.3** Images for corn silage diets and cow manure and anaerobically digested cow manure. Corn silage diets (a), the coarsely milled corn silage diets supplemented with additional protein feed (b), solid cow manure materials after removing the liquid part by mechanical packing (c), anaerobically digested cow manure obtained from the physical separation of slurry after anaerobic digestion process for biogas production (d).

Generally, cow dung is directly introduced as a fertilizer in the soil because it has a large number of macronutrients that are helpful in plant growth. Figure 5.3 shows corn silage diets, cow manure, and anaerobically digested cow manure. But currently, this manure is used to produce methane-rich biogas. Cow dung contains a fraction of lignocellulosic biomass, which is an alternative for producing bioethanol (Yan et al., 2018).

### 5.2.3 Kitchen Food Waste

Kitchen waste is extensively obtained organic matter. It is degradable along with microbial infestation and mainly consists of starch and lignocellulose. An example is rice straw that is a primary substrate used to produce ethanol (Yano et al., 2009). It possesses a large amount of cellulose, protein, and lignin, which exhibits its potential for bioethanol production. Food waste includes vegetable oils, which are also used for the production of biofuels, for example, olive oil which is rich in lignin and cellulose. Bioethanol extracted from kitchen waste is an attractive and sustainable renewable energy source, used in vehicles to replace gasoline.

## 5.3 CELLULOSIC BIOMETHANOL PRODUCTION

Fossil fuels are the foremost requirement in the development of the industrial sector and growth of motorization which has resulted in their depletion. Sustainable methods and economically viable sources are primary needs to replace fossil fuels. This is not only to enhance the production of fuel, but also to mitigate the impact of rising fuel prices worldwide, and reduce greenhouse emissions during the entire cycle by adopting conventional sources to produce biofuel like biomethanol (Sahota et al., 2016; Suresh et al., 2018a & 2018b). Methanol has been produced through thermochemical processes, for instance the auto thermal gasification process, using fossil fuels such as natural gas, coal, and oil products (naphtha, gasoline), but due to increases in greenhouse gas emissions and the continuous depletionof fossil fuels, the focus has been shifted towards utilizing biochemical methods. The economic crisis and the increase in waste generation not only stimulated interest in the production of biomethanol, but also how efficiently the process can minimize greenhouse emissions has become a prime target. On the other hand, an increase in municipal and industrial wastes has become an issue of concern requiring a sustainable and stable solution for waste valorization (Iaquaniello et al., 2018).

Ethanol from biomass plays a crucial role in the transportation sector; it not only has a high octane number which can be mixed with petroleum without any modifications in engine designs, but it also has a low cetane number which is beneficial for diesel engines. Bioethanol can be efficiently produced from carbonylation of biomethanol (Marie-Rose et al., 2011). Biomethanol is referred to as renewable methanol since the methanol derived from fossil fuels using thermochemical methods is identical to the one which is derived from conventional methods (IRENA, IEA-ETSAP, 2013). Biomass is primarily produced from carbon-based compounds including biogas from landfills, waste streams, sewage, black liquor from the paper and pulp industry, bagasse from the sugarcane industry, wood, grass straws, edible feedstocks like corn (maize), and wheat. Biomass is the raw material that is the prime source of carbon or lignocellulose or starch-based feedstock. Biomethanol from thermochemical methods has a similar process which is used for production from fossil fuels, including coal.

$$2H_2 + CO \rightarrow CH_3OH \text{ water gas shift reaction} \quad (5.1)$$

In this chemical process, syngas is produced using water gas shift reaction and the required chemical ratio of $H_2/CO$ should ideally be two, but this process needs a sufficient amount of $H_2$ to convert the biomass carbon into biomethanol and it requires extra production costs, which increases the biomethanol price to above that of fossil fuels (Demirbas, 2008). Although the emissions from this method have been reduced to 25–40% as compared withfossil fuels, since the process is similar, it still releases gases like $CO_2$ into the atmosphere, resulting in the greenhouse effect (IRENA, IEA-ETSAP, 2013).

Biomass is a potential fuel but the method of technology used can alter the efficiency of the production of biomethanol. Progressive and efficient results have been obtained using the biological method. Globally the production of methanol is about

45 million tonnes per year from fossil fuels, mainly natural gas and coal (IRENA, IEA-ETSAP, 2013). It implies that to fulfill the requirements for fuel worldwide in larger proportions, biomass in abundance is required to produce biofuel. Remains from agricultural waste, the food industry, and municipal waste are often used in biorefineries worldwide (Agrawal et al., 2020). The feedstock used in the first generation mainly consists of starch-rich crops including corn, potato, rice, sugar cane, and wheat, which are part of the food chain. The misuse of these starch-rich crops in large amounts not only has adverse effects on theland and on water services, but impacts the ecological system as well. This initiated the development of second-generation biomass from solid waste, crop residues, cotton linters, black liquor from paper and pulp industries, sludge, livestock manure, and energy crops including switchgrass and miscanthus (Nair et al., 2017).

The use of microorganisms and enzymatic hydrolysis to produce bioethanol is an ancient art but these are now referred to as biorefineries or biochemical factories (Kumar and Verma, 2021). Biomethanol from biomass is a more beneficial biofuel than from fossil fuels because it has low pollution emissions. In the process, feedstock including raw grass-like *Salix schwerinii*, *Panicum virgatum* reduces the emissions by up to 60% but requires pretreatment for the breakdown of the rigid structure of lignocellulose (Chen et al., 2021). The most promising and potential source for the production of biofuels is cellulosic biomass. It reduces carbon emissions and does not affect the ecological system. It is a renewable source as well as less expensive than first generation feedstock including starch and sugar, moreoverit is abundantly available in earth's biosphere. Cellulose is majorly present in agricultural waste and industrial waste. Livestock manure and sludge contain polysaccharides, proteins, and other organic materials that have a good proportion of cellulosic biomass (Champagne, 2008). Cellulose has a complex structure in close linkage with hemicellulose and lignin. It is a long chain of glucose units that are linked by β-1, 4 glucosidic linkages, which is the available polymer for renewable energy generation (Mishra et al., 2018). Lignocellulosic biomass is composed of cellulose (40–50%), hemicellulose (25–30%), and lignin (15–20% and a small amount of protein, pectin, and ash (Chaturvedi and Verma, 2013; Baruah et al., 2018). Lignin is a three-dimensional, cross-linked, phenolic polymer located in the internal wall and attaches to the hemicellulose with covalent linkages and covalent cross-linkages providing the plant's strength and resistance to microbial degradation (Chen et al., 2021). It has a complex hierarchical and recalcitrant nature which resists breakdown (Baruah et al., 2018). In order to obtain fermentable sugars including glucose and xylose from cellulosic biomass, it needs pretreatment. Methods used for delignification of cellulosic biomass into glucose are chemical and enzymatic hydrolysis which converts cellulose to glucose (Champagne, 2008). Although chemically used biological degradation of cellulose requires a few chemicals, nonetheless later in the process it needs detoxification before fermentation (Khan et al., 2016). The use of the enzymatic action of enzymes including cellulase, proteases, and amylases have shown environmentally friendly results.

Cellulose possesses advantageous properties including biocompatibility, stereo regularity, hydrophilicity, and reactive hydroxyl groups and it also serves as a renewable resource for fuels (Baruah et al., 2018). Cellulose is a main structural and integral part of lignocellulosic biomass which is a linear polysaccharide polymer and consists of D-glucose subunits linked by β-(1, 4)-glycosidic bonds (Baruah et al., 2018). Cellulase is a multiple enzyme system consisting of endo-1, 4- β -d-glucanases and exo-1, 4- β -d-glucanases along with cellobiase (β -d-glucosideglucano hydrolase) (Khan et al., 2016). Cellulase is responsible for the hydrolysis of the β-1, 4-glucosidic bonds in cellulose and it is an inducible enzyme produced from a variety of microorganisms including bacteria, actinobacteria, and yeasts. Hemicellulose is the second major component of lignocellulosic biomass that consists of short chains of different polysaccharides such as xylan, galactomannan, glucuronoxylan, arabinoxylan, glucomannan, and xyloglucan that are held together by β-(1,4)- and/or β-(1,3)-glycosidic bonds. In contrast to cellulose, hemicellulose is readily degradable into monosaccharides due to its low degree of polymerization and noncrystalline nature (Baruah et al., 2018). Mostly, the species of aspergillus produces cellulase under many variations in the conditions, that is why this particular species has great dominance in the process. The majority of enzymes are produced by microorganisms, whereas some are soil bacteria belonging to genera of *Bacillus*, *Paenibacillus*, *Streptomyces*, *Hymenbacter* (Mihajlovski et al., 2020). Among these, members of *Hymenobacter* CKS3 genera have great potential in this field. When bioethanol is produced by incorporating *Hymenobacter* CKS3 strain in the hydrolysis of waste bread, it produces amylases, cellulases, and pectinases and all these enzymes significantly contribute to the release of sugars from waste bread, which is further fermented to produce bioethanol. Results from utilizing *Hymenobacter* give a higher concentration of bioethanol (Mihajlovski et al., 2020). However,there is much more to discover in this field that can lead towards fulfilling the potential of *Hymenobacter* for the production of bioethanol.

There are various biological technologies for lignocellulosic biomass which are already in use to optimize the production of bioethanol. For the production of bioethanol, the following scheme is mostly used by various technologies:

* Pretreatment for enzyme hydrolysis
* Enzyme hydrolysis for the conversion of cellulose to sugars.
* Fermentation using microorganisms to produce bioethanol.
* Distillation of bioethanol using multistage distillation units.

### 5.3.1 Pretreatment of Lignocellulosic Biomass and Feedstock Preparation

The primary aim of the pretreatment process is to reduce the recalcitration of the structure of lignocellulose which results in breakage and disruption of lignin sheath of cellulose. There are several factors which affect the pretreatment including the crystalline structure of cellulose, the degree of lignification, and the structural heterogeneity and complexity of cell wall constituents which is the cause of its recalcitrance, although it completely depends upon the feedstock which is the source of the cellulose (Baruah et al., 2018; Suresh and Sudhakar, 2013; Suresh et al., 2017). This is the initial as well as the costliest stage as it consumes a large amount of energy for the cleaning, milling, and grinding of the biomass. In the preparation of feedstock, biomass material is firstly cleaned out if it includes wood, raw grass straw, stover, or agricultural waste to remove the impurities present. Following the proper cleaning of the biomass, it is used for the pretreatment which is an important step for the cellulose conversion process. The foremost purpose of pretreatment is to convert biomass into cellulose, hemicellulose, and lignin (Amin et al., 2017; Yu et al., 2018). Pretreatment maximizes the production of fermentable sugars, and reducing the amount of enzyme inhibitors formed during pretreatment is still a challenge (Kumari and Singh, 2018). Pretreatment brings various changes in the feedstock including physical as well as chemical. Methods used can be physical, chemical, mechanical, physicochemical, microaerobic, and biological and this also depends upon the feedstock used (Aftab et al., 2019). Physical methods include screening, grinding, milling, chipping, freezing, ultrasonic treatment, and different types of irradiation which are utilized for the accessibility of biomass. This brings out the necessary changes in the biomass, including increase in the surface area of lignocellulose biomass, and changes in the structure of biomass so that the cellulose present can be made easily available for enzymes. Physical methods need a large amount of energy which affects the feasibility as well as the cost-effectiveness, whereas chemical methods require a large amount of expensive chemicals (Aftab et al., 2019). However, most of the methods use acid, alkali treatment, or the neutralization step prior to the hydrolysis. Even pentoses and hexoses such as furfural, 5- hydroxymethylfurfural, and phenolic compounds formed by the delignification and aliphatic acids (acetic, formic, and levulinic acid) formed during acidic pretreatment act as fermentation inhibitors (Nair et al., 2017). Strong acids are also used including sulfuric acid and phosphoric acid for the sulfur contamination of the feedstock (UoI, 2021). Pretreatment using weak acids is also employed in some cases which causes biomass hydrolysis, and further breakdown of xylose to furfural occurs.This type of treatment is favored by high temperatures. There are two types of acid treatments which can be given; one is a short treatment at higher temperature, whereas the other has a long duration of 30–90 minutes and temperatures less than 120°C (Duff et al., 1994).

Alkaline treatment has many advantages which are based on the solubilization of lignin in the alkali solution using alkali reagents including hydroxides like sodium, potassium, calcium, and ammonium. Sodium hydroxide has shown effective results. In this alkali treatment, saponification reaction takes place which causes cleavage of the intermolecular ester linkages between hemicelluloses and lignin. This results in solubilization of lignin and hemicellulose fragments in the alkali solution and brings the cellulose in the interaction of enzymes. It changes the lignocellulosic structure via cellulose swelling that leads to a reduction in crystallinity and degree of polymerization thereby increasing internal surface area. In addition, the removal of acetyl groups and uronic acid substitutions in hemicelluloses during alkali pretreatment also increases the accessibility of the carbohydrates to enzymatic hydrolysis.

After the complete examination and delignification, cellulose is more accessible to the cellulase, which further degrades it into glucose. As cellulose is a polymer that is soluble in water unless at extremely low or high pH, or soluble in solvents including ionic liquids and N-methylmorpholine-N-oxide levels, it is important to maintain the pH of the biomass (Baruah et al., 2018).

### 5.3.2 Enzymes for Biofuel Production

Many enzymes obtained from different microorganisms and plants play a crucial role in forming bioethanol from liquid wastes. They can change the waste into value-added material. Recent research has focused on enzymatic treatments for the conversion of waste. Enzymes can be used to reduce food waste by processing it into nonfood products. Some major enzymes are explained here.

#### 5.3.2.1 Cellulolytic Enzymes

The cellulolytic enzymes are used to treat sewage sludges from drinking and pulping. Duff et al. (1995), investigated the probability of hydrolyzing the high cellulosic sludge yielded from paper and pulp analysis to generate a source of renewable energy like ethanol. About 50–60 kg of primary sludge generates 1000 kg of semiliquid (pulp) stuff. Hence, bioethanol generation is an interesting method to decrease the proportion of sewage sludge disposed of and handled while producing a saleable product (Shoemaker, 1986). It also centralizes the conversion of cellulose substrate having the least value from the drinking method and fiber reuse to fermentable sugars. The enzymes used include a mixture of b-glucosidase, cellulose, and cellobiohydrolase. Considering all the facts, it is noted here that the existence of surfactant increases the rate of hydrolysis, particularly in the initial phase of any reaction.

#### 5.3.2.2 Amylases

These are mainly polysaccharide hydrolases known for starch fermentation and saccharification together, and the operation of food wastewaters containing starch. Coleman, (1990) reported that with rice-processing liquids, amylases can produce alcohol. The report also attests that amylases can increase the secondary sludge operation by decreasing the time duration of treatment. Venugopal et al. (1989) explained an exciting approach of glucoamylase and α-amylase in the manufacturing of biodegradable and photodegradable plastic. The procedure involves converting food waste containing starch (potato and cheese) by industrial food processing to absolutely degradable, nontoxic plastic. Initially, it is used to disintegrate the (long) starchy molecules into small fragments.

#### 5.3.2.3 Proteases

These are the group of hydrolases broadly employed in food industries to process meat and fish waste. Protease can easily dissolve the protein in the waste stream, resulting in retrievable dry solids or liquid concentrate having nutritional value for livestock or fish. The hydrolysis of an inexplicable protein by multistep treatment is carried by proteases through which the enzyme, primarily adsorbed on the solid substrates, disintegrates the polypeptide chains, which are loosely connected through the surface. Therefore, the closely packed core dissolution occurs at a steady rate, which depends on the diffusion of enzymes to core particles and surface-active sites (Blaschek, 1992).

#### 5.3.2.4 Pectinases

Pectinesterase from *Clostridium thermos* sulfurogenes and pectinlyase from *Clostridium* can be used for degrading protein (a water-soluble substance that binds adjacent cell walls in plants) (Karam and Nicell, 1997). Apple pomace, which is a food processing waste, can be degraded to produce butanol. L-galactonolactone oxidase enzyme obtained from yeast can be utilized to bio-convert galactose yields from the hydrolysis of the lactose in L-ascorbic acid, which is an essential synthetic product.

#### 5.3.2.5 Lactases

Lactases give rise to value-added products from the processing of dairy wastes. The essence of protein is segregated from saturated solids, which incorporate a higher level of lactose whose transformation and degradation are taken up through enzymatic action. A large quantity of whey is manufactured yearly, and hence, several valuable treatments could have a remarkably positive effect on the environment (Saha et al., 2017).

### 5.3.3 Enzyme Hydrolysis

Enzymatic hydrolysis represents an effective method used to convert cellulose into processed biomass. Under moderate environment, with the pH range between 4.5–5 and temperature approx. 40–50°C, the cellulose is converted into glucose by the cellulase enzyme. The preliminary investigation of lignocellulosic biomass constitutes a significant role in the efficacy of hydrolysis. Such pretreatment involves the solubility of hemicellulose, enzymatic-loading, the time-span for hydrolysis, and lignin removal. The rate of hydrolysis is affected by the crystalline structure of the enzyme (cellulose). Henceforth, the existing lignin and hemicellulose bond with cellulose and manage the hydrolysis procedure. The efficiency of hydrolysis can be enhanced by the polymers or non-ionic surfactants like PEG (polyethylene glycol) because they can reduce enzyme loading by altering the properties of cellulose (Börjesson et al., 2007). Li et al. (2019) suggested that adsorbed PEGs at lignocellulose is the outcome of hydrophobic PEGs and the share of hydrogen of lignin in lignocellulose. Thus, higher enzyme accumulation is accessible for cellulose degradation. Moreover, the inclusion of polyethylene glycol is proved further to expedite enzymatic hydrolysis (Cheng et al., 2020). Ostadioo et al. (2019), reported that the insertion of PEG in enzymatic hydrolysis increases the sugar yields. A different approach to enhancing bioethanol production via enzymatic hydrolysis is to upgrade the method by increasing the concentration of substrate, which influences the hydrolysis rate to optimize the glucose yield in the hydrolysate. Vasić et al. (2021) reported the ingenious enzymatic protocols accepting xylanase from *Thermomyces lanuginosus*, facilitating the hydrolysis of hemicellulose, where different substrates with varied concentrations are used, like sugarcane bagasse biomass from wheat straw and xylans from birchwood and oats spelt. Furthermore, the effective conversion of sugars in bioethanol can be attained by developing processed parameters, that is, hydrolyzation time, enzyme loading, and shaking speed (Edeh, 2020). Following are the factors influencing enzymatic hydrolysis:

a. Crystallinity of cellulose,
b. The particle size of ligno-cellulosic biomass,
c. The pore volume of ligno-cellulosic biomass,
d. Accessible surface area.

### 5.3.4 Fermentation

This is a natural mechanism in which microorganisms such as yeast, fungus, or bacteria convert the monomeric units of sugars acquired during the hydrolysis stage into ethanol, acids, and gases. Yeast, particularly *Saccharomyces cerevisiae*, is by far the most widely utilized microbe owing to its high ethanol production and tolerance limitations (Suresh and Chandrasekhar, 2009, Suresh et al., 2018a; Zabed et al., 2014), though currently, the bacterium *Zymomonas mobilis* is also being considered for this purpose. Batch, fed-batch, and continuous fermentation are the most common methods for producing bioethanol (Kiuchi et al., 2015). Equation 5.2 represents the process of fermentation.

$$C_6H_{12}O_6 + \text{yeast} \rightarrow 2C_2H_5OH + 2CO_2 \tag{5.2}$$

In order to increase bioethanol production in aerobic conditions, several critical parameters are important to evaluate its interaction like aeration, mass and heat transfer, power consumption, suitability for on-line monitoring, control of various environmental and other operating parameters (Suresh et al., 2009a; Suresh et al., 2009b; Suresh et al., 2011; Suresh, 2016).

### 5.3.5 Distillation

Bioethanol identified as a renewable fuel has become a potent alternative to petroleum gasoline fuel. The distillation technique is widely used for refining biofuels like bioethanol, but this process needs excessive heat input and energy, and thereby the process decreases its carbon-neutral value. The earlier utilized distillation technique in distillation units uses atmospheric pressure, while nowadays vacuum distillation and other configurations are employed. The process utilizes the substrate feed which is then fed to the distillation units under vacuum conditions, thereby producing ethanol-rich streams having a yield of ethanol around 40 to 50 wt%. Further, these streams are subjected to atmospheric rectification columns, in which hydrous ethanol (approximately 93 wt. %) is produced. The steam needed to be utilized can be regulated or saved in the reboilers using the double-effect distillation technique because varied temperature levels can be made available in column condensers and reboilers, thus providing thermal integration of columns (Kiuchi et al., 2015).

Distillation is widely used for the purification of bioethanol, in spite of having higher costs, as it is a sustainable method for purification. The fermentation mixture with a larger proportion of water contains a small amount of bioethanol which requires purification. The fraction of bioethanol is between 15 and 20% which needs to be concentrated up to 95% of ethanol (Yang et al., 2012). The distillation process first evaporates and then condenses the liquid mixture which results in the separation of both components since the boiling point of bioethanol at normal temperature is 79°C and that of water is 100°C. For the greater efficiency of the process, a large amount of latent vaporization to the reboiler to vaporize the liquid mixture is used, and the removal of latent heat of condensation from the vapor by the condenser to condense vapor to liquid. That is why this process consumes a large amount of energy, despite the advantage of the ease of operation. The energy recovery from the bottom products enhances energy efficiency keeping the energy consumption the same. Instead of having greater efficiency, it requires significant improvements for cost reduction as well as for the reduction in greenhouse emissions because the latent heat recovery technology has great potential to make this process of bioethanol more sustainable.

## 5.4 CONCLUSIONS AND FUTURE RECOMMENDATIONS

Bioethanol has great potential, and lignocellulosic biomass is identified as the most suitable renewable raw material for sustainably producing bioethanol. Cellulosic biomass includes waste ingredients having bulk availability with low cost as it contains large amounts of carbohydrates. The conversion of lignocellulosic biomass into ethanol happens to be a complicated process that limits its usage to produce biofuels at a large scale. Another challenge is the lower yield of ethanol, which is mainly attributed to less yield of fermentable sugars based on the nature of the feedstock, particularly its properties like viscosity, moisture content, and the inhibitors utilized in the process. Feedstock greatly differs with regard to its nature, source, availability, and the characteristics of lignocellulosic biomass. The reason for low fermentable sugars yield is mainly because of insufficient biodegradation of lignocellulosic materials during the process of hydrolysis. Some important recommendations are listed below:

1. Optimization of biofuel production techniques helps in energy input reduction during the process and hence enhances the feedstock availability. Another approach is thermal integration which effectively decreases the amount of steam and water utilities, thereby optimizing the energy input required during bioethanol production.
2. Use of techniques like pinch technology, double distillation, self-heat recuperation technology, and distillation in bioethanol production can be optimized thereby reducing production costs.
3. Enzymatic hydrolysis needs to be optimized by adopting novel enzymes having high efficiency and attributing low cost of biofuel production. Apart from this, upcoming technologies for handling biomass solids will be beneficial.
4. Integrations of technologies shows promising results for example. simultaneous saccharification and fermentation (SSF) and pinch technology which produces higher yields of bioethanol. Hence, process integration remains the critical parameter which requires implementation and further improvements.

## REFERENCES

Aftab, M.N., Iqbal, I., Riaz, F., Karadag, A., and Tabatabaei, M., (2019). *Different Pretreatment Methods of Lignocellulosic Biomass for Use in Biofuel Production.* Biomass for Bioenergy-Recent Trends and Future Challenges.

Agrawal, K., Bhatt, A., Chaturvedi, V., and Verma, P. (2020). Bioremediation: An Effective Technology Toward a Sustainable Environment via the Remediation of Emerging Environmental Pollutants. In *Emerging Technologies in Environmental Bioremediation* (pp. 165–196). Elsevier.

Amin, F.R., Khalid, H., Zhang, H., u Rahman, S., Zhang, R., Liu, G., and Chen, C. (2017). Pretreatment methods of lignocellulosic biomass for anaerobic digestion. *AMB Express*, 7(1), pp. 1–12.

Anyanwu, R., Rodriguez, C., Durrant, A., and Olabi, A.G., (2018) Micro-macroalgae properties and applications. In *Reference Module in Materials Science and Materials Engineering*. Elsevier BV. https://doi.org/10.1016/B978-0-12-803581-8.09259-6

Azhar, S.H.M., Abdulla, R., Jambo, S.A., Marbawi, H., Gansau, J.A., Faik, A.A.M., and Rodrigues, K.F. (2017). Yeasts in sustainable bioethanol production: A review. *Biochemistry and Biophysics Reports*, 10, pp. 52–61. https://doi.org/10.1016/j.bbrep.2017.03.003

Baruah, J., Nath, B.K., Sharma, R., Kumar, S., Deka, R.C., Baruah, D.C., and Kalita, E., (2018). Recent trends in the pretreatment of lignocellulosic biomass for value-added products. *Frontiers in Energy Research*, 6, p. 141.

Bhardwaj, N., Agrawal, K., Kumar, B., and Verma, P. (2021). Role of Enzymes in Deconstruction of Waste Biomass for Sustainable Generation of Value-added Products. In: H. Thatoi, S. Mohapatra, and S. K. Das (eds.), *Bioprospecting of Enzymes in Industry, Health care and Sustainable Environment* (pp. 219–250). Springer, Singapore.

Blaschek, H.P. (1992). Food processing industry more environmentally friendly. *Trends in Food Science & Technology*, 3(May), pp. 107–10.

Börjesson, J., Peterson, R., and Tjerneld, F. (2007). Enhanced enzymatic conversion of softwood lignocellulose by poly (Ethylene Glycol) addition. *Enzyme and Microbial Technology*, 40(4), pp. 754–762.

Bušić, A., Marđetko, N., Kundas, S., Morzak, G., Belskaya, H., Ivančić Šantek, M., Komes, D., Novak, S., and Šantek, B. (2018). Bioethanol production from renewable raw materials and its separation and purification: A review. *Food Technology and Biotechnology*, 56(3), pp. 289–311. https://doi.org/10.17113/ftb.56.03.18.5546

Champagne, P. (2008). Bioethanol from agricultural waste residues. *Environmental Progress*, 27(1), pp. 51–57.

Chaturvedi, V. and Verma, P. (2013). An overview of key pretreatment processes employed for bioconversion of lignocellulosic biomass into biofuels and value added products. *Biotech*, 3(5), pp. 415–431. https://doi.org/10.1007/s13205-013-0167-8

Chen, J., Wang, X., Zhang, B., Yang, Y., Song, Y., Zhang, F., Liu, B., Zhou, Y., Yi, Y., Shan, Y., and Lü, X., (2021). Integrating enzymatic hydrolysis into subcritical water pretreatment optimization for bioethanol production from wheat straw. *Science of the Total Environment*, 770, p. 145321.

Cheng, M.H., Kadhum, H.J., Murthy, G.S., Dien, B.S., and Singh, V. (2020). High solids loading biorefinery for the production of cellulosic sugars from bioenergy sorghum. *Bioresource Technology*, 318, p. 124051. https://doi.org/10.1016/j.biortech.2020.124051.

Coleman, R. (1990). Biodegradable plastics from potato waste double savings to environment. *Agricultural Engineering*, 71(6), p. 20È2.

Demirbas, A. (2009). Biofuels securing the planet's future energy needs. *Energy Conversion and Management*, 50(9), pp. 2239–2249. https://doi.org/10.1016/j.enconman.2009.05.010

Demirbas, A. (2008). Biomethanol production from organic waste materials. *Energy Sources, Part A*, 30(6), pp. 565–572.

Dimian, A.C., Bildea, C.S., and Kiss, A.A. (2014). Process intensification. *Computer Aided Chemical Engineering*, 35, pp. 397–448. Elsevier. https://doi.org/10.1016/B978-0-444-62700-1.00010-3

Duff, Sheldon J.B., Moritz, John W., and Andersen, Kari L. (1994). Simultaneous hydrolysis and fermentation of pulp mill primary clarifier sludge. *The Canadian Journal of Chemical Engineering*, 72(6), pp. 1013–1020.

Duff, Sheldon J.B., Moritz, John W., and Casavant, Tracy E. (1995). Effect of surfactant and particle size reduction on hydrolysis of deinking sludge and nonrecyclable newsprint. *Biotechnology and Bioengineering*, 45(3), pp. 239–244.

Edeh, I. (2020). Bioethanol production: An overview. *Bioethanol Technologies, Freddie Inambao, IntechOpen*. https://doi.org/10.5772/intechopen.94895.

Ethanol Fuel basics (EFB) (2021). Alternative fuel data center, U.S Energy Department. https://afdc.energy.gov/fuels/ethanol_fuel_basics.html

Goswami, R.K., Agrawal, K., and Verma, P. (2021). Multifaceted role of microalgae for municipal wastewater treatment: A futuristic outlook toward wastewater management. *Clean Soil Air Water*, 1–18.

Grandview Research (GVR) (2020). Ethanol Market Size, Share & Trends Analysis Report By Source (Second Generation, Grain-based), By Purity (Denatured, Undenatured), By Application (Beverages, Fuel & Fuel Additives), And Segment Forecasts, 2020 – 2027, Published Date: June, 2020. https://www.grandviewresearch.com/industry-analysis/ethanol-market

Haldar, D., Sen, D., and Gayen, K. (2016). A review on the production of fermentable sugars from lignocellulosic biomass through conventional and enzymatic route—a comparison. *International Journal of Green Energy*, 13(12), pp. 1232–1253. https://doi.org/10.1080/15435075.2016.1181075

Hoekman, S.K. (2009). Biofuels in the U.S. – Challenges and opportunities. *Renewable Energy*, 34(1), pp. 14–22. https://doi.org/10.1016/j.renene.2008.04.030

Iaquaniello, G., Centi, G., Salladini, A., and Palo, E. (2018). 4: Waste as a Source of Carbon for Methanol Production. In: A. Basile and F. Dalena (eds.), *Methanol* (pp. 95–111). Science and Engineering Elsevier.

IRENA, IEA-ETSAP (2013). Production of Biomethanol Technological Brief. IEA-ETSAP and IRENA© Technology Brief I08. Accessed at: https://irena.org/-/media/Files/IRENA/Agency/Publication/2013/IRENA-ETSAP-Tech-Brief-I08-Production_of_Biomethanol.pdf

Kamyotra, J.S. and Bhardwaj, R.M. (2011). Municipal wastewater management in India. India infrastructure report, pp. 1–439. Oxford University Press, New Delhi.

Karam, J. and Nicell, J.A. (1997). Potential applications of enzymes in waste treatment. *Journal of Chemical Technology & Biotechnology: International Research in Process, Environmental and Clean Technology*, 69(2), pp. 141–153.

Khan, M.N., Luna, I.Z., Islam, M.M., Sharmeen, S., Salem, K.S., Rashid, T.U., Zaman, A., Haque, P., and Rahman, M.M. (2016). Cellulase in Waste Management Applications. In: V.-J. Gupta (ed.), *New and Future Developments in Microbial Biotechnology and Bioengineering* (pp. 237–256). Elsevier. https://doi.org/10.1016/B978-0-444-63507-5.00021-6

Kiuchi, T., Yoshida, M., and Kato, Y., (2015). Energy saving bioethanol distillation process with self-heat recuperation technology. *Journal of the Japan Petroleum Institute*, 58(3), pp. 135–140.

Kuila, A., Sharma, V., Garlapati, V.K., Singh, A., Roy, L., and Banerjee, R. (2017). Present status on Enzymatic Hydrolysis of Lignocellulosic Biomass for Bioethanol Production. In: L.K. Singh and G. Chaudhary (eds.), *Advances in Biofeedstocks and Biofuels*. https://doi.org/10.1002/9781119117322.ch4

Kumar, B., Bhardwaj, N., Agrawal, K., Chaturvedi, V., and Verma, P. (2020a). Current perspective on pretreatment technologies using lignocellulosic biomass: An emerging biorefinery concept, *Fuel Processing Technology*, 199, p. 106244. https://doi.org/10.1016/j.fuproc.2019.106244

Kumar, B., Bhardwaj, N., Agrawal, K., and Verma, P. (2020b). Bioethanol Production: Generation-Based Comparative Status Measurements. In: *Biofuel Production Technologies: Critical Analysis for Sustainability* (pp. 155–201). Springer, Singapore.

Kumar, B. and Verma, P. (2020). Biomass-based biorefineries: An important architype towards a circular economy. *Fuel*, 288, p. 119622. Elsevier.

Kumari, D. and Singh, R. (2018). Pretreatment of lignocellulosic wastes for biofuel production: A critical review. *Renewable and Sustainable Energy Reviews*, 90, pp. 877–891.

Li, H., Wang, C., Xiao, W., Yang, Y., Hu, P., Dai, Y., and Jiang, Z., (2019). Dissecting the effect of polyethylene glycol on the enzymatic hydrolysis of diverse lignocellulose. *International Journal of Biological Macromolecules*, 131, pp. 676–681. https://doi.org/10.1016/j.ijbiomac.2019.03.131.

Maps and Data (M & D) (2021). Global Ethanol Production by Country or Region, Alternative Fuels Data Center, U.S Department of Energy. https://afdc.energy.gov/data/10331

Marie-Rose, S.C., Chornet, E., Lynch, D., and Lavoie, J.M. (2011). From biomass-rich residues into fuels and green chemicals via gasification and catalytic synthesis. *WIT Transactions on Ecology and the Environment*, 143, pp. 123–132.

Mihajlovski, K., Rajilić-Stojanović, M., and Dimitrijević-Branković, S., (2020). Enzymatic hydrolysis of waste bread by newly isolated *Hymenobacter* sp. CKS3: Statistical optimization and bioethanol production. *Renewable Energy*, 152, pp. 627–633.

Mishra, S., Singh, P.K., Dash, S., and Pattnaik, R., (2018). Microbial pretreatment of lignocellulosic biomass for enhanced biomethanation and waste management. *3 Biotech*, 8(11), pp. 1–12.

Mondala, A., Liang, K., Toghiani, H., Hernandez, R., and French, T., (2009). Biodiesel production by in situ transesterification of municipal primary and secondary sludges. *Bioresource Technology*, 100(3), pp. 1203–1210. http://dx.doi.org/10.1016/j.biortech.2008.08.020.

Nair, R.B., Lennartsson, P.R., and Taherzadeh, M.J. (2017). Bioethanol Production from Agricultural and Municipal Wastes. In: J. Wong, R.D. Tyagi, and A. Pandey (eds.), *Solid Waste Management* (pp. 157–190). Elsevier, USA.

Ostadjoo, S., Hammerer, F., Dietrich, K., Dumont, M.J., Friscic, T., and Auclair, K. (2019). Efficient enzymatic hydrolysis of biomass hemicellulose in the absence of bulk water. *Molecules*, 24(23), p. 4206.

Renewable Fuels Association (RFA) (2021). Annual Fuel Ethanol Production. https://ethanolrfa.org/statistics/annual-ethanol-production/

Saha, K., Sikder, J., Chakraborty, S., da Silva, S.S., and dos Santos, J.C. (2017). Membranes as a tool to support biorefineries: Applications in enzymatic hydrolysis, fermentation and dehydration for bioethanol production. *Renewable and Sustainable Energy Reviews*, 74, pp. 873–890. http://dx.doi.org/10.1016/j.rser.2017.03.015.

Sahota, S., Menaria, K., Suresh, S., Arisutha, S., Singh, D.S., and Shah, G. (2016). Biological pretreatment of water hyacinth (*Eichhornia Crassipies*) for biofuel production-A review. *Journal of Bioenergy and Biofuels*, 2(2), p. 97.

Shoemaker, S. (1986). The Use of Enzymes for Water Management in the Food Industry. In: S.K. Harlander and T.P. Labuza (eds.), *Biotechnology in Food Processing* (pp. 259–267). Nayes Publications, Park Ridge, NJ.

Suresh, S. (2016). Mixing in Shake Flask Bioreactor. In: *Encyclopedia of Industrial Biotechnology: Bioprocess, Bio-separation, and Cell Technology* (pp. 1–16). John Wiley & Sons, Inc., USA.

Suresh, S. and Chandrasekhar, G. (2009). Production of bioethanol from cashew waste. *Petroleum Conservation Research Association*, pp. 16–17.

Suresh, S., Sakthivel, S., Prasanna, V., and Arisutha, S. (2018a). Biofuel/Bioenergy-Technical and Economic Viability in India. In *Biorefining of Biomass to Biofuels-Opportunities and Perception*, 4, pp. 343–359. Springler-Verlag, Germany.

Suresh, S., Kumar, A., Shukla, A., Singh, R., and Krishna, C.M. eds. (2017). *Biofuels and Bioenergy, Springer Proceedings in Energy*. Springer International, 2017, pp. 1–197. (ISBN: 978-3-319-47255-3)

Suresh, S., Srivastava, V.C., and Mishra, I.M., (2009a). Critical analysis of engineering aspects of shaken flask bioreactors. *Critical Reviews in Biotechnology*, 29(4), pp. 255–278.

Suresh, S., Srivastava, V.C., and Mishra, I.M., (2009b). Techniques for oxygen transfer measurement in bioreactors: A review. *Journal of Chemical Technology & Biotechnology: International Research in Process, Environmental & Clean Technology*, 84(8), pp. 1091–1103.

Suresh, S., Srivastava, V.C., and Mishra, I.M. (2011) "Oxygen Mass Transfer in Bioreactors" comprehensive. *Biotechnology*, 2, pp. 947–956. (ISBN: 978-0-08-088504-9)

Suresh, S., Srivastava, V.C., Sakthivel, S., and Arisutha, S. (2018b). Kinetic Modeling of Ethanol Production for Substrate-Microbe System. In: *Biorefining of Biomass to Biofuels-Opportunities and Perception*, 4 (pp. 361–372). Springler-Verlag, Germany.

Suresh, S. and Sudhakar, K. (2013). *Global Scenario in Environment and Energy*, BS Publisher (India), Pvt. Ltd. pp. 1–424, 2013. (ISBN: 978-81-7800-286-6)

Tan, K.T., Lee, K.T., and Mohamed, A.R. (2008). Role of energy policy in renewable energy accomplishment: The case of second-generation bioethanol. *Energy Policy*, 36(9), pp. 3360–3365. https://doi.org/10.1016/j.enpol.2008.05.016

UoI (2021). The Future of Ethanol-Cellulosic, University of Illinois Extension. https://web.extension.illinois.edu/ethanol/cellulosic.cfm, accessed on 07.11.2021

Vasić, K., Knez, Ž., and Leitgeb, M. (2021). Bioethanol production by enzymatic hydrolysis from different lignocellulosic sources. *Molecules*, 26(3), p. 753.

Venugopal, V., Alur, M.D., and Nerkar, D.P. (1989). Solubilization of fish proteins using immobilized microbial cells. *Biotechnology and Bioengineering*, 33(9), pp. 1098–1103.

Yan, Q., Liu, X., Wang, Y., Li, H., Li, Z., Zhou, L., Qu, Y., Li, Z., and Bao, X. (2018). Cow manure as a lignocellulosic substrate for fungal cellulase expression and bioethanol production. *AMB Express*, *8*(1), pp. 1–12. https://doi.org/10.1186/s13568-018-0720-2

Yang, Y., Boots, K., and Zhang, D. (2012). A sustainable ethanol distillation system. *Sustainability*, 4, pp. 92–105. https://doi.org/10.3390/su4010092

Yano, S., Inoue, H., Tanapongpipat, S., Fujimoto, S., Minowa, T., Sawayama, S., Imou, K., and Yokoyama, S. (2009). Potential of ethanol production from major agricultural residues in Southeast Asia. *International Energy Journal*, 10(4), pp. 209–214.

Yu, H.T., Chen, B.Y., Li, B.Y., Tseng, M.C., Han, C.C., and Shyu, S.G. (2018). Efficient pretreatment of lignocellulosic biomass with high recovery of solid lignin and fermentable sugars using Fenton reaction in a mixed solvent. *Biotechnology for Biofuels*, 11(1), pp. 1–11.

Zabed, H., Faruq, G., Sahu, J.N., Azirun, M.S., Hashim, R., and Nasrulhaq Boyce, A. (2014). Bioethanol production from fermentable sugar juice. *The Scientific World Journal*, 2014(957102), pp. 1–11. https://doi.org/10.1155/2014/957102

Zeghlouli, J., Christophe, G., Guendouz, A., El Modafar, C., Belkamel, A., Michaud, P., and Delattre, C. 2021. Optimization of bioethanol production from enzymatic treatment of argan pulp feedstock. *Molecules*, 26(9), p. 2516. https://doi.org/10.3390/molecules26092516

# 6 The Role of Pectinases in Waste Valorization

*Cecil Antony*
National Institute of Technology Calicut, Kozhikode, India

*Praveen Kumar Ghodke*
National Institute of Technology Calicut, Kozhikode, India

*Saravanakumar Thiyagarajan*
Michigan State University, East Lancing, USA

**CONTENTS**



DOI: 10.1201/9781003187684-6

## 6.1 INTRODUCTION

Pectinases are enzymes that depolymerize pectin: a component of middle lamella that cements cells together. Pectin, a heteropolysaccharide, forms a matrix with hemicellulose and cellulose and other polymers (Venkatesh & Umesh-Kumar, 2005). Pectin is a complex polysaccharide functionally and structurally and it has various industrial applications as stabilizing and gelling agents in the food industry. Galacturonic acid is the essential component of pectin that includes homogalacturonan, xylogalacturonan, rhamnogalacturonan I and rhamnogalacturonan II. More than 60 transferases are involved in the biosynthesis of pectin. Recently, researchers manipulated pectin biosynthesis pathways to obtain desired agronomical properties required for the production of biofuel.

Since the 1930s pectinases have been used commercially for the preparation of juice and by wine industries to enhance the extraction of juice from the fruits. Now pectinases are one of the major enzymes that have been gaining attention globally for their biocatalytic features and hold more than 20% of the commercial food market worldwide (Nakkeeran, Umesh-Kumar, & Subramanian, 2011; Jayani et al., 2005). Pectinases are the combination of three enzymes namely, Pectinesterase (PE), Pectate lyase (PL), and Polygalacturonase (PG) that degrade different pectic substances (Siddiqui, Pande, & Arif, 2012). Based on the properties of pectinases they are grouped into two categories: (1) Acid pectinases, and (2) Alkaline pectinases that have been used in different commercial applications. Acid pectinases from the fungus *Aspergillus niger* are widely used by the wine and fruit juice industries. Alkaline pectinases have their importance in the pretreatment of wastewater containing pectic materials from juice industries, retting and degumming in a process that is mandatory for the separation of fibers in the textile industry, oil extraction, and paper making among others (Hoondal et al., 2002).

Despite the abundant presence of pectinases in plants, only microbial sources have the potential to solve the industrial demands. This is mainly due to less generation time, easy scale-up, diverse microbial sources, and ease of recombinant DNA (rDNA) technology implementation. Till now, pectinases from different microbial sources like bacteria, yeast, and fungus have been successfully identified for industrial application (Amin et al., 2017a, 2017b; Godinez et al., 2001). The microbial fermentation process is the only cost-effective way to produce pectinases with the catalytic stability and in the high volumes required for industrial applications. Different types of microbial fermentations are used according to the application of pectinases such as solid-state fermentation (SSF), submerged fermentation (SmF), and immobilized microbial culture. Optimization is one of the key practices that have to be followed for improving the production and activity of the pectinases. In recenttimes,traditional optimization methods have been slowly transformed with genetic engineering-based approaches due to their quick results, and this could further advance waste valorization in the near future (Hoondal et al., 2002).

The process of recycling, reuse, transformation, or composting of waste materials to value-added products, fuels or chemicals is termed waste valorization. Pectinases are widely implemented in the valorization of food and agricultural waste. The volumes of both wastes vary by supply, produce and geographical location, post-harvest, and consumption levels. Food and agricultural waste cannot be directly fed into wastewater treatment plants due to their high biological oxygen demand. This is because of the high quantities of carbohydrates, proteins, and lipid contents in the food and agricultural wastes. Indeed their rich biopolymers could be transformed into valuable products through various chemical reactions catalyzed by enzymes (e.g., acylation, hydrolysis, phosphorylation, oxidation, deamination, hydrogenation, and transesterification, etc.) (Kumar & Verma, 2020). The conventional chemical-based pectin depolymerization approaches have a few drawbacks like reaction specificity, high energy input, chemical catalysis, and undesired by-product formation. Therefore, enzyme-catalyzed biotransformations could overcome the limitations of chemical transformations in the waste valorization process, thereby increasing the efficiency of biotransformation. Effluents from juice and wine industries, food, and agricultural waste contain high amounts of pectic materials that could be treated with pectinolytic enzyme cocktails for the degradation of pectin polymers to improvised waste treatment (Asgher et al., 2016).

## 6.2 PECTIN AND ITS STRUCTURE

Pectin is a polymer of D-galacturnoic acid units that are linked by α-1,4 glycosidic bonds. This polymer comprises 70% of the total cell wall polysaccharides and is of three types namely, homogalacturonan (HG), rhamnogalacturonan I (RGI), and rhamnogalacturonan II (RGII) as shown in Figure 6.1.

### 6.2.1 Homogalacturonan (HG)

This is a linear polymer of repeating D-galacturnoic acid (D-GalA) units that may be acetylated at O-2 or O-3 positions or methyl esterified at the C-6 position. The level of methylated carboxyl groups varies in different pectic substances, for instance,

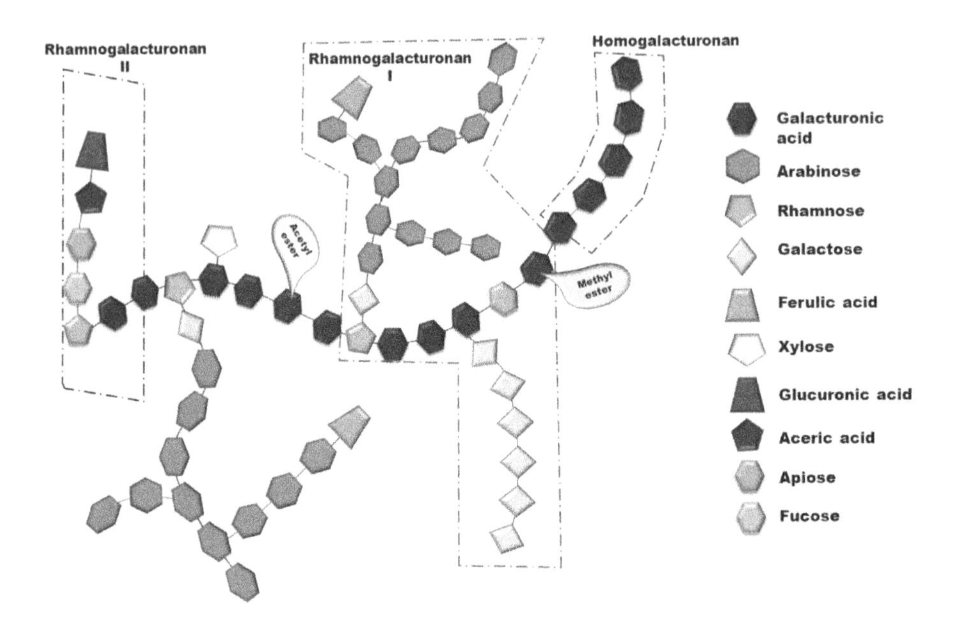


**FIGURE 6.1** Structure of pectin.

75% carboxyl groups methylated were found in pectin whereas in pectinic acid this is much lesser. On the contrary, polygalaturonic acid or pectic acid have none of its carboxyl groups methylated (Jayani et al., 2005). HG reported to be present in stretches of ~100 GalA units in length and even shorter HGs have been found scattered among the other cell wall pectic polysaccharides (Nakamura et al., 2002; Yapo et al., 2007). The structure of pectic substances becomes further complicated when HGs are added with other complex polysaccharides such as, xyloalaturonan (XGA), rhamnogalacturonan II (RG-II), rhamnogalacturonan I (RG-I), and apiogalacturonan (AP) (Mohnen, 2008).

### 6.2.2 Rhamnogalacturonan I (RGI)

This is a polysaccharide with repeating rhamnose (Rha) and GalA disaccharide units $[\text{-}\alpha\text{-D-GalA-1,2-}\alpha\text{-L-Rha-1-4}]_n$ (Willats et al., 2006). This repeating disaccharide backbone of RGIs shows high complexity and cell-specific expression with type and number of monosaccharides, oligosaccharides, and branching oligosaccharides attached to the backbone. The reason for such a high structural complexity is not yet decoded. The GalA side chains can be acetylated and the backbone disaccharide residues carry sugars like arabinose, xylose and galactose. The side chain branching contains α-1,5-linked-L-arabinan with 2- and 3- linked arabinose or arabinan and β-(1,4)-linked-D-galactans with a higher degree of polymerization ranging to the maximum of 47 (Nakamura et al., 2002). The presence of α-L-fucose, β-D-galactose and methylated- β-D-galactose attached to the side chains have also been reported (O'Neill & York, 2003).

### 6.2.3 Rhamnogalacturonan II (RGII)

This is considered to be a highly complex pectic polysaccharide among the pectic substances. RGII with the conserved structure across various plant species has repeating units of a minimum of eight HGs. The backbone with α-1,4-linked α-D-GalA units comprises 12 different sugars with 20 different linkages. RGIIs are the complex side chains of RGIs where both are referred to as hairy regions of the pectic substances. RGIIs exist as dimers cross-linked by 1,2-boratediol ester between apiosyl residues in the side chain A of two RGII molecules (O'Neill, Ishii, Albersheim, & Darvill, 2004). The classification of pectic substances have been provided by the American Chemical Society which divides pectic substances into four major types (Alkorta et al., 1998) as given below;

1. Protopectin: a water-insoluble polysaccharide available in undamaged plant tissues.
2. Pectic acid: a soluble polymer of galacturonans that hardly contains methoxyl groups.
3. Pectinic acids: a polygalacturonan chain containing 1–75% of methylated galacturonate residues.
4. Pectin: polymethyl galacturonate that has 75% of its CO-groups esterified with methanol. This is the pectic substance that provides rigidity to the cell wall when combined with cellulose.

## 6.3 DEPOLYMERIZATION OF PECTIN SUBSTANCES

There are different methods available to perform the depolymerization of pectin substances which include chemical, physical, and enzymatic processes.

### 6.3.1 Chemical Method

In the context of stability, pectins are more stable around pH = 3.5 whereas extreme pH either high or low would induce the removal of neutral sugars, acetyl, and methoxyl groups with backbone cleavage (Schols & Voragen, 2002). Reports are available on the base-catalyzed breakdown of glycosidic linkages next to esterified GalA through β-elimination reactions. This happens while pectin is in a weakly acidic or neutral condition when subjected to heating, and the breakdown is directly proportional to its degree of methylation (Renard & Thibault, 1996). On the contrary, the rate of β-elimination reaction has been reported to be increased by the heat per se than the demethoxylation process at alkaline conditions (Kirtchev et al., 1989). Degradation of pectin through β-elimination was observed while the concentration of cations in the pectin solution increased (Sajjaanantakul et al., n.d.). Copper has been also used as a metallic catalyst to catalyze the depolymerization of pectin in the presence of free radicals (Elboutachfaiti et al., 2008). However, chemical methods are often prone to certain drawbacks such as sample coloration and formation of undesired by-products.

### 6.3.2 Physical Method

The depolymerization of pectin can be induced by high pressure, photolysis, ultrasonication, and radiation. High hydrostatic pressure treatments do not have any effect on pectin depolymerization. When high pressure combined with temperature-induced demethoxylation and inhibited β-elimination reactions (Chen et al., 2015). This is the key reason for high pressure sterilized vegetables and fruits not losing their texture integrity as β-elimination reactions account for the thermal softening. On the contrary, dynamic high pressure is capable of degrading pectin as was experimentally demonstrated by researchers (Chen et al., 2012). The authors used a dynamic high-pressure microfluidization (DHPM) technique to degrade pectin that resulted in better viscosity with the reduction in average molecular weight (Mol. Wt) and particle size. In addition, the amount of reducing sugars increased with the increase in DHPM pressure conditions without β-elimination or demethoxylation reactions.

A combination of $TiO_2$/UV light has been used for the photolysis-induced degradation of pectin. This method generates undamaged or complete pectin that could be utilized by the food and pharmaceutical industries. The molecular size of pectin could be controlled by the exposure duration to UV light and pH conditions (Buranaosot et al., 2010). Ultrasonication improved the optical properties with better viscosity and reduced Mol.Wt due to cavitation, increase in temperature, and OH radicals (Leighton, 1995). Radiation is known to degrade biological polymers and is commonly used to control the microbial pathogens and spoilage of packed food (Harish Prashanth & Tharanathan, 2007). The ionizing radiations cause the free radical-induced cleavage of glycosidic bonds of polysaccharides, thereby degrading pectin. Irradiation of pectin in aqueous solutions causes a vivid decrease in viscosity and Mol.Wt which confirms pectin degradation (Ayyad et al., 1990).

### 6.3.3 Enzyme Method

Enzymatic methods are preferred over the chemical and physical methods because of targeted depolymerization of pectin, as shown in Figure 6.2. and prevention of the formation of undesired by-products. Pectin depolymerization is carried out by a variety of enzymes namely, polygalacturonase, pectate lyase, pectinesterase, β-galactosidase, arabiosidase, and pectin acetylesterase. The enzymes can be classified into two main categories based on their action: (i) HG degrading, and (ii) RG, and side-chain degrading. The former includes pectin acetylesterase, pectin methylesterase, pectate lyase, and polygalacturonase, whereas the latter incudes rhamnogalacturonanan lyase, rhamnogalacturonanan galactohydrolase, rhamnogalacturonase, and rhamnogalacturonanan rhamnohydrolase. A few enzymes catalyze the side-chains degradation, like α-arabinosidases and β-galatosidases.

## 6.4 PECTINOLYTIC ENZYMES

The enzymes that degrade pectin or pectic material through de-esterification and depolymerization reactions are characterized as pectinolytic enzymes. Based on the mode of action different pectinolytic enzymes have been described in Table 6.1.

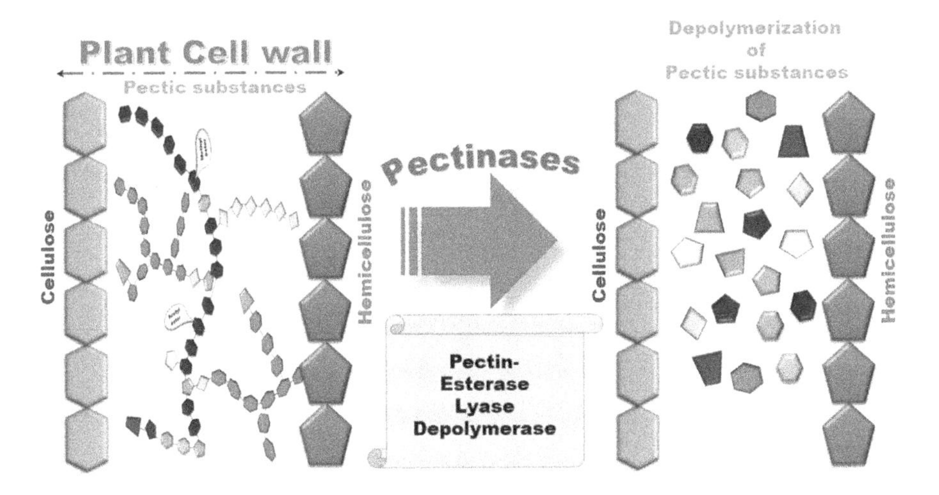


**FIGURE 6.2** Degradation of pectic substances by pectinases or pectinolytic enzymes.

Pectinases are widespread in this biosphere and they are produced by a variety of living organisms (i.e., plants, microbes like bacteria, fungi, and yeast) (Gummadi & Panda, 2003; Whitaker, 1990). Pectinolytic enzymes have a different role in different species, for instance, in plants they are mainly required for fruit ripening, whereas microbes especially plant pathogens make use of these enzymes to break cell walls for cell entry. Microbial decomposers also produce pectinolytic enzymes to break down cell wall/structure and followed by the action of other enzymes the organic matter is finally mineralized into carbon dioxide, nitrogen, methane, and others. Overall pectinases fall into three classes:

1. PPases: solubilize pectin from polymerized protopectin.
2. Esterases: act by removing acetyl and methoxyl groups from pectin and form poly-D-GalA.
3. Depolymerizing: break α-(1,4) glycosidic bonds in poly-GalA.

## 6.5 BIOCHEMICAL PROPERTIES OF PECTINASES

### 6.5.1 Protopectinases (PPases)

Unripe fruits have a higher content of insoluble protopectin and by the action of PPases this insoluble polymer is converted into soluble pectin. This conversion during the ripening process is mainly responsible for fruit tissue softening. PPases are further categorized into two types namely, (i) A type and (ii) B-type. The A-type PPases react with the poly-D-GalA region of protopectin whereas B-type PPases react with the polysaccharide chains that link poly-D-GalA with the cell wall components (Sakai, Sakamoto, Hallaert, & Vandamme, 1993). PPases enzyme activity is measured by the amount of release of pectic substances from the substrate protopectin determined by carbazole sulphuric method (Dische, 1947). There are 3 A-type PPases (F, L and S) all have a Mol.Wt of ~30kDa. The optimum pH range varies with

**TABLE 6.1**
**Pectinolytic Enzymes Involved in Pectin Depolymerization**

| S. No | Enzyme | EC Number | Mode of Action | Reference |
|---|---|---|---|---|
| 1 | Protopectinases (PPases) | 3.2.1.99 | Catalyzes the degradation of the insoluble protopectin to soluble pectin | Tapre and Jain (2014) |
| 2 | Pectin methyl esterases (PME) | 3.1.1.11 | Catalyzes the de-esterification of the methoxyl group of pectin forming pectic acid and methanol | Kashyap et al. (2001) |
| 3 | Pectin acetyl esterases (PAE) | 3.1.1.- | Catalyzes the hydrolysis of (acetylester) pectin forming pectic acid and acetate | Shevchik and Hugouvieux-Cotte-Pattat (1997) |
| 4 | Polymethylgalacturonases (PMG) | | Catalyzes the hydrolytic cleavage of α-1,4-glycosidic bonds in pectin backbone, that are highly esterified pectin, forming 6-methyl-D-galacturonate | Jayani et al. (2005) |
| 5 | Polygalacturonases (PG) | Endo-PG 3.2.1.15 Exo-PG 3.2.1.67 | Catalyzes the hydrolysis of α-1,4-glycosidic linkages in polygalacturonic acid producing D-GalA | Coutinho and Henrissat (1999) |
| 6 | Pectate lyases (PGL) | endo-PGL 4.2.2.2 exo-PGL 4.2.2.9 | Catalyzes the cleavage of glycosidic linkages preferentially on polygalacturonic acid forming an unsaturated product Δ-4,5-D-galacturonate | Rombouts and Pilnik (1980); Pitt (1988) |
| 7 | Pectin lyases (PL) | all PLs are endo-PLs 4.2.2.10 | Catalyzes the cleavage of high esterified pectin, producing unsaturated methyloligogalacturonates through transelimination of glycosidic linkages | Sinitsyna et al. (2007) |
| 8 | Rhamnogalacturonan Rhamnohydrolases | 3.2.1.40 | Catalyzes the hydrolytic cleavage of rhamnogalacturonan chain at nonreducing end forming rhamnose | (Mutter et al. (1994) |
| 9 | Rhamnogalcturonan Galacturonohydrolases | 3.2.1.- | Catalyzes the hydrolytic cleavage of the rhamnogalacturonan chain at nonreducing end forming monogalacturonate | Mutter et al. (1998) |
| 10 | Rhamnogalacturonan Hydrolases | - | Catalyzes the random hydrolysis of rhamnogalacturonan chain forming oligogalacturonates | Mutter et al. (1998) |
| 11 | Rhamnogalacturonan Lyases | 4.2.2.- | Catalyzes the random transelimination of the rhamnose-galcturonate linkage to produce rhamnogalacturonan chain | Mutter et al. (1996) |

*(Continued)*

**TABLE 6.1 (CONTINUED)**
**Pectinolytic Enzymes Involved in Pectin Depolymerization**

| S. No | Enzyme | EC Number | Mode of Action | Reference |
|---|---|---|---|---|
| 12 | Rhamnogalacturonan Acetylesterases | 3.1.1.- | Catalyzes hydrolytic cleavage of acetyl groups from rhamnogalacturonan chain | Searle-Van Leeuwen et al. (1992) |
| 13 | Xylogalacturonan Hydrolase | 3.2.1.- | Catalyzes the hydrolytic cleavage of glycosidic linkages between two galacturonate residues in xylose-substituted rhamnogalacturonan chain forming xylose galacturonate dimers | Vlugt-Bergmans et al. (2000) |
| 14 | α-arabinosidases or α-L-arabinofuranosidase | 3.2.1.55 | Catalyzes the hydrolysis of terminal non-reducing alpha-L-arabinofuranoside residues in alpha-L-arabinosides | Kaji and Yoshihara (1971) |
| 15 | β-galatosidases | 3.2.1.23 | Catalyzes the hydrolysis of terminal non-reducing beta-D-galactose residues | Kuo and Wells (1978) |

each A-type PPases, F is 2.0–8.0 and of L and S are 2.0–7.0. Each PPases have a different isoelectric point F = 5.0; L = 8.4–8.5; S = 7.6–7.8 with an optimum pH of 5.0 and temperature range 50–60°C (Sakai, 1999).

PPases B-type are C and T have a Mol. Wt range from 30 to 55 kDa and isoelectric point C = 9.0 and T = 8.1. The optimum pH range of C is 5.0–9.0 and T is 2.0–6.0 with an optimum temperature of 60°C (Sakai, 1992).

### 6.5.2 Polygalacturonases (PG)

Polygalacturonate substances are hydrolyzed by polygalacturonases that catalyze the hydrolysis of α-1,4-glycosidic linkages. There are two types of PGs performing different breakdown reactions (i) Exo-PGs and (ii) Endo-PGs. The Exo-PGs action on polygalacturonases results in mono- and di-galacturonates whereas the Endo-PGs beak the backbone into oligogalacturonates. Measuring the reducing sugar formed due to the hydrolytic action of the enzymes could be used to ascertain the enzyme activity (Kant et al., 2013; Rebello et al., 2017). The reducing sugars released by the enzyme action could readily be measured by the addition of 3,5-dinitrosalicyclic reagent (Miller, 1959) or arsenomolybdate–copper reagent. Yet another method that depends on the viscosity of the sample upon enzyme action is less familiar due to its lesser sensitivity to Exo-PGs (Rexová-Benková & Markoviĉ, 1976).

The cup-plate method is also used to measure the PGs activity where the substrate containing cups are cut from agar after solidification and added with a solution containing the enzymes. This is followed by an incubation period to allow the enzyme to act over its substrate that can be finally stained with iodine to visualize the zones of degradation for measuring the enzyme activity (Dingle, Reid, & Solomons,1953). The optimum pH range for both Exo-PGs and Endo-PGs is 2.6–6.0 (± 0.3). They have a Mol. Wt. approximately 38–65 kDa and the optimum temperature ranges from 30 to 50°C (Mutter et al., 1998; Jayani et al., 2005).

Plants, plant pathogenic nematodes, bacteria, yeast, and fungi have been reported to produce PGs (Niture, 2008) but *Aspergillus* is considered to be a significant source due to its versatility in scaling up the production. There are many Exo-PGs and Endo-PGs isolated and characterized from *Aspergillus* (Nakkeeran et al., 2011; Nagai et al., 2000; Sakamoto et al., 2002; Zhou et al., 2015).

### 6.5.3 Pectin Esterases (PEs)

The acetyl and methoxyl residues of pectin are removed by the action of pectin esterases that gives an end-product poly-GalA. PEs are of two classes namely (i) pectin acetylesterase (PAE) and (ii) pectin methylesterase (PME). Both act on pectin; the former hydrolyzes the acetyl esters in pectin forming pectic acid and acetate, while the latter releases the methanol-transforming pectin to pectic acid (Pedrolli et al., 2009; Micheli, 2001). PMEs are produced by plants, bacteria, and fungi.The advantage of microbial PMEs, especially from fungi, is the elimination of methyl groups through multichain action in a randomized manner. Plant-based PEs in comparison to their fungal counterparts act either next to carboxyl residue or at the non-reducing end in a non-randomized manner resulting in blocks of deesterified GalA units (Forster, 1988). The Mol.Wt of PEs lies in the range 30–50 kDa with a PI from 4.0 to 8.0 (Christensen et al., 2002). The optimum pH differs between 3.8 and 9.0 (±0.3) and the optimum temperature range between 40 and 60°C. Ruthenium red binding to pectin is used as an assay to ascertain the enzyme activity. This assay utilizes the increased affinity of ruthenium red to pectin that has reduced methyl esters (Downie et al., 1998). PMEs can be further classified into types I and II based on the PRO domain (an N-terminal extension region resembles PME inhibitor). The type I has a Mol.Wt about 52–105 kDa while type II has only 27–45 kDa. Type II has a high similarity with plant pathogenic microbes (Tian et al., 2006; Micheli, 2001).

### 6.5.4 Polygalacturonase (PGs)

The polygalacturonic acid backbone α-(1,4)-glycosidic bonds are hydrolysis by PGs releasing D-GalA. The PGs (i) polymethylgalacturonate PMG and (ii) polygalacturonase (PGs) can act either Exo or Endo. The Endo PG and PGM act on the substrate by randomly cleaving it whereas the Exo PG and PMG act on the substrates by cleaving at the substrate non-reducing end thereby releasing mono- or di-galacturonates (Alkorta et al., 1998; Kashyap et al., 2001; Jayani et al., 2005; Rombouts & Pilnik, 1980; Vincken et al., 2003; Willats et al., 2006). Highly esterified pectin substances are the substrates for PMGs, the polymer backbone α-(1,4) glycosidic linkages are hydrolyzed releasing 6-methyl-D-GalA. The PMG enzyme activity can be determined by measuring the reduced sugars formed as reaction end products or by measuring the drop in substrate viscosity. Pectic acid and pectate derivatives barely react with PMGs while highly esterified pectin is found to be the apt substrate. Among the microbial sources of PMGs *Aspergillus* was reported to be the principal producer with the optimal pH range 4.0–7.0 (±0.3) (Koller & Neukom, 1967). PGs and PMGs have an optimal temperature about 3.0–7.0 (±0.3) and an optimal temperature of 40–60°C (Pedrolli et al., 2009).

### 6.5.5 Pectin Lyases (PLs)

Pectin lyase catalyzes pectin degradation by transelimination of glycosidic linkages yielding unsaturated methyl-oligogalacturonates. Until now only Endo-PLs have been described while Exo-PLs have not yet been identified or isolated. PLs cleave the glycosidic linkages at the fourth carbon position and simultaneously eliminate the hydrogen at the fifth carbon position giving rise to the formation of unsaturated products (i.e., di-, tri-, and tetra-galacturonates from methyl-oligogalacturonates). PL enzyme activity can be measured by an increase in unsaturated oligo-galacturonates that results in absorbance at 235 due to the formation of Δ 4:5 double bonds produced at the non-reducing ends of unsaturated products (Liao et al., 1999). The PL enzyme activity can also be measured by thiobarbituric acid method or reducing sugar method (Nedjma et al., 2001). PL does not have an obligatory requirement of $Ca^{2+}$ ions for the enzyme activity. The Mol. Wt of PL ranges from ~30–40 kDa with PI between 7.0 and 11.0. The optimal pH of PL ranges from 7.5 to 11.0 in the alkaline scale and the optimal temperature ranges from 40 to 50°C (Jayani et al., 2005).

### 6.5.6 Polygalacturonate Lyase (PGLs)

PGLs also act similarly to PLs by catalyzing the transelimination reaction to yield Δ 4:5 unsaturated oligo-galacturonates. The end products of transelimination reaction can be used to determine the enzyme activity as discussed in PLs. The sources of PGLs are only pathogenic microbes and they have an obligatory requirement of $Ca^{2+}$ ions for their activity. The Mol. Wt of PGLs differs from ~30–40 kDa with an alkaline optimum pH range of 8.0–10.0 (±0.3). The optimal temperature of PGLs is between 30 and 40°C. PGLs are further classified into Endo-and Exo PGLs and have varied activity on their substrates. Endo-PGLs enzyme activity decreases with a decrease in substrate chain length, often shows low activity upon binding bi- and tri-galacturonates while Exo-PGLs have no such preference in substrate length.

## 6.6 PURIFICATION OF MICROBIAL PECTINASES

Microbial pectinases have gained importance over plants, insects, nematode, and protozoan origin due to their easy scale-up, wide distribution, high growth rate, less complexity, and ease of gene manipulation. Microbes that produce pectinases are either plant pathogens or decomposers; the former require such enzymes to invade host cells crossing the complex cell wall barrier (Hoondal et al., 2002) and the latter play an important role in biomineraliztion. Saprophytic fungi are the largest producer of pectinases among the available microbial pectinase producers (Gummadi & Panda, 2003). From the microbial sources of pectinases, one can obtain different forms of enzymes (i.e., varied $K_m$, temperature, pH, etc.) as they are naturally selected for a particular type of environmental condition or plant species. The selection pressure decides and tailors the microbial genes (especially the enzyme genes) that have to be expressed alone or in combination for a successful cell invasion thereby crossing the plant cell wall barrier and their components. Hence microbial pathogens and decomposers secrete various cell-wall-degrading enzymes together to establish infection or mineralize the organic matter (Naessens & Vandamme, 2003; Singh et al., 2015).

Microorganisms synthesize pectinolytic enzymes with different biochemical characteristics and modes of action (Gummadi & Panda, 2003). In general, both prokaryotic and eukaryotic microorganisms have been reported to produce pectic enzymes, with prokaryotic microorganisms producing alkaline pectinases and eukaryotic fungi producing acid pectinases (Hoondal et al., 2002; Kashyap et al., 2001). Even though bacteria, yeast, and fungi produce pectinases, the majority of industrially significant pectinase production have been found to be from fungal sources (Beg et al., 2000; Reid & Ricard, 2000).

### 6.6.1 Production Techniques for Pectinases

#### 6.6.1.1 Fermentation Strategies

Solid-state (SSF) and submerged fermentation (SmF) are frequently used methods that have been recognized for large-scale production of pectinases as shown in Figure 6.3. However, more than 90% of pectinases produced industrially have been

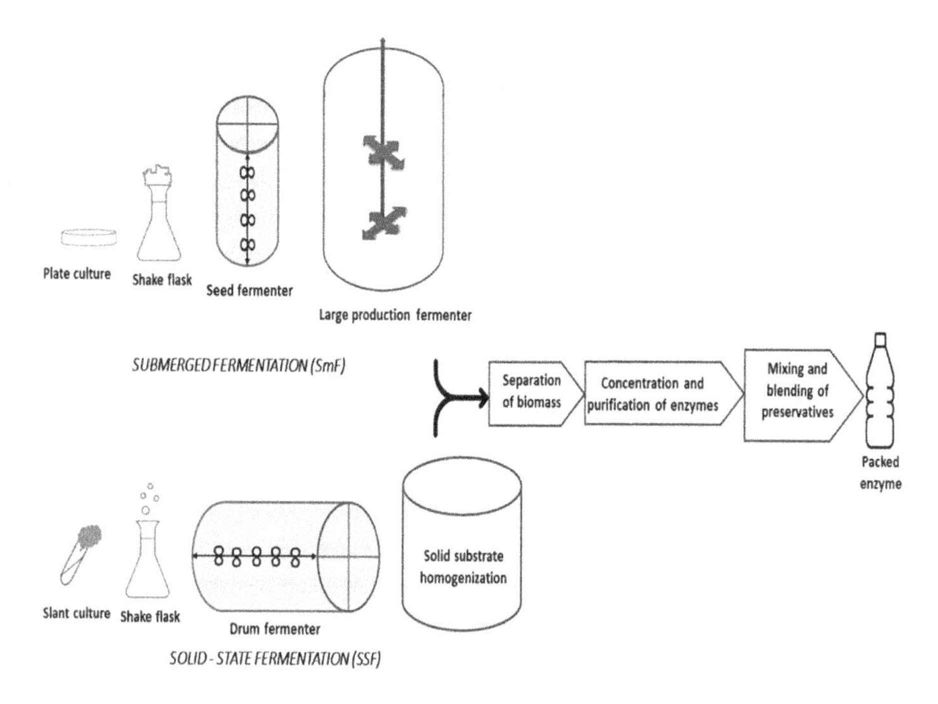


**FIGURE 6.3** Pectinase production by SmF and SSF fermentation.

obtained by submerged fermentation (Pedrolli & Carmona, 2010). In recent decades trends are toward the production of industrially relevant enzymes using SSF technologies. SSF has been used since ancient times by reproducing microbial processes like composting and ensiling. These natural processes can be utilized in a controlled manner for industrial purposes (Couto & Sanromán, 2006). The main advantage of SSF is that it does not require mechanical energy as it is a static process in comparison to SmF which is non-static and requires large amounts of energy to meet its oxygen demands (Viniegra-González et al., 2003). Enzyme production in SmF techniques is generally carried out in stirred tank reactors aerobically using batch or fed-batch systems. SmF requires more water requirement and generates lot of industrial effluents. In the SSF, microbial growth and products formation are conducted without water, and yet another advantage of SSF is it does not require aseptic conditions (Castilho et al., 2000).

#### 6.6.1.2 Submerged Fermentation

In the SmF technique, the type of strain used dictates the yield of the enzyme. Various other parameters that contribute to the enzyme yield in SmF are nutritional requirements such as substrate concentration, pH balance, temperature, incubation time, aeration, agitation, and extraction methods. In the case of microbial pectinolytic enzymes, the production level relies on the activation of genes, thus gene inducers play a key role in induction. Thereby the production can be controlled by the apt usage of inducers. As mentioned earlier, some bacterial cultures produce alkaline pectinases, and yeast culture results in acidic pectinases, so there are certain exemptions in deciphering the biosynthesis of alkaline pectinases by fungal strains using SmF (often connected with soft rot, and food spoilage). Based on the different genetic and environmental backgrounds microbes can produce several types of pectinases (Alimardani-Theuil et al., 2011).

#### 6.6.1.3 Solid-state Fermentation

SSF is attracting many researchers in recent decades due to the following reasons.

Low-cost agricultural waste residues can be used for microbial growth, downstream processing is simple, and there is less intensive energy input, a high-level yield of microbial metabolites, and less catabolic repression (CCR) of microbes. Because it resembles the natural habitat in moist conditions, SSF appears a promising tool for enzyme extraction from fungi. Certain controlled environmental conditions in SSF for bacteria and yeast also produce pectinases with high water activity which results in a good yield of enzymes during extraction (Ahlawat et al., 2009). Fungal spores show high resistance to dehydration and have higher germination rates even after long time periods of lyophilization in SSF when compared to SmF (Kawano et al., 1999). The other disadvantages of SSF are maintaining reproducibility due to heterogeneous fermentation conditions and hurdles in process scale-up. The downstream processes or product purification is quite complicated due to the usage of a mixture of organic growth substrates (McMillan et al., 1992). A comparison of SmF and SSF is given in Table 6.2.

**TABLE 6.2**
**Comparison of Submerged Fermentation and Solid-state Fermentation**

| S. No | Parameters | Submerged Fermentation | Solid-state Fermentation |
|---|---|---|---|
| 1 | Culture inoculum | Optimal for the growth of fungi | Optimal for the growth of bacteria |
| 2 | Inoculum size | Large | Small |
| 3 | Shaking condition | Static | Agitated |
| 4 | Oxygenation method | Supply by diffusion | Supply by aeration |
| 5 | Utilization of water | Usage is limited | Continuous usage of water |
| 6 | Scope of scale-up | Scale-up is restricted | Easier to scale up |
| 7 | Aseptic conditions | Contamination is lesser | Susceptible to contaminants |
| 8 | Size of the reactors | Smaller in size | Larger |
| 9 | Downstream processing | Easier separation methods | Multiple steps involved in separation |
| 10 | Capital cost | Less investment | High |
| 11 | Requirement of energy | Low | High |
| 12 | Maintenance | Difficult | Easy |

#### 6.6.1.4 Immobilization of Cell Culture

Microorganisms and their immobilization on an inert polymeric solid carrier provide more benefits like reuse, increased stability, easy separation of products, and biomass from microbes (Moreno-García et al., 2018). Different types of natural polymers like chitin, carrageenan, alginate, and chitosan, and synthetic polymers like polyethylene glycol and polyacrylamide are being used as a support material. In contrast to synthetic polymers, naturally occurring biopolymers are used as immobilization matrices due to their biocompatible characteristics (Angelim et al., 2013). In a study researchers used immobilized *A. niger* cells on polyurethane support material to compare pectinase production by SSF and SmF. The results showed that there was an increase in the enzyme yield in SSF in comparison to SmF as there was a high rate of biomass production without promoting catabolite repression (Diaz-Godinez et al., 2001). But still, implementation of immobilization technology for industrial production of enzymes and metabolites is hindered.

### 6.6.2 The Role of Genetic Engineering in Microbial Pectinase Production

The advent of rDNA technology has given rise to a modern era of biotechnological tools like gene cloning, expression, and mutation with which desired enzyme traits could be engineered at an industrial scale. rDNA technology is helpful in the overexpression of targeted proteins and enzymes. Advanced techniques like site-directed mutagenesis are helpful in the development of robust enzymes with increased pH stability, thermal stability, and catalytic activity.

Genetic engineering has reduced the cost of enzyme production greatly which increases the profits of bioindustries. Pectinolytic enzymes have been overexpressed successfully in homologous as well as heterologous expression systems. For bacterial expression systems, *E. coli* species are used for intracellular production, whereas *Bacillus subtilis* are being used in the secretory expression. Yeast expression systems provide superior yield with proper protein folding ability. Two industrial workhorses that were used for three decades are *Saccharomyces cerevisiae* and *Pichia pastoris*, with their ability to manipulate the constitutive expression and over-expression of selective inducers. Most of these expression systems are targeted and secretory thus facilitating the separation, concentration, and purification processes at an industrial scale (Abdulrachman et al., 2017). Interestingly, so far 97 pectinase genes have been mapped and 60 novel genes have been recognized from the whole genome of *A. niger* and with this genetic information, researchers were able to successfully produce pectinases on a commercial scale (Pel et al., 2007). By harnessing the homologous recombination technique scientists have been able to genetically engineer and generate pure enzymes devoid of side effects, similar to classical genetics.

### 6.6.3 Fermentation Media Optimizations for Pectinase Production

Pectinase production depends on the composition of the growth medium, carbon/nitrogen sources, moisture, cultivation conditions, aeration, strain type, inducer type, pH, and temperature. The agricultural industry harbors a vast amount of biomass waste which contains all the vital nutritional requirements required by the microbes for the biosynthesis of pectinases (Amin et al., 2017b). Screening of agricultural and industrial waste fuels the demand for the availability of substrates used in fermentative processes. Particularly, food and agro-waste like fruit pomaces, rice bran, orange peels, wheat bran, and sugarcane bagasse are utilized for the production of the enzyme at the industrial level (Munir et al., 2015).

These agricultural wastes serve as an important substrate for pectinases production as they contain large amounts of pectin as well as other nutrients (da Silva et al., 2005). Other synthetically defined media, such as starch, sugar, polysaccharides, and certain intricate compositions like peptone, yeast extract, malt extract, and so on, have met the nutritional requirements of microbes and therefore play a crucial role in pectinase expression (Jahan et al., 2017). In order to achieve a better yield in pectinase production, the fermentation media should be formulated by combining both agro-waste and synthetic media. Several research groups are working on media optimization through statistically significant methods for formulating synthetic media as well as utilizing agro-wastes in order to achieve efficient pectinase production. Bridging cutting-edge skills like genetic engineering with statistical optimizations could increase the desired levels of enzyme expression with less usage of substrate, cost, energy, and time (Zou et al., 2013).

### 6.6.4 Factors Affecting Pectinase Production

Along with microbial sources like wild type, recombinant, and mutagenized, various other parameters can influence the production of pectinase including pH, temperature, and metal ions. Because of cell permeability utilization surfactants like Tween 20 and Tween 80 leads to increased enzyme production, whereas enzyme production is inhibited by the presence of detergents like SDS (Phutela et al., 2005).

#### 6.6.4.1 Role of the Substrate in Pectinase Production

The presence of substrate in the culture medium is based on the organisms used for inoculation. The most-used substrates are potato dextrose agar, yeast extract, wheat bran, dextrose agar, and grape pomace. In some cultures, maximum yield was obtained with wheat bran, while the usage of glucose leads to least production in pectinases, thus glucose acts as a repressor. Yeast extract has the highest output of pectinase among the nitrogen sources for pectinase synthesis. Glycine, urea, and ammonium nitrate limit pectinase synthesis, whereas wheat bran, peptone, ammonium chloride, and yeast extract increase pectinase production (Kashyap et al., 2003).

#### 6.6.4.2 Effect of pH and Temperature on Enzyme Production

All PGases have an acidic optimal pH of 3.3–7 with an exception of Exo PGase produced by *Fusarium oxysporum* and Endo PGase by *Bacillus licheniformis*. Salts conceal the carboxylic charged groups involved in enzyme-substrate identification; hence the optimal pH of pectinases is determined by the salt content in the medium (Laratta et al., 2008). The optimal temperature for the biosynthesis of pectinase by *Bacillus subtilis* was discovered to be 37°C. The optimum temperature for pectate lyase in the standard test was 70°C. *Streptomyces sp* pectinase is most active at 60°C. The optimal temperature for pectinase activity in fungi is 50°C, while the temperature range for yeast is 40–60°C (Favela-Torres et al., 2006).

#### 6.6.4.3 Effect of Metal Ions on the Activity of Pectinases During Production

Because thiol groups may be implicated in the enzyme's active site, the metal ions $Hg^{2+}$, $Zn^{2+}$, and $Mg^{2+}$ inhibit enzyme production. $Mn^{2+}$ increases pectinase activity, while $Li^{2+}$, $Fe^{2+}$, and $Rb^{2+}$ have no effect. At high metal ion concentrations, enzyme synthesis is inhibited by a protein secretion barrier into the extracellular medium. $Ca^{2+}$ is required for bacterial pectinase to be active, whereas for fungal pectinase $Ca^{2+}$ is not needed (Saad et al., 2007).

## 6.7 APPLICATION OF PECTINASES IN WASTE VALORIZATION

### 6.7.1 Food and Agro-waste

When raw materials are processed for obtaining the required products, there is always a non-specific or processing by-products formation that gets accumulated as waste. The food industry has different types of wastes that are residual materials from vegetables, fruits, and animal meat. It is noted that there is a significant increase in the emission of greenhouse gases at different phases of food processing. This adds more than 30% of greenhouse gases from Europe alone, according to a study (Garnett, 2011). Food, drug, and cosmetic industries implement various extraction procedures for the recovery of nutraceuticals, small molecules, vitamins, bioactive compounds, and other plant-based chemicals. The leftover biomaterials after extraction of desired products form the food and agro-waste that have high organic content within them (Russ & Meyer-Pittroff, 2004). Pectinases have multiple applications in the valorization of food and agro-waste as summarized in Table 6.3.

**TABLE 6.3**
**Application of Pectinase**

| S. No | Industrial Sector | Application | Function |
|---|---|---|---|
| 1 | Textile | Bioscouring and degumming | Increases the absorbing ability of fabrics and decomposition of non-cellulosic parts |
| 2 | Pulp and paper | Pectin depolymerization | Biobleaching and softening of cell wall |
| 3 | Animal feed | Ruminant feed production | Low viscosity feed |
| 4 | Oil industry | Preparation and extraction of oil from seeds | Biocompatible enzymatic process |
| 5 | Food and beverages | Fruit juice extraction, clarification, and gelling agent | Reduction of viscosity in juices, stabilizing the property of some food |
| 6 | Biofuel | Pretreatment of biomass | Enhances the enzyme activity of cellulases |

#### 6.7.1.1 Food and Agro-waste Valorization Methods

##### *6.7.1.1.1 Composting*

This method is used for the aerobic degradation of food and agro-waste through controlled conditions of temperature and pH that would give rise to an end product termed as compost. Different aerobic microbes act on the target waste thereby utilizing available carbon (C) I and nitrogen (N) to build their biomass with the generation of heat. The resulting end product that is, the compost has less of both C and N content, and this resembles the process of biomineralization and humification that transfers the organic content back to soil (Antizar-Ladislao et al., 2006). Multiple microorganisms have been reported for their involvement at different degrees of composting based on the waste quality such as C/N ratio, humidity, and temperature (Hassen et al., 2001; Tuomela et al., 2000). Composting has two main stages: the first stage involves microbial action on the biodegradable waste and the formation of organic substance, and the second stage comprises the formation of stable organic substances to form the humus (Adani et al., 1999).

The composting process passes through different temperatures from mesophilic to thermophilic ranging from 40 to 70°C and this eradicates weeds and pathogens at the completion (Tuomela et al., 2000; Schaub & Leonard, 1996). This method of waste valorization has been practiced by farmers over centuries mainly for enriching the soil nutrients and disinfecting the soil (Aladjadjiyan, 2007). In a recent study, researchers used a consortium of microbes to compost vegetable waste and found the *Streptomyces sp* Al-Dhabi 30 promoted the composting process by efficient production of pectinases along with cellulases, proteases, amylases, and chitinases (Al-Dhabi et al., 2019). Green waste contains approximately 75% of cellulosic and hemicellulosic material that is mainly degraded by pectinases and xylanases (Zieminski et al., 2012). A study reported the increase of pectinases during the thermophilic phase of the composting that played a major role in the biochemical transformation of green waste (Lu Zhang, 2019).

#### 6.7.1.1.2 *Fermentation*

Carbon, nitrogen, and phosphorus can be utilized from food and agro-waste to produce products of great commercial value such as biodegradable plastics, biofuels, and other valuable products. Biodegradable polylactic acid, a derivative of lactic acid, could be a better replacement for packaging and fibres than conventional petroleum-based polymers that are nonbiodegradable. The production of lactic acid through fermentation of food and agro-waste is an economically significant way of waste valorization (Tuck, Pérez, Horváth, Sheldon, & Poliakoff, 2012). Carbohydrates present in the food and agro-waste biomass can be fermented to biofuels thereby supporting the fuel demands in a sustainable manner (Augustine & Roy, 1981; Kumar & Verma, 2021a). The common principle behind the biotransformation of food and agro-waste is the hydrolysis of cellulose and hemicellulose biomasses (Kumar & Verma, 2021b). Hydrolysis of cellulose a homopolymer yields glucose whereas hemicellulose results in pentose sugars most probably xylose and arabinose. Apart from this, the pectic substances release arabinose, polygalacturnose, and rhamnose among others that make the fermentation process inefficient in converting the carbohydrates to valuable products.

A large number of fermentation processes are efficiently carried for the production of citric acid, lactic acid, ethanol, and industrially significant enzymes. Recent reports are available on pectinase pretreatment of waste biomasses containing pectic substances that have resulted in increased yield. However, pectin dictates the rate of carbohydrate hydrolysis by preventing the access of the enzyme to the vulnerable substrate regions. Pretreatment of sugar beet pulp with pectinases in combination with cellulases synergistically catalyzed tepolymerizationion of lignocellulosic and pectic substances that significantly improved the bioethanol production (Rezić et al., 2013). Similarly, pretreatment of alfalfa fiber with pectinases and cellulases increased the lactic acid production (Sreenath et al., 2001).

#### 6.7.1.1.3 *Anaerobic Digestion*

This method is mainly used for the production of biogas from wastewater and organic waste. The breakdown of organic material is carried out in the absence of oxygen, that is, in anaerobic conditions that result in the evolution of carbon dioxide ($CO_2$) and methane ($CH_4$) and minimal amounts of ammonium ($NH_3$) and hydrogen sulfide ($H_2S$) as shown in Figure 6.4. The biogas composition depends upon the quantity of carbohydrates, protein, and fat present in the waste material. Food and agro-waste containing starch have a high degree of crystallinity that prevents breakdown by hydrolysis, hence, a pretreatment step before subjecting to anaerobic digestion might resolve the degradability issues (Kumar et al., 2020; Chaturvedi & Verma, 2013). Although different pretreatment methods like chemical or thermal are available, only biological methods deal with either microbial or enzymatic (pectinase) treatment. Prior to anaerobic digestion hydrolysis of organic waste by pectinases in combination with other enzymes for 24 hours was reported to be highly effective in methane production (Zieminski et al., 2012).

### 6.7.1.2 Recycling of Wastepaper

De-inking requires hazardous chemicals that are neither safe for humans nor the environment. It is one of the major tasks in wastepaper recycling that is now being

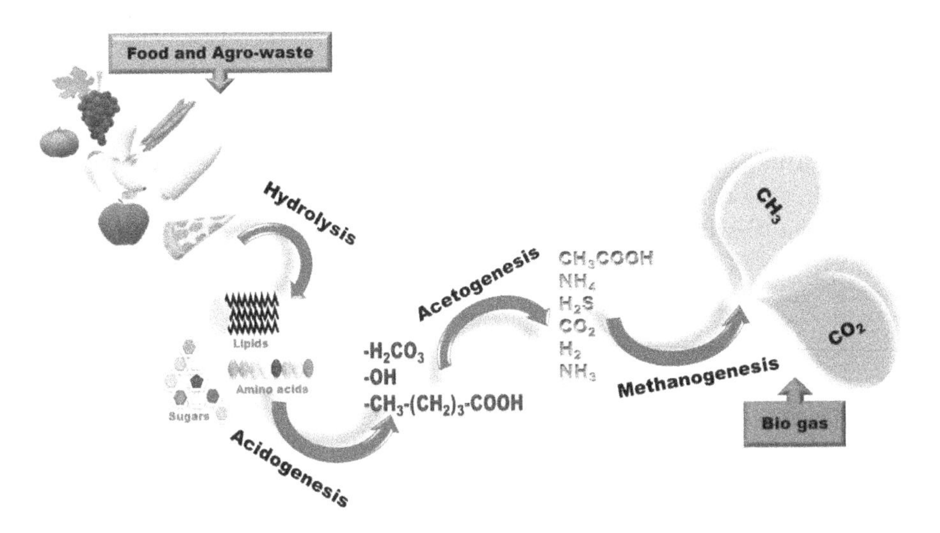


FIGURE 6.4 Valorization of food and agro-waste.

replaced by enzymes causing less pollution and energy saving without compromising the performance. Cellulases, hemicellulases, ligninolytic enzymes, and pectinases are used in the de-inking process. The enzymes alter the surface of the fiber or bonds near the ink allowing the release of ink molecules by floatation or washing (Puneet et al., 2010; Xu et al., 2011; Kumar et al., 2021). The usage of enzymes in de-inking is safe for the environment rather than using hazardous chemicals that would increase the cost of wastewater treatment.

#### 6.7.1.3 Oil Extraction

The quality of oil has a great influence on extraction and storage conditions. An economic and widely used oil extraction process is the pressing method. The arrival of centrifugation methods has advanced the oil extraction process to the next level with a significant increase in oil yield. The plant cell vacuole and the cytoplasm are the regions that contain oil, of which the vacuole has the maximum storage of more than 75%. The oil present in the cytoplasm is more than 20% and occurs in dispersed droplets as colloids that are hard to extract by the existing conventional methods. In order to extract oil present in the plant cell vacuoles and cytoplasm (as colloids) the plant cell wall structure has to be disintegrated. Therefore, the usage of cell-wall-degrading enzymes would be advisable as they could disintegrate the cell wall components such as cellulose, hemicellulose, and pectic substances without compromising the oil yield (Ranalli, 1997).

In general, hexane has been widely used as a solvent for oil extraction that is harmful and carcinogenic (Kashyap et al., 2001). Pectinases have slowly replaced the hexane extraction method as they do not have any detrimental effects on humans and indeed improve the oil extraction. For instance, the use of pectinases in lemon oil extraction has significantly increased the yield by de-emulsifying pectic substances (Mohnen, 2008). Extraction of vegetable oils using pectinases in combination with other cell-wall-degrading enzymes have not only improved the oil yield but also the vitamin E and polyphenol content of the oils (Iconomou et al., 2010; Kashyap et al., 2001).

#### 6.7.1.4 Animal Feed

Food and agro-wastes that contain high amounts of pectic substances along with other nonbiodegradable fibers prevent nutrient uptake. As the nutrients are blocked by these nonbiodegradable fibers as well, they are responsible for the increase in feces quantity. Upon pretreating, the wastes with pectinases in combination with other enzyme cocktails result in a decrease in the viscosity of feed that can facilitate the nutrient absorption or releases nutrients from nonbiodegradable fibers (Jayani et al., 2005). Supplementing enzyme cocktails along with feed improved the digestion and energy intake that altogether increased the weight of the animal (Ghorai et al., n.d.; Petersen, 2001).

#### 6.7.1.5 Wastewater Treatment

One of the major problems in the wastewater treatment process is the resistance of pectin substances to degradation during activated sludge management. In order to overcome this issue researchers have used alkaline pectinase producing *Bacillus* sp (GR612) that produced endopectate lyase. The enzyme is secreted by the bacteria when exposed to an alkaline medium of pH = 10 and can degrade pectic substances. *Bacillus* sp (GR612) has been successfully used to treat wastewater from the citrus processing industry (Tanabe et al., 1987). Similarly, immobilized pectinase from *Aspergillus ibericus* has been used to treat pectin-containing industrial wastewater and showed 94% efficiency in pectin degradation (Mahesh et al., 2016).

## 6.8 CONCLUSION

Pectinases from different sources could be utilized for various commercial as well as valorization purpose. The key issues in exploiting pectinases in waste valorization is their stability during the process. Globally several researchers are working on improving the stability through implementing genetic engineering strategies using rDNA technology. Among the acidic and alkaline pectinases, the latter has issues with its implementation on an industrial scale. Considering the importance of pectinases in waste valorization, the enzyme requires more improvements or upgrades related to temperature and pH. This is possible as the majority of pectinases used both in industry or for waste valorization purposes are of microbial origin. Given the smaller genome size of microbes that could be easily manipulated through rDNA technology and results could be ascertained in a shorter duration. Once a promising enzyme engineering results from a pilot setup, it is easy to scale up the enzyme production or to use the microbe directly in ex-situ waste valorization in an established containment facility. Altogether the valorization of waste would boost waste management and assist the sustainable development of economies globally. New strategies have to be devised to counter the growing biodegradable wastes by converting them into useful chemicals or bioenergy. This would have a positive impact on the economy and environment of developing countries in particular.

## REFERENCES

Abdulrachman, D., Thongkred, P., Kocharin, K., Nakpathom, M., Somboon, B., Narumol, N., Champreda, V., Eurwilaichitr, L., Suwanto, A., Nimchua, T., & Chantasingh, D. (2017). Heterologous expression of Aspergillus aculeatus endo-polygalacturonase in *Pichia pastoris* by high cell density fermentation and its application in textile scouring. *BMC Biotechnology*, *17*(1), 1–9. https://doi.org/10.1186/S12896-017-0334-9

Adani, F., Genevini, P., Gasperi, F., & Tambon, F. (1999). Composting and humification. *Compost Science & Utilization*, *7*(1), 24–33.

Ahlawat, S., Dhiman, S. S., Battan, B., Mandhan, R. P., & Sharma, J. (2009). Pectinase production by *Bacillus subtilis* and its potential application in biopreparation of cotton and micropoly fabric. *Process Biochemistry*, *44*(5), 521–526. https://doi.org/10.1016/J.PROCBIO.2009.01.003

Aladjadjiyan, S. S. N. V. (2007). Composting of food and agricultural wastes. In V. Oreopoulou & W. Russ (Eds.), *Utilization of by-products and treatment of waste in the food industry* (pp. 283–301). Boston, MA: Springer. https://doi.org/10.1007/978-0-387-35766-9_15

Al-Dhabi, N. A., Esmail, G. A., Mohammed Ghilan, A. K., & Valan Arasu, M. (2019). Composting of vegetable waste using microbial consortium and biocontrol efficacy of *Streptomyces* sp. Al-Dhabi 30 isolated from the Saudi Arabian environment for sustainable agriculture. *Sustainability*, *11*(23), 6845. https://doi.org/10.3390/su11236845

Alfonso Ranalli, G. D. M. (1997). Characterization of olive oil produced with a new enzyme processing aid. *Journal of the American Oil Chemists' Society*, *74*, 1105–1113.

Alimardani-Theuil, P., Gainvors-Claisse, A., & Duchiron, F. (2011). Yeasts: An attractive source of pectinases—From gene expression to potential applications: A review. *Process Biochemistry*, *46*(8), 1525–1537. https://doi.org/10.1016/J.PROCBIO.2011.05.010

Alkorta, I., Gabirsu, C., Lhama, M. J., & Serra, J. L. (1998). Industrial applications of pectic enzymes: A review. *Process Biochemistry*, *33*, 21–28.

Amin, F., Bhatti, H. N., Bilal, M., & Asgher, M. (2017a). Multiple parameter optimizations for enhanced biosynthesis of exo-polygalacturonase enzyme and its application in fruit juice clarification. *International Journal of Food Engineering*, *13*(2). https://doi.org/10.1515/IJFE-2016-0256

Amin, F., Bhatti, H. N., Bilal, M., & Asgher, M. (2017b). Purification, kinetic, and thermodynamic characteristics of an exopolygalacturonase from *Penicillium notatum* with industrial perspective. *Applied Biochemistry and Biotechnology*, *183*, 426–443.

Angelim, A. L., Costa, S. P., Farias, B. C. S., Aquino, L. F., & Melo, V. M. M. (2013). An innovative bioremediation strategy using a bacterial consortium entrapped in chitosan beads. *Journal of Environmental Management*, *127*, 10–17. https://doi.org/10.1016/J.JENVMAN.2013.04.014

Antizar-Ladislao, B., Lopez-Real, J., & Beck, A. J. (2006). Investigation of organic matter dynamics during in-vessel composting of an aged coal–tar contaminated soil using fluorescence excitation–emission spectroscopy. *Chemosphere*, *64*(5), 839–847.

Asgher, M., Khan, S. W., & Bilal, M. (2016). Optimization of lignocellulolytic enzyme production by *Pleurotus eryngii* WC 888 utilizing agro industrial residues and bio-ethanol production. *Romanian Biotechnological Letters*, *21*(1), 11133–11143.

Augustine, B., & Roy, G. K. (1981). Recovery of chemicals from agricultural residuesd—A critical review. *Journal of the Institution of Engineers*, *62*, 36–41.

Ayyad, K., Hassanien, F., & Ragab, M. (1990). The effect of $\gamma$ irradiation on the structure of pectin. *Food/Nahrung*, *34*, 465–468.

Beg, Q. K., Bhushan, B., Kapoor, M., & Hoondal, G. S. (2000). Production and characterization of thermostable xylanase and pectinase from *Streptomyces* sp. QG-11-3. *Journal of Industrial Microbiology and Biotechnology*, *24*(6), 396–402. https://doi.org/10.1038/SJ.JIM.7000010

Buranaosot, J., Soonthornchareonnon, N., Hosoyama, S., Linhardt, R. J., & Toida, T. (2010). Partial depolymerization of pectin by a photochemical reaction. *Carbohydrate Research*, *345*, 1205–1210.

Castilho, L. R., Medronho, R. A., & Alves, T. L. M. (2000). Production and extraction of pectinases obtained by solid state fermentation of agroindustrial residues with *Aspergillus niger*. *Bioresource Technology*, *71*(1), 45–50. https://doi.org/10.1016/S0960-8524(99)00058-9

Chaturvedi, V., & Verma, P. (2013). An overview of key pretreatment processes employed for bioconversion of lignocellulosic biomass into biofuels and value added products. *Biotech*, 3(5), 415–431.

Chen, J., Liang, R.-H., Liu, W., Liu, C.-M., Li, T., Tu, Z.-C., & Wan, J. (2012). Degradation of high-methoxyl pectin by dynamic high pressure microfluidization and its mechanism. *Food Hydrocolloids*, *28*, 121–129.

Chen, Jun, Liu, Wei, Liu, Cheng-Mei, Li, T., Liang, R. H., & Luo, S. J. (2015). Pectin modifications: A review. *Critical Reviews in Food Science and Nutrition*, *55*(12), 1684–1698. https://doi.org/10.1080/10408398.2012.718722

Christensen, T. M., Nielsen, J. E., Kreiberg, J. D., Rasmussen, P., & Mikkelsen, J. (2002). Pectin methyl esterase from orange fruit: Characterization and localization by in-situ hybridization and immunohistochemistry. *Planta*, *206*, 493–503.

Coutinho, P. M., & Henrissat, B. (1999). Carbohydrate-active enzymes: An integrated database approach. In H. J. Gilbert, B. Svensson, Gideon Davies, & Bernard Henrissat (Eds.), *Recent advances in carbohydrate bioengineering* (pp. 3–12). Cambridge: The Royal Society of Chemistry.

Couto, S. R., & Sanromán, M. Á. (2006). Application of solid-state fermentation to food industry—A review. *Journal of Food Engineering*, *76*(3), 291–302. https://doi.org/10.1016/J.JFOODENG.2005.05.022

da Silva, E. G., Borges, M. D. F., Medina, C., Piccoli, R. H., & Schwan, R. F. (2005). Pectinolytic enzymes secreted by yeasts from tropical fruits. *FEMS Yeast Research*, *5*(9), 859–865. https://doi.org/10.1016/J.FEMSYR.2005.02.006

Diaz-Godinez, G., Soriano-Santos, J., Augur, C., & Viniegra-González, G. (2001). Exopectinases produced by *Aspergillus niger* in solid-state and submerged fermentation: A comparative study. *Journal of Industrial Microbiology & Biotechnology*, *26*(5), 271–275. https://doi.org/10.1038/SJ.JIM.7000113

Dingle, J., Reid, W. W., & Solomons, G. L. (1953). The enzymatic degradation of pectin and other polysaccharides. II. Application of the 'Cup-plate' assay to the estimation of enzymes. *Journal of the Science of Food and Agriculture*, *4*, 149–155.

Dische, Z. (1947). A new specific color reaction of hexuronic acids. *Journal of Biological Chemistry*, *167*, 189–198.

Downie, B., Dirk, L. M. A., Hadfield, K. A., Wilkins, T. A., Bennet, A. B., & Bradford, K. J. (1998). A gel diffusion assay for quantification of pectin methylesterase activity. *Analytical Biochemistry*, *264*, 149–157.

Elboutachfaiti, R., Delattre, C., Michaud, P., Courtois, B., & Courtois, J. (2008). Oligogalacturonans production by free radical depolymerization of polygalacturonan. *International Journal of Biological Macromolecules*, *43*, 257–261.

Favela-Torres, E., Volke-Sepúlveda, T., & Viniegra-González, G. (2006). *Production of Hydrolytic Depolymerising Pectinases*. Undefined.

Forster, H. (1988). Pectinesterase from *Phytophthora infestans*. *Methods in Enzymology*, *161*, 355–357.

Garnett, T. (2011). Where are the best opportunities for reducing greenhouse gas emissions in the food system (including the food chain)? *Food Policy*, *36*, S23–S32. https://doi.org/10.1016/j.foodpol.2010.10.010

Ghorai, S, Banik, S. P., Verma, D., Chowdhury, S., Mukherjee, S., & Khowala, S. (n.d.). Fungal biotechnology in food and feed processing. *Food Research International*, *42*, 577–587.

Godinez, G. D., Soriano-Santos, J., Augur, C., & Viniegra-González, G. (2001). Exopectinases produced by *Aspergillus niger* in solid state and submerged fermentation: A comparative study. *Journal of Industrial Microbiology and Biotechnology*, *26*, 271–275.

Gummadi, S. N., & Panda, T. (2003). Purification and biochemical properties of microbial pectinases – A review. *Process Biochemistry*, *38*, 987–996.

Harish Prashanth, K. V., & Tharanathan, R. N. (2007). Chitin/chitosan: Modifications and their unlimited application potential—an overview. *Trends in Food Science & Technology*, *18*(3), 117–131. https://doi.org/10.1016/J.TIFS.2006.10.022

Hassen, A., Belguith, K., Jedidi, N., Cherif, A., Cherif, M., & Boudabous, A. (2001). Microbial characterization during composting of municipal solid waste. *Bioresource Technology*, *80*(3), 217–225.

Hoondal, G., Tiwari, R., Tewari, R., Dahiya, N. B. Q. K., & Beg, Q. (2002). Microbial alkaline pectinases and their industrial applications: A review. *Applied Microbiology and Biotechnology*, *59*(4–5), 409–418. https://doi.org/10.1007/S00253-002-1061-1

Iconomou, D., Arapoglou, D., & Israilides, C. (2010). Improvement of phenolic antioxidants and quality characteristics of virgin olive oil with the addition of enzymes and nitrogen during olive paste processing. *Grasas Aceites*, *61*, 303–311.

Jahan, N., Shahid, F., Aman, A., Mujahid, T. Y., & Qader, S. A. U. (2017). Utilization of agro waste pectin for the production of industrially important polygalacturonase. *Heliyon*, *3*(6), 330. https://doi.org/10.1016/J.HELIYON.2017.E00330

Jayani, R. S., Saxena, S., & Gupta, R. (2005). Microbial pectinolytic enzymes: A review. *Process Biochemistry*, *40*, 2931–2944.

Kaji, A., & Yoshihara, O. (1971). Properties of purified -L-arabinofuranosidase from *Corticium rolfsii. Biochim Biophys Acta*, *250*(2), 367–371. https://doi.org/10.1016/0005-2744(71)90193-8

Kant, Shashi, Vohra, Anuja, & Gupta, R. (2013). Purification and physicochemical properties of polygalacturonase from *Aspergillus niger* MTCC 3323. *Protein Expression and Purification*, *87*(1), 11–16. https://doi.org/10.1016/j.pep.2012.09.014

Kashyap, D. R., Soni, S. K., & Tewari, R. (2003). Enhanced production of pectinase by *Bacillus* sp. DT7 using solid state fermentation. *Bioresource Technology*, *88*(3), 251–254. https://doi.org/10.1016/S0960-8524(02)00206-7

Kashyap, D. R., Vohra, P. K., Chopra, S., & Tewari, R. (2001). Application of pectinases in the commercial sector: A review. *Bioresource Technology*, *77*, 215–227.

Kawano, C. Y., Chellegatti, M. A. D. S. C., Said, S., & Fonseca, M. J. V. (1999). Comparative study of intracellular and extracellular pectinases produced by *Penicillium frequentans*. *Biotechnology and Applied Biochemistry*, *29*(2), 133–140.

Kirtchev, N., Panchev, I., & Kratchanov, C. (1989). Kinetics of acid catalysed de-esterification of pectin in a heterogeneous medium. *International Journal of Food Science & Technology*, *24*, 479–486.

Koller, A. & Neukom, H. (1967). Onteruschimgen uber den pektolytischen enzyme von *Aspergillus niger. Mitt Geb Debensmittelunter Hug*, *58*, 512.

Kumar, B., Agrawal, K., & Verma, P. (2021). Current perspective and advances of microbe assisted electrochemical system as a sustainable approach for mitigating toxic dyes and heavy metals from wastewater. *ASCE's Journal of Hazardous, Toxic, and Radioactive Waste*, *25*(2), 04020082.

Kumar, B., Bhardwaj, N., Agrawal, K., Chaturvedi, V., & Verma, P. (2020). Current perspective on pretreatment technologies using lignocellulosic biomass: An emerging biorefinery concept. *Fuel Processing Technology*, *199*, 106244.

Kumar, B., & Verma, P. (2020). Enzyme mediated multi-product process: A concept of bio-based refinery. *Industrial Crops and Products*, *154*, 112607.

Kumar, B., & Verma, P. (2021a). Biomass-based biorefineries: An important architype towards a circular economy. *Fuel*, *288*, 119622. Elsevier.

Kumar, B., Verma, P. (2021b) Techno-economic assessment of biomass-based integrated biorefinery for energy and value-added product. In: *Biorefineries: A step towards renewable and clean energy* (pp. 581–616). Singapore: Springer.

Kuo, C. H., & Wells, W. W. (1978). beta-Galactosidase from rat mammary gland. Its purification, properties, and role in the biosynthesis of 6beta-O-D-galactopyranosyl myoinositol. *Journal of Biological Chemistry*, *253*(10), 3550–3556.

Laratta, B., Masi, L. De, Minasi, P., & Giovane, A. (2008). Pectin methylesterase in *Citrus bergamia* R.: Purification, biochemical characterisation and sequence of the exon related to the enzyme active site. *Food Chemistry*, *110*(4), 829–837. https://doi.org/10.1016/J.FOODCHEM.2008.02.065

Leighton, T. (1995). Bubble population phenomena in acoustic cavitation. *Ultrasonics Sonochemistry*, *2*, S123–S136.

Liao, C.-H., Revear, L., Hotchkiss, A., & Savary, B. (1999). Genetic and biochemical characterization of an exo polygalacturonase and a pectate lyase from *Yersinia enterolitica. Canadian Journal of Microbiology*, *45*, 396–403.

Lu Zhang, X. S. (2019). Changes in physical, chemical, and microbiological properties during the two-stage composting of green waste due to the addition of β-cyclodextrin. *Compost Science & Utilization*, *27*(1), 46–60. https://doi.org/10.1080/1065657X.2019.1585304

Mahesh, M., Arivizhivendhan, K. V., Maharaja, P., Boopathy, R., Hamsavathani, V., & Sekaran, G. (2016). Production, purification and immobilization of pectinase from *Aspergillus ibericus* onto functionalized nanoporous activated carbon (FNAC) and its application on treatment of pectin containing wastewater. *Journal of Molecular Catalysis B: Enzymatic*, *133*, 43–54. https://doi.org/10.1016/J.MOLCATB.2016.07.012

McMillan, G. P., Johnston, D. J., & Pérombelon, M. C. M. (1992). Purification to homogeneity of extracellular polygalacturonase and isoenzymes of pectate lyase of Erwinia carotovora subsp. atroseptica by column chromatography. *Journal of Applied Bacteriology*, *73*(1), 83–86. https://doi.org/10.1111/J.1365-2672.1992.TB04974.X

Micheli, F. (2001). Pectin methylesterases: Cell wall enzymes with important roles in plant physiology. *Trends in Plant Science*, *6*(9), 414–419. https://doi.org/10.1016/s1360-1385(01)02045-3

Miller, G. L. (1959). Use of dinitrosalicylic acid reagent for determination of reducing sugars. *Analytical Chemistry*, *31*, 426–428.

Mohnen, D. (2008). Pectin structure and biosynthesis. *Current Opinion in Plant Biology*, *11*, 266–277. https://doi.org/10.1016/j.pbi.2008.03.006

Moreno-García, J., García-Martínez, T., Mauricio, J. C., & Moreno, J. (2018). Yeast immobilization systems for alcoholic wine fermentations: Actual trends and future perspectives. *Frontiers in Microbiology*, *9* (FEB), 241. https://doi.org/10.3389/FMICB.2018.00241

Munir, N., Asghar, M., Tahir, I. M., Riaz, M., Bilal, M., & Shah, S. M. A. (2015). Utilization of agro-wastes for production of ligninolytic enzymes in liquid state fermentation by Phanerochaete chrysosporium-Ibl-03. Undefined.

Mutter, M., Beldman, G., Schols, H. A., & Voragen, A. G. J. (1994). Rhamnogalacturonan α-L-Rhamnopyranohydrolase: A novel enzyme specific for the terminal nonreducing rhamnosyl unit in rhamnogalacturonan regions of pectin. *Plant Physiology*, *106*, 241–250.

Mutter, M., Colquhoun, I. J., Schols, H. A., Beldman, G., & Voragen, A. G. J. (1996). Rhamnogalacturonase B from *Aspergillus aculeatus* is a rhamnogalacturonan α-l-rhamnopyranosyl-(1→4)-α-d-galactopyranosyluronide lyase. *Plant Physiology*, *110*(1), 73–77. http://www.jstor.org/stable/4276966

Mutter, M., Renard, C. M. G. C., Beldman, G., Schols, H. A., & Voragen, A. G. (1998). Mode of action of RG-hydrolase and RG-lyase toward rhamnogalacturonan oligomers: Characterization of degradation products using RG-rhamnohydrolase and RG-galacturonohydrolase. *Carbohydrate Research*, *311*, 155–164.

Naessens, M., & Vandamme, E. J. (2003). Multiple forms of microbial enzymes. *Biotechnology Letters*, *25*, 1119–1124.

Nagai, M., Katsuragi, T., Terashita, T., Yoshikawa, K., & Sakai, T. (2000). Purification and characterization of an endo-polygalacturonase from *Aspergillus awamori*. *Bioscience, Biotechnology, and Biochemistry*, *64*(8), 1729–1732.

Nakamura, A., Furuta, H., Maeda, H., Takao, T., & Nagamatsu, Y. (2002). Structural studies by stepwise enzymatic degradation of the main backbone of soybean soluble polysaccharides consisting of galacturonan and rhamnogalacturonan. *Bioscience, Biotechnology, and Biochemistry*, *66*, 1301–1313.

Nakkeeran, E., Umesh-Kumar, S., & Subramanian, R. (2011). *Aspergillus carbonarius* polygalacturonases purified by integrated membrane process and affinity precipitation for apple juice production. *Bioresource Technology*, *102*(3), 3293–3297.

Nedjma, M., Hoffmann, N., & Belari, A. (2001). Selective and sensitive detection of pectin lyase activity using a colorimetric test: Application to the screening of micro-organisms possessing pectin lyase activity. *Analytical Biochemistry*, *291*, 290–296.

Niture, S. (2008). Comparative biochemical and structural character izations of fungal polygalacturonases. *Biologia*, *63*(1), 1–19.

O'Neill, M. A., Ishii, T., Albersheim, P., & Darvill, A. G. (2004). Rhamnogalacturonan II: Structure and function of a borate cross-linked cell wall pectic polysaccharide. *Annual Review of Plant Biology*, *55*, 109–139.

O'Neill, M. A., & York, W. S. (2003). The composition and structure of plant primary cell walls. In R. J. K. C. Ithaca (Ed.), *The plant cell wall* (pp. 1–54). New York: Blackwell Publishing/CRC Press.

Pedrolli, D. B., & Carmona, E. C. (2010). Purification and characterization of the exopolygalacturonase produced by *Aspergillus giganteus* in submerged cultures. *Journal of Industrial Microbiology & Biotechnology*, *37*(6), 567–573. https://doi.org/10.1007/S10295-010-0702-0

Pedrolli, D. B., Monteiro, A. C., & Carmona, E. G. and Carmona, E. (2009). Pectin and pectinases: Production, characterization and industrial application of microbial pectinolytic enzymes. *The Open Biotechnology Journal*, *3*, 9–18.

Pel, H. J., de Winde, J. H., Archer, D. B., Dyer, P. S., Hofmann, G., Schaap, P. J., Turner, G., de Vries, R. P., Albang, R., Albermann, K., Andersen, M. R., Bendtsen, J. D., Benen, J. A. E., van den Berg, M., Breestraat, S., Caddick, M. X., Contreras, R., Cornell, M., Coutinho, P. M., ... Stam, H. (2007). Genome sequencing and analysis of the versatile cell factory *Aspergillus niger* CBS 513.88. *Nature Biotechnology*, *25*(2), 221–231. https://doi.org/10.1038/nbt1282

Petersen, S. (2001). Enzymes to upgrade plant nutrients. *Feed Mix*, *9*, 12–15.

Phutela, U., Dhuna, V., Sandhu, S., & Chadha, B. S. (2005). Pectinase and polygalacturonase production by a thermophilic *Aspergillus fumigatus* isolated from decomposting orange peels. *Brazilian Journal of Microbiology*, *36*(1), 63–69. https://doi.org/10.1590/S1517-83822005000100013

Pitt, D. (1988). Pectin lyase from *Phoma medicaginis* var. pinodella. *Methods in Enzymology*, *161*, 350–354.

Puneet, P., Bhardwaj, N. K., & Singh, A. K. (2010). Enzymatic deinking of office waste paper: An overview. *Ippta Journal*, *22*, 83–88.

Rebello, Sharrel, Anju, Mohandas, Aneesh, Embalil Mathachan, Sindhu, R., Binod, P., & Pandey, A. (2017). Recent advancements in the production and application of microbial pectinases: An overview. *Reviews in Environmental Science and Bio/Technology*, *16*, 381–394.

Reid, I., & Ricard, M. (2000). Pectinase in papermaking: Solving retention problems in mechanical pulps bleached with hydrogen peroxide. *Enzyme and Microbial Technology*, *26*(2–4), 115–123. https://doi.org/10.1016/S0141-0229(99)00131-3

Renard, C. M. G. C., & Thibault, J.-F. (1996). Degradation of pectins in alkaline conditions: Kinetics of demethylation. *Carbohydrate Research, 286*, 139–150.

Rexová-Benková, L., & Markoviĉ, O. (1976). Pectic enzymes. *Advances in Carbohydrate Chemistry and Biochemistry, 33*, 323–385.

Rezić, Tonči, Oros, Damir, Marković, Iva, Kracher, Daniel, Ludwig, Roland, & Santek, B. (2013). Integrated hydrolyzation and fermentation of sugar beet pulp to bioethanol. *Journal of Microbiology and Biotechnology, 23*(9), 1244–1252. https://doi.org/10.4014/jmb.1210.10013

Rombouts, F. M., & Pilnik, W. (1980). Pectic enzymes. In A. H. Rose (Ed.), *Microbial enzymes and bioconversions* (pp. 5: 227–272). London: Academic Press.

Russ, W., & Meyer-Pittroff, R. (2004). Utilizing waste products from the food production and processing industries. *Critical Reviews in Food Science and Nutrition, 44*(1), 57–62.

Saad, N., Briand, M., Gardarin, C., Briand, Y., & Michaud, P. (2007). Production, purification and characterization of an endopolygalacturonase from *Mucor rouxii* NRRL 1894. *Enzyme and Microbial Technology, 41*(6–7), 800–805. https://doi.org/10.1016/J.ENZMICTEC.2007.07.012

Sajjaanantakul, T., Van Buren, J. P., & Downing, D. L. (n.d.). Effect of cations on heat degradation of chelator-soluble carrot pectin. *Carbohydrate Polymers, 20*, 207–214.

Sakai, T. (1992). Degradation of pectins. In G. Winkelmann (Ed.), *Microbial degradation of natural products* (pp. 57–81). Weinheim: VCH.

Sakai, T. (1999). Research on Protopectinase: A new aspect of research on Pectolytic enzymes: Development of a novel microbial function and its utilization in the bioindustry. *Memoirs of the Faculty of Agriculture of Kinki University 32*, 1–19.

Sakai, T., Sakamoto, T., Hallaert, J., & Vandamme, E. (1993). Pectin, pectinase and protopectinase: Production, properties and applications. *Advances in Applied Microbiology, 39*, 213–294.

Sakamoto, T., Bonnin, E., Quemener, B., & Thibault, J. F. (2002). Purification and characterization of two exo-polygalacturonases from *Aspergillus niger* able to degrade xylogalacturonan and acetylated homogalacturonan. *Biochimica et Biophysica Acta, 1572*(1), 10–18.

Schaub, S. M., & Leonard, J. J. (1996). Composting: An alternative waste management option for food processing industries. *Trends in Food Science & Technology, 7*(8), 263–268.

Schols, H. A., & Voragen, A. G. J. (2002). The chemical structure of pectins. In G. B. Seymour & J. P. Knox (Eds.), *Pectins and their manipulation*. USA: Wiley-Blackwell, CRC Press.

Searle-Van Leeuwen, M. J. F., Van Den Broek, L. A. M., Schols, H. A., Beldman, G., & Voragen, A. G. J. (1992). Rhamnogalacturonan acetylesterase: A novel enzyme from *Aspergillus aculeatus*, specific for the deacetylation of hairy (ramified) regions of pectins. *Applied Microbiology and Biotechnology, 38*, 347–349.

Shevchik, V. E., & Hugouvieux-Cotte-Pattat, N. (1997). Identification of a bacterial pectin acetyl esterase in *Erwinia chysanthemi*. *Molecular Microbiology, 24*(6), 1285–1301.

Siddiqui, M. A., Pande, V., & Arif, M. (2012). Production, purification, and characterization of Polygalacturonase from *Rhizomucor pusillus* isolated from decomposting orange peels. *Journal Enzyme Research, 2012*, 138634.

Singh, A., Kaur, A., Dua, A., & Mahajan, R. (2015). An efficient and improved methodology for the screening of industrially valuable xylanopectino-cellulolytic microbes. *Enzyme Research*. https://doi.org/10.1155/2015/725281

Sinitsyna, O. A., Fedorova, E. A., & Semenova, M. V., Gusakov, A. V., Sokolova, L. M., Bubnova, T. M., Okunev, O. N., Chulkin, A. M., Vavilova, E. A., Vinetsky, Y. P. & Sinitsyn, A. P. (2007). Isolation and characterization of extracellular pectin lyase from Penicillium canescens. *Biochemistry, 72*(5), 565–571.

Sreenath, Hassan K., Moldes, Ana B., Koegel, R. G., & Straub, R. J. (2001). Lactic acid production from agriculture residues. *Biotechnology Letters, 23*, 179–184. https://doi.org/10.1023/A:1005651117831

Tanabe, H., Yoshihara, Y., Tamura, K., Kobayashi, Y., Akamatsu, I., Niyomwan, N., & Footrakul, P. (1987). Pretreatment of pectic wastewater from orange canning process by an alkalophilic *Bacillus* sp. *Journal of Fermentation Technology*, *65*, 243–246.

Tapre, A. R., & Jain, R. K. (2014). Pectinases; Enzymes for fruit processing industry. *International Food Research Journal*, *21*, 447–453.

Tian, Guo-Wei, Chen, Min-Huei, Zaltsman, Adi, & Citovsky, V. (2006). Pollen-specific pectin methylesterase involved in pollen tube growth. *Developmental Biology*, *294*(1), 83–91. https://doi.org/10.1016/j.ydbio.2006.02.026

Tuck, C. O., Pérez, E., Horváth, I. T., Sheldon, R. A., & Poliakoff, M. (2012). Valorization of biomass: Deriving more value from waste. *Science*, *337*(6095), 695–699.

Tuomela, M., Vikman, M., Hatakka, A., & Itävaara, M. (2000). Biodegradation of lignin in a compost environment: A review. *Bioresource Technology*, *72*(2), 169–183.

Venkatesh, K. S., & Umesh-Kumar, S. (2005). Production of pectinases and utilization in food processing. In Robert E. Levin, K. Shetty, G. Paliyath, & A. Pometto (Eds.), *Food biotechnology* (pp. 329–348). London: Taylor and Francis.

Vincken, J. P., Schols, H. A., Oomen, R. J., McCann, M. C., Ulvskov, P., Voragen, A. G., & Visser, R. G. (2003). If homogalacturonan were a side chain of rhamnogalacturonan I: Implications for cell wall architecture. *Plant Physiology*, *132*, 1781–1789.

Viniegra-González, G., Favela-Torres, E., Aguilar, C. N., Rómero-Gomez, S. de J., Díaz-Godínez, G., & Augur, C. (2003). Advantages of fungal enzyme production in solid state over liquid fermentation systems. *Biochemical Engineering Journal*, *13*(2–3), 157–167. https://doi.org/10.1016/S1369-703X(02)00128-6

Vlugt-Bergmans, C. J. B., Meeuwsen, P. J. A., Voragen, A. G. J., & Van Ooyen, A. J. J. (2000). Endo-xylogalacturonan hydrolase, a novel pectinolytic enzyme. *Applied and Environmental Microbiology*, *66*(1), 36–41.

Whitaker, J. R. (1990). Microbial Pectinolytic enzymes. In W. M. Fogarty & C. T. Kelly (Eds.), *Microbial enzymes and biotechnology* (2nd Edition, pp. 133–176). Dordrecht: Springer.

Willats, W. G. T., Knox, P., & Mikkelsen, J. D. (2006). Pectin: New insights into an old polymer are starting to gel. *Trends in Food Science & Technology*, *17*, 97–104.

Xu, Q. H., Wang, Y. P., Qin, M. H., Fu, Y. J., Li, Z. Q., Zhang, F. S., & Li, J. H. (2011). Fibre surface characterization of old newprint pulp deinked by combining hemicellulase with laccase-mediated system. *Bioresource Technology*, *102*, 6536–6540.

Yapo, B. M., Lerouge, P., Thibault, J.-F., & Ralet, M. C. (2007). Pectins from citrus peel cell walls contain homogalacturonans homogenous with respect to molar mass, rhamnogalacturonan I and rhamnogalacturonan II. *Carbohydrate Polymers*, *69*, 426–435.

Zhou, H., Li, X., Guo, M., Xu, Q., Cao, Y., Qiao, D., Cao, Y., & Xu, H. (2015). Secretory expression and characterization of an acidic endo-polygalac turonase from *Aspergillus niger* SC323 in *Saccharomyces cerevisiae*. *Journal of Microbiology and Biotechnology*, *25*(7), 999–1006.

Zieminski, K., Romanowska, I., & Kowalska, M. (2012). Enzymatic pretreatment of lignocellulosic wastes to improve biogas production. *Waste Management*, *32*(6), 1131–1137. https://doi.org/10.1016/j.wasman.2012.01.016

Zou, M., Li, X., Shi, W., Guo, F., Zhao, J., & Qu, Y. (2013). Improved production of alkaline polygalacturonate lyase by homologous overexpression pelA in *Bacillus subtilis*. *Process Biochemistry*, *48*(8), 1143–1150. https://doi.org/10.1016/J.PROCBIO.2013.05.023

# 7 Recent Advances in Enzyme-assisted Hydrolysis of Waste Biomass to Value-added Products

*Ria Majumdar, Umesh Mishra, Biswanath Bhunia, and Muthusivaramapandian Muthuraj*
National Institute of Technology Agartala, India

## CONTENTS



DOI: 10.1201/9781003187684-7

## 7.1 INTRODUCTION

According to a UN (United Nations) report, the global population may reach 9.7 billion in 2050 from 7.7 billion, with a growth rate of ~2 billion people in the next three decades (Global-Agriculture, 2009; UN, 2019). With this demographic expansion and growing luxury requirements, demand for every precursor material is increasing. The fulfillment of growing demands primarily relies on petroleum-based conventional resources. For example, 85% of the global energy demand depends on fossil fuel resources (Singh & Dwevendi, 2019), which augments the $CO_2$ emissions and raises global warming alarms. An ~21% rise in $CO_2$ emissions is also expected in 2040 from $35.6 \times 10^9$ tons of $CO_2$ emissions in 2020 (Conti et al., 2016), which would increase the earth's temperature further (Li et al., 2017; Van Nes et al., 2015). These environmental and geopolitical concerns force the world to choose cost-effective, eco-friendly, non-toxic, and sustainable resources.

Biomass is the best alternative source for sustainable and cleaner environmental applications to overcome environmental and socioeconomic problems. Additionally, waste biomass utilization has gained attention because of the abundance of biomass, its non-competitiveness with food, and the avoidance of improper disposal. Wastes are currently an emerging issue worldwide and will keep growing with society and industry development. A World Bank report states that global biomass waste may increase by 70% from the current amount by 2050 (Waste, 2018). Furthermore, the attempts to transform waste biomass into various valuable materials give tremendous direct and indirect advantages to the environment, help minimize or avoid dependency on fossil fuel-based materials, and add revenue. Lignocellulosic biomasses (LCB) are the largest reservoir of renewable feedstock on earth and consist of potentially fermentable carbohydrates (Mtui & Nakamura, 2007; Agrawal & Verma, 2020a). The LCB are generated as wastes from agricultural residues obtained after crop harvesting and from food processing industries that rely on crops and agricultural products (Qi et al., 2005; Rodríguez et al., 2008; Roig et al., 2006; Bhardwaj et al., 2020; Nair et al., 2022a). Due to their availability, renewability and nutrient-rich nature, there has been a great interest in utilizing such waste biomasses for value-addition.

This chapter discusses the merits and demerits of various enzymatic hydrolysis treatment methodologies and summarizes the socioeconomic and environmental advantages of transforming waste into different high-value, bio-based components. The chapter gives a general idea of different recoverable value-added products and, concurrently, the role of enzymes in the recovery process, along with opportunities and prospects of such valorization. Therefore, this chapter provides guidelines for researchers, industrialists, policymakers, stakeholders, and other companies to understand the current status, progress, limitations, merits, demerits, and future pathways for waste biomass transformation for effective value-addition by enzymatic hydrolysis.

## 7.2 CURRENT SCENARIO OF WASTE BIOMASS POTENTIAL

As reported by the National Renewable Energy Laboratory (NREL), biomass material represents organic matter derived from any living beings, that is, plants and animals (NREL, 2012). Biomass is abundantly available in nature (Ahorsu et al., 2018) and is most commonly used for energy or fuel production (Alayi, Sobhani, & Najafi, 2020). Therefore, it excludes fossil fuels derived from the biomass of dead organisms that take millions of years to convert (Gopal & Sathiyagnanam, 2018; Shamel et al., 2014). Biomass is a stored reservoir of solar energy and inorganic $CO_2$ that plants and trees capture through photosynthesis and convert into plant growth molecules such as cellulose, hemicellulose, and lignin. Therefore, it encompasses a large amount of organic matter and is considered the most important food and fiber production source. Alternatively, wastes are defined as any rejected, abandoned, discarded, unwanted, or surplus matter, whether or not intended for sale or recovery, recycling, or purification by a different method, unlike the processing method of the primary raw materials (EPA, 2019). It is a by-product of zero or comparatively little economic value. Several types of biomass wastes include municipal solid wastes (MSWs), agricultural residues, wood residues, and wastes resulting from logging, aquatic plants and algae (Ozturk et al., 2017; Goswami et al., 2021; Goswami et al., 2022a).

Assessment of biomass potential provides a general overview of the quantity and quality of biomass, plantation, production, harvestings, waste generation, management, conversion techniques, components, marketing strategies, and other associated activities (Dessie et al., 2020). According to the International Energy Agency (IEA), eight different precautions must be taken before the utilization of biomass for energy generation. Biomass employed for energy generation can be categorized as agricultural residues, energy crops, municipal wastes, and industrial wastes (from food industries, oil industries, etc.). With massive agricultural land availability in India, many agricultural residues are generated, and the overall production capacity of agricultural residues reaches 511 metric tons per year. Table 7.1. illustrates previously reported estimations of biomass potential. Maximum biomass potential among agricultural residues was reported for paddy husk (169 Mt per year), followed by wheat and cotton. Among the energy crops cultivated as biomass feedstock, sunflower, soybean, mustard, bamboo, and sugar cane are the predominant sources. A study showed that India has a vast potential for power generation from biomass (17,538 MW), bagasse (5000 MW), and wastes (2707 MW). Unfortunately, the estimation of biomass potential shows considerable variation due to seasonal and regional variations or different definitions of biomass (Long et al., 2013; Thrän et al., 2010). For example, tires made from either natural or synthetic rubbers do not fall in the category of

**TABLE 7.1**
**Biomass Potential in India**

| Biomass Category | Year | Estimation | References |
|---|---|---|---|
| Biomass | 1999 | 2 to 17 t/ha/yr | Sudha and Ravindranath (1999) |
| Biomass | 2003 | 6.6 t/ha/yr | Sudha et al. (2003) |
| Fuel wood | 2003 | 82 t/yr | Sudha et al. (2003) |
| Lignocellulose | 2018 | 200 billion t/yr | De Bhowmick et al. (2018) |
| Crop residue | 2014 | 686 t/yr | Hiloidhari (2014) |
| Agricultural residue | 2015 | 611 t/yr | Cardoen et al. (2015) |
| Switch grass | 2020 | 260 t/yr | Usmani (2020) |
| Mix crops | 2020 | 230 t/yr | Usmani (2020) |
| Miscanthus | 2020 | 770 t/yr | Usmani (2020) |
| Solid waste | 2018 | 100,000 t/day | Fiksel and Lal (2018) |
| Agricultural residue waste | 2020 | 500 t/yr | Kapoor et al. (2020) |
| Industrial waste | 2012 | 7.2 t/yr | Dixit and Srivastava (2016) |
| Livestock waste | 2017 | 2600 t/yr | Kaur et al. (2017) |

biomass, but for political purposes, they may be included under biomaterials (Toklu, 2017). However, the biomass potential is enormous for both developed and developing countries.

## 7.3 ENVIRONMENTAL IMPACTS OF WASTE BIOMASS

With the growing population, the amount of waste biomass generation is rapidly increasing globally, and needs to be utilized with the advanced and proper technology. Such waste biomasses have various adverse impacts on the environment. On the other hand, waste biomass requires a huge amount of financial support for its management. Major issues associated with waste biomass arise during its generation, collection, and transportation. From research data, about 100,000 metric tons solid wastes are generated in India per day (Fiksel & Lal, 2018). The majority of such solid waste materials are dumped in landfills or in household areas by the municipal corporation (Misra & Pandey, 2005). For example, the municipal corporation of greater Mumbai collects 6,000 t/day of waste (Annepu, 2012); however, they have only four landfills (dumping sites), one of which has operated since 1927. Therefore, a significant portion of these solid wastes is dried and burnt to minimize dumping grounds or used as household fuel feedstocks. This led to a rise in GHG emissions and global warming. Various gases are emitted during the incineration of solid wastes, such as methane ($CH_4$) from the dumping sites, $CH_4$ and nitrous oxide ($N_2O$) from the combustion of agricultural wastes, and carbon dioxide ($CO_2$) and CO (carbon monoxide) from fuel burning. Additionally, leachate generated from wastes after their partial degradation may pollute soil and water resources. It also affects the health condition of workers, transporter, and residents near the dumping area.

For commercial purposes, the waste biomass needs to be utilized to minimize landfills, incineration, GHG emission, and water pollution. In India, plastic wastes are collected by waste pickers, and 70% of those plastics are recycled and converted into valuable resources. Composting plastics is not a promising way of recycling as it may take a considerable time in relation to the rate of waste generation. Organic wastes create foul odors after their decomposition and may serve as a habitat for various pests and insects. Hence, recycling and bioremediation of organic wastes can be performed by their segregation, biological decomposition, and stabilization (Sharholy et al., 2008). However, organic waste materials can also be recycled through the composting process, where bioconversion is done under moist and open-air conditions. The end product is well known as compost that is rich in nutrients and has high agricultural value. The volume of waste can be reduced by 40–50% by the composting method (Sharholy et al., 2008), which can be carried out manually or mechanically. Moreover, vermicompost is a highly nutrient-rich soil fertilizer made by stabilization of organic matter with the help of both aerobic microbes and earthworms (Bundela et al., 2010). It is also known as a biofertilizer. Biomethanation of anaerobic digestion is another method of waste recycling where anaerobic bacteria decompose the organic matter from waste material buried under the ground. It is a natural process that usually occurs in landfills and is slower than aerobic decomposition. Biomethanation of waste recycling leads to energy recovery by the generation of biogas. In 2000, the Ministry of Environment and Forests (MoEF) of India issued regulations for the proper handling and management of MSW, ensuring appropriate collection techniques, segregation, transportation facility, processing and pretreatment techniques, and disposal to a safe site. Nowadays, various NGOs and private sectors are working on waste minimization to make our environment healthy and safe.

## 7.4 STRUCTURE, COMPOSITION, AND PROPERTIES OF BIOMASS WASTE

The only renewable, natural carbon source to be used as a substitute for petroleum sources is biomaterial. Biomass sources include trees, vegetation, water, and land-based organisms, dead biomass, MSW, agricultural residues, livestock manure, forestry biomass, industrial sludge, algae, grass, wood, and biosolids (Champagne, 2007; Goswami et al., 2022b; Goswami et al., 2022c). LCB wastes are composed of biopolymers such as cellulose (40–50%), hemicellulose (25–35%), and lignin (15–20%) (Mtui, 2009; S. Sun et al., 2016; Kumar & Verma, 2020, Nair et al., 2022b) that are firmly combined via hydrogen and covalent bonding. This may lead to the formation of a complex biomass structure that may act as a recalcitrant for value-addition. Lignin, a phenolic polymeric component, is considered the inhibitor in the conversion process of biomass surrounding crystalline cellulose and pentose sugar. The structure of cellulose, hemicellulose, and lignin is $(C_6H_{10}O_5)_n$, $(C_5H_8O_4)_m$, and $[C_9H_{10}O_3(OCH_3)_{0.9\text{-}1.7}]_x$ respectively. In addition to these three biopolymeric components, biomass is also composed of a minor amount of pectin, extractive, ash, chlorophyll, protein, and others (Akhtar et al., 2016). Table 7.2 illustrates variations in the composition of different biomass materials. The composition of these significant components may differ with the category and source of feedstock.

**TABLE 7.2**
**Composition of Various Type of Waste Biomass**

| Waste Biomass | Cellulose (%) | Hemicellulose (%) | Lignin (%) | Ash | Moisture | References |
|---|---|---|---|---|---|---|
| Sugarcane bagasse | 40–50 | 30–35 | 20–30 | - | - | Cardona et al. (2010) |
| Corn stover | 38.4 | 23 | 20.2 | - | - | Wan and Li (2010) |
| Elephant grass | 31.3 | 29 | 15.8 | 8.7 | - | Scholl et al. (2015) |
| Rice husk | 37.1 | 29.4 | 24.14 | - | - | Kalita et al., (2015) |
| Rice straw | 32.15 | 28.0 | 19.64 | 11.33 | - | Shawky et al. (2011) |
| Corn stalks | 29.80 | 33.30 | 16.65 | 8.50 | - | Shawky et al. (2011) |
| Corn stover | 36.8 | 21.5 | 17.5 | 9.3 | - | Lin et al. (2010) |
| Soybean hull | 28.6 ± 0.7 | 20.0 ± 0.5 | 13.1 ± 0.2 | 0.2 ± 0.1 | - | Qing et al. (2017) |
| Soybean straw | 42.3 ± 0.9 | 16.7 ± 0.4 | 21.6 ± 0.5 | 0.2 ± 0.1 | - | Qing et al. (2017) |
| Newspaper | 49.3 | 12.2 | 19.18 | 1.5 | 3 | Guerfali et al. (2015) |
| Office papers | 78.6 | 4.7 | 1.2 | 9 | 3.2 | Guerfali et al. (2015) |
| Mature bamboo | 47.4 | 27.2 | 15.1 | - | - | Li et al. (2015) |
| Bamboo | 40.4 | 21.6 | 25.9 | 2.8 | 9.3 | Xiao et al. (2014) |
| Poplar | 51.3 | - | 23.4 | - | - | Meng et al. (2016) |
| Switch grass | - | 25.5 | - | - | - | Meng et al. (2016) |
| Bermuda grass | 25.59 | 15.88 | 19.33 | 6.6 | - | Wang et al. (2010) |
| Corncob | 30.0 ± 0.5 | 34.0 ± 1.0 | 18.4 ± 0.3 | 1.6 ± 0.2 | - | Xu et al., (2017) |
| Sorghum stalk | 38.2 ± 0.2 | 33.0 ± 0.1 | 9.9 ± 0.1 | 3.1 ± 0.2 | - | Xu et al. (2017) |
| Corn stover | 30.0 ± 0.1 | 26.1 ± 0.1 | 11.0 ± 0.1 | 4.9 ± 0.2 | - | Xu et al. (2017) |

Lignocellulosic biomass (LCB) wastes may be grouped into different categories such as waste papers, grasses, agricultural residues (straw, stalks, nutshells, husks, peels, nonfood seeds, domestic wastes, cobs, etc.), wood residues (paper mill residues and sawdust), food industry residues, and municipal solid wastes (MSW) (Cardona et al., 2010; Bhardwaj and Verma 2020). Agricultural wastes contribute the most prominent portion to the LCB wastes for the production of biodegradable polymer and include potato peels (Xie et al., 2020), fruit seeds (Santana et al., 2018), coconut shells (Nunes et al., 2020), fruit peels (Bashir et al., 2018), and so on. The limitation of any feedstock is the structural complexity that primarily depends on the nature and composition of the biomass. Cellulose, a linear polysaccharide, is an essential component for value-addition and consists of subunits of glucose linked by β-(1,4)-glycosidic bonds (Kumar et al., 2017). It is insoluble in normal water but slightly or fully soluble in highly acidic and alkaline water. Nonetheless, cellulose is also soluble in solvents such as ionic liquids (ILs) and N-methylmorpholine-N-oxide (NMMO). It possesses various beneficial properties like biocompatibility, hydrophilicity, stereoregularity, and hydroxyl groups. It also serves as a versatile reservoir for value-added materials, for example, nanofibers, biocomposites, medicine, fertilizers, bioethanol, enzyme recovery, chemicals and nutrients (Jedvert & Heinze, 2017; Kumar et al., 2020a; 2020b).

## 7.5 SIGNIFICANCE OF ENZYMATIC HYDROLYSIS TREATMENT

Lignocellulosic biomass does not have monosaccharides available for bioconversion to different valuable products. Instead, such biomasses contain polysaccharides (cellulose, hemicellulose) which need to break down to simple sugars for fermentation either by acid or enzymatic breakdown. Enzyme-assisted hydrolysis is considered a promising method of obtaining fermentable sugars from LCBs, but has lower chances of accessibility due to its complex structure limiting cellulose-to-ethanol production (Cardona et al., 2010; Bhardwaj et al., 2021). Several factors such as high-lignin content, high crystallinity and degree of polymerization, protection of cellulose by lignin, hemicellulose-cellulose coatings, accessibility of low surface area of cellulose to an enzyme, toxicity of derivatives of lignin to microorganisms, the strength of fibers, among others, are responsible for the recalcitrance of biomass (Arantes & Saddler, 2010; Zhang et al., 2007). Thus, pretreatment technologies are an essential step before the enzymatic conversion of biomass. Pretreatment reduces the crystallinity of cellulose, removes the matrices, and increases biomass porosity, thereby accelerating the hydrolysis process (Keller et al., 2003).

An ideal pretreatment method should be eco-friendly and economical, using non-toxic chemicals, handy equipment, and simple procedures. It should not be so harsh as to reduce total yield nor so weak as to limit the accessibility of enzymes for breaking down the cellulose. A combination of physical and chemical pretreatments provides an innovative way to treat LCBs where hemicellulose is appropriately dissolved in the process, and the structure of lignin can be transformed. In this manner, cellulose accessibility can be improved. For example, steam explosion is considered as one of the essential physicochemical treatments where lignin conversion and hemicellulose degradation occurs due to intense heat and pressure. It may improve the workability of enzyme-assisted hydrolysis, minimize generation of inhibitory elements and thus make possible complete liquefaction of xylan, hemicellulose, galactan, glucan, arabinan, and mannan (Jeoh & Agblevor, 2001; Sun & Cheng, 2002).

Various pretreatment methods such as pulsed electric field, ionizing and non-ionizing radiation, ultrasonic, and high-pressure treatments are emerging as the most valuable techniques for the valorization of waste biomass by cellulose conversion (Hassan, Williams, & Jaiswal, 2018). But, their massive consumption of energy limits their use due to high final production costs. Therefore, the adoption of cheaper sources, energy-saving technologies, and eco-friendly advanced procedures have emerged in the last decades. Extensive research has been implemented to investigate the effect of pretreatment technologies in biomass composition; however, each method has its pros and cons. Among all the physical, physicochemical, and chemical pretreatments, biological treatment prior to enzymatic hydrolysis is the most economical and eco-friendly conversion technique to obtain the required products. There is no formation of inhibitors during such a pretreatment process, which supports faster biomass synthesis. Various microbes such as fungi, bacteria, and enzymes are used in the biological conversion of LCBs. Biological treatment by fungi is an emerging technique for better digestibility of cellulose. White-rot, brown-rot, and soft-rot fungi are generally used for delignification and hemicellulose removal from LCBs (Chen et al., 2010; Alma et al. 2021; Agrawal and Verma 2021). However, as well as lignin removal, there are several other biological pretreatment applications such as removing antimicrobial components (Wan & Li, 2012).

Enzymatic hydrolysis has several advantages over acid hydrolysis treatment as it generates negligible amounts of toxic products and also consumes a minimal amount of energy, thereby making the cellulose conversion process perfect and cost-effective cellulose conversion process (Bhatia et al., 2017; Galbe & Zacchi, 2002; Sindhu et al., 2016). The improvement of enzyme-assisted hydrolysis can be achieved by optimizing the concentration of substrates, dose of cellulose, enzyme mixture, surfactants' role, and enzyme recycling. However, the main limitation with LCBs is their high processing costs and treatment processes because of their structural complexity. The pores of the LCB structure are blocked by lignin and hemicellulose content, thereby limiting the enzymatic degradation process. Therefore, the efficiency of enzymatic saccharification depends on the nature of the cellulose and if it can be enhanced by the removal of impurities (wax, pectin, lignin, hemicellulose, chlorophyll, etc.,) (Wyman et al., 2005). Nonetheless, utilization of dry substrates increases the inhibitory components that lead to the reduction of sugar yield at a certain level. The development of cost-effective technologies for value-addition depends on various government policies. Processing and treatment of such biomasses are thus aimed at nanobiotechnology and genetic engineering with advanced methods.

## 7.6 DIFFERENT TYPES OF ENZYMES AND THEIR ROLE IN NOVEL WASTE CONVERSION STRATEGY

Enzymes are proteins that act as biocatalysts. They are used to speed up the reaction process a million times faster than it would have been without it. Therefore, they can be used for various purposes. Cellulose hydrolysis can be performed using various microbial cellulolytic enzymes (Cardona et al., 2010) and ligninolytic enzymes (Biko et al., 2020). For this purpose, commercially available cellulose enzymes are generally used for high-yield sugar production (Lynd et al., 1996). Enzymes used for delignification or lignin removal from the LCBs are known as ligninolytic enzymes (Baruah et al., 2018), and help in boosting the digestibility of LCBs, thereby reducing the cost of the whole conversion process. Some bacteria, fungi, and insects can produce enzymes that can generate cellulolytic, hemicellulolytic, and ligninolytic enzymes. Therefore, various oxidative and hydrolytic enzymes are also prepared for commercial purposes instead of whole organisms. The ligninolytic enzymes can be further classified into two families, namely peroxidase (lignin peroxidase (LiP), manganese peroxidase (MnP), and versatile peroxidase (VP)) and phenol oxidase (laccase) (Brinton et al., 2015; Agrawal et al., 2020; Agrawal & Verma, 2020b). Table 7.3 lists various types of enzymes involved in the hydrolysis of biomaterials.

Though most of the research work has been performed using commercial enzymes, the high cost is a major drawback for the application of enzymes in the hydrolysis process. The literature illustrates that procurement of enzymes amounts to almost 40% of the overall cost of bioethanol production (Arora et al., 2015). On the other hand, the pulp-paper and textile industries release a higher concentration of

**TABLE 7.3**
**Role of Various Types of Enzymes in Bioconversion**

| Feedstock | Enzyme | Type & Role | Source | Preparation Conditions | Effect on Biomass Feedstock | References |
|---|---|---|---|---|---|---|
| Industrial effluent | Lignin peroxidase (LiP) and Manganese peroxidase (MnP) | Lignino-lytic enzyme (detoxifica-tion) | *Pseudomonas aeruginosa* and *Serratia marcescens* | - | Able to degrade and detoxify lignin by 44% and 49% respectively. 70% and 75% decolori-zation of textile effluent was observed | Bholay et al. (2012) |
| Corn stover | MnP and laccase | Lignino-lytic enzyme (delignifica-tion) | *Ceriporiopsis subvermi spora* | Solid-state cultivation for 42d | Delignifica-tion (for cellulose accessibility) – 39.2% lignin removal and 57–67% glucose yield increase after 18-42d of pretreatment | Wan and Li (2010) |
| Corn stover | Xylanase | Cellulolytic enzyme | *Ceriporiopsis subvermi-spora* | Solid-state cultivation | Overall glucose yield of 72% was obtained | Wan and Li (2010) |
| Apple pomace<br>Coffee silverskin | Laccase | Lignino-lytic enzyme (lignin removal) | *Pleurotus ostreatus* | Commercial enzyme | Saccharifica-tion yield = 83%<br>Saccharifica-tion yield = 73% | Giacobbe et al. (2018) |
| Bamboo (*Dendrocalamus*) | Cellulase | Cellulolytic enzyme (cellulose conversion) | - | Hydrolysis conducted in a shaking incubator at 50°C, 200 rpm, 120min | Glucose yield was increased from 15.7% to 75.7% | Xiao et al. (2014) |
| Rice straw | Laccase | Lignino-lytic enzyme (lignin removal) | *Tyromyces chioneus* | Incubation period 4h, Incubation temp: 30°C, in a rotary shaker at 150 rpm. | Removal efficiency of phenolic compounds = 49.8%<br>Saccharification yield was observed up to 74.2% | Dhiman et al. (2015) |
| Willow tree | | | | | Removal efficiency of phenolic compounds = 32.6%<br>Saccharifica-tion yield was observed up to 63.6% | |

*(Continued)*

**TABLE 7.3 (CONTINUED)**
**Role of Various Types of Enzymes in Bioconversion**

| Feedstock | Enzyme | Type & Role | Source | Preparation Conditions | Effect on Biomass Feedstock | References |
|---|---|---|---|---|---|---|
| Wheat straw | Laccase | Lignino-lytic enzyme (lignin removal) | *Trametes versicolor* | - | Released higher glucose concentration up to 2.3 g/L as compared to control | Heap et al. (2014) |
| Rice husk | Cellulase and hemicellu-lase | Cellulolytic enzyme (cellulose conversion) | *Phanero chaete chryso sporium* | Commercial enzyme (MTCC) – Solid state fermentation - 30±1°C, for 5 days | Maximum amount of reducing sugar released was 485mg/g of the substrate | Saratale et al. (2014) |
| Banana stalk, sugarcane bagasse | Cellulase | Cellulolytic enzyme (Cellulose conversion) | *Trichoderma viride, Aspergillus niger, Fusarium oxysporum* | The culture was incubated for 5-7 days at 28°C | Highest cellulose content observe in banana stalks = 36.23%<br>Highest hemicellu-lose content in sugarcane bagasse = 24.82%<br>*Trichoderma viride* is most efficient.<br>Yield cellulose conversion of banana stalk, sugarcane bagasse and paddy straw = 55.28, 51.59 and 53.70% respectively | Kadarmoidheen et al. (2012) |
| Wood (*Eucalypt-us globulus*) | Laccase | Lignin-olytic enzyme (lignin removal) | *Mycelio phthora thermophila* | 45°C for 72h in thermostate shaker at 170 rpm | 50% lignin removal was observed | Rico et al. (2014) |
| Rice straw | Laccase | Lignin-olytic enzyme (detoxifi-cation) | *Coltricia perennis* | Incubated for 13 days at 30°C and pH was adjusted to 5.0 | Removal of 37%, 45%, 76% and 68% of phenolic components at pH 3.5, 4, 4.5 and 5 respectively | Kalyani et al. (2012) |

| | | | | | | |
|---|---|---|---|---|---|---|
| Elephant grass | Cellulase and xylanase | Cellulolytic enzyme | *Penicillium echinulatum* | Solid state cultivation | Enzyme recovery (cellulase and xylanase | Scholl et al. (2015) |
| News papers | Cellulase and β-glucosidase | Cellulolytic enzyme | *Trichoderma reesei* and *Aspergillus niger* | Incubated for 5 days, 30 °C, at rotary shaker (160rpms) | Improved the digestibility by 45% | Guerfali et al. (2015) |
| Poplar Switch grass | Cellulase and β-glucosidase | Cellulolytic enzyme | *Trichoderma reeseil* and Novozyme | Incubation under continuous agitation at 50°C, 150 rpm, 3 days, antimycotic and antibiotic conditions | Switch grass has better glucose yield before and after pretreatment. Complete hemi-cellulose removal | Meng et al. (2016) |
| Sugarcane bagasse | Xylanase and β-glucosidase | Cellulolytic enzyme | *Aspergillus fumigatus* | Solid-state fermentation- the culture was incubated at 50°C for 12h and 24h at 150 rpm | 10 times more glucose production after saccharification | (Dos Santos et al. (2019) |
| Banana waste | Mix species | Cellulolytic and ligninolytic enzyme | *Pleurotus ostreatus* and *P.sajor-caju* | Solid-state fermentation – growth measurement for 40d, | Enzyme recovery (cellulolytic and ligninolytic enzyme) – Maximum activity was observed between 10 and 20 days | Reddy et al. (2003) |

pollutants into the environment and such waste disposal is an emerging issue. Therefore, enzyme recovery from waste biomass (naturally available and cheap) and their application in the conversion process may be an effective and nontoxic alternative path in its sector (Dessie et al., 2018). Fungi are considered the main cellulose enzyme-producing organisms. However, some bacteria and actinomycetes have also been reported to yield cellulase activity. Different types of enzymes are detailed in Table 7.3.

### 7.6.1 Cellulolytic Enzymes

Different microbes like bacteria, fungi, and actinomycetes can produce cellulolytic enzymes, where fungi are the main cellulase-producing organisms. Cellulase and hemicellulase production from *Penicillium, Trichoderma, Aspergillus*, and *Botrytis* has extensively been exploited from various agricultural waste biomass past decades. These cellulolytic enzymes hydrolyze the cellulose content of LCB to simple fermentable monosaccharides like glucose (Kadarmoidheen et al., 2012).

### 7.6.2 Ligninolytic Enzymes

#### 7.6.2.1 Laccases

Laccase (benzenediol: oxygen oxidoreductase) is an essential class of ligninolytic enzyme that has the potential ability for oxidation. It belongs to blue copper-containing oxidase and is therefore known as blue copper proteins. Yoshida first discovered it in 1883 from a Japanese lacquer tree called *Rhus vernicifera*. Laccase is mainly found in white-rot fungi. In 1896, both Laborde and Bertrand first reported the presence of laccase in fungi (Peisach et al., 1966; Thurston, 1994). Fungal laccase has a high ability in redox reaction compared to plant or bacterial laccase, therefore, high potential application in the biotechnological field. The molecular weight of laccase varies between 60 and 80 kDa.

Meanwhile, fungal laccase is also involved in the oxidation of toxic phenolic compounds and aromatic amines in the presence of compounds with low molecular weight, which are well known as mediators or enhancers. Due to high specificity, eco-sustainability, and proficient catalytic activity (Giacobbe et al., 2018), laccase is very efficient in delignification and modification, which eventually intensify the yield of both saccharifications as well as the fermentation process by altering the lignin porosity and hydrophobicity (Fillat et al., 2017; Piscitelli et al., 2011; Agrawal and Verma, 2020b; Agrawal and Verma, 2020a). Several fungal and bacterial laccases are used alone or combined with a mediator for delignification and/or detoxification of pretreated feedstocks (Fillat et al., 2017). Even though high detoxification was obtained, contrasting outcomes on delignification have been observed because of the grafting phenomenon (covalent coupling between phenolic radicals and aromatic lignin fibers) (Moreno et al., 2016; Oliva-Taravilla et al., 2015).

There are diverse sources of laccase-producing organisms like bacteria, fungi, insects, and plants. Recently, laccase has been used in a wide range of applications: in food industries for organic synthesis; in pulp and paper industries for bioremediation such as detoxification and decoloration; in wine factories for removal of phenolic compounds; in pharmaceutical industries for the biosynthesis of complex compounds; and in wastewater treatment centers for dye removal, especially from detergent and washing powders. The utilization of laccase in the biotechnological field can be enhanced by introducing a laccase-mediator system. This system can oxidize non-phenolic compounds that are either difficult or not able to be oxidized by the enzyme alone. The efficiency and potentiality of a laccase-mediator system on the industrial scale have widely been studied for the modification and removal of lignin from untreated as well as pretreated wood residues, improvement of the hydrolysis process, and analysis of chemical modifications (Rico et al., 2014).

#### 7.6.2.2 Lignin Peroxidase (LiP)

Lignin peroxidase was first discovered in a white-rot fungi named *Phanerochaete chrysosporium*, which grows in a liquid nitrogen medium. It belongs to a family of extracellular heme proteins that produces during the secondary metabolism of those fungi. The molecular mass of LiP ranges between 35 and 40 kDa. LiP can oxidize $H_2O_2$-dependent aromatic lignin compounds, eventually resulting in aryl cation radical formation. Like laccase, LiP also has a wide range of applications due to its high potential in delignification, detoxification, and bioconversion to energy and various chemicals (C. A. Reddy & D'Souza, 1994). Nowadays, several research works have reported the production of LiP from white-rot fungi such as *Trametes versicolor, Phlebia flavido-alba, Phanerochaete chrysosporium, Bjerkandera* sp. strains *BOS55,Gloeophyllum trabeum, Trametes trogii, Phlebia tremesllosa*, among others, (Baruah et al., 2018).

#### 7.6.2.3 Manganese Peroxidase (MnP)

Manganese peroxidase (MnP) belongs to the second family of extracellular peroxidase, that is, *P. chrysosporium* and other white-rot fungi. It catalyzes Mn-dependent reactions. Even though MnP is a heme glycoprotein like LiP, they are different in their activity. LiP oxidizes non-phenolic compounds of lignin whereas MnP oxidizes both phenolic and non-phenolic parts of lignin. It also catalyzes $Mn^{2+}$ and $Mn^{3+}$ that further oxidizes the phenolic ring to phenoxy radicals leading to the decomposition of compounds (Brown & Chang, 2014).

#### 7.6.2.4 Versatile Peroxidase (VP)

Versatile peroxidase (VP) oxidizes phenolic as well as non-phenolic lignin compounds. It has a similar catalytic activity to both LiP and MnP, that is, it catalyzes high redox reactions of phenolic compounds like LiP, and oxidizes Mn group compounds like MnP (Abdel-Hamid et al., 2013; Zavarzina et al., 2018).

## 7.7 ENVIRONMENTAL AND SOCIOECONOMIC BENEFITS OF REUSING WASTE MATERIALS

Generally, the transformation of waste materials provides various advantages such as social, environmental, and economic value. These procedures can mitigate (a) improper disposal, (b) improper usage of fossil fuels which are very limited, and (c) generation of environmental pollution. The socioeconomic benefits are as follows:

### 7.7.1 Eco-friendliness

Production of valuable products from waste biomass material provides dual environmental benefits: (i) prevention of environmental pollution by replacing fossil fuel-based materials, and (ii) utilization of wastes that could have been released into the ecosystem. Additionally, biomass wastes are readily biodegradable by any microorganisms, and are comparatively less toxic. For example, biofuel showed much less risk of pollution (from various research data), whereas oil refinery industries emit high quantities of various wastes which detrimentally affect our environment (Nyashina et al., 2019).

### 7.7.2 Sustainable and Abundant

From ancient times, fossil fuel has been the primary source of energy formed by natural processes, that is, by anaerobic decomposition of buried dead organisms, and it takes millions of years in formation, sometimes 650 million years (Mann et al., 2003). During the combustion process such organisms emit energy, which is stored during photosynthesis or food-feed combination. In contrast, biomass materials can be generated within a short period (six months to 12 years) in a sustainable manner. They are abundantly available in nature. Biomass abundancy is derived from the quantity available and its renewability, diversity, and distinctive characteristics that cannot also be attributed to fossil fuels.

### 7.7.3 Cost-effectiveness

The majority of the processing steps to convert biomass materials into desired products require less manpower, energy, and revenue than petrochemical resources (Culler, 2016). Furthermore, second generation waste biomass (agricultural residues) is much less expensive than the first-generation edible raw materials. Therefore, utilization of waste material is a cost-effective process that adds value to a country's revenue.

### 7.7.4 Land Management

Waste materials generated domestically or industrially are usually disposed of in municipal landfill areas, or industries' dumping grounds. This process may affect the environment by leachate or other harmful chemicals from the wastes. Therefore, utilization of such wastes first harnesses land management, and second, helps produce various value-added products. Alternatively, the transformation of wastes can replace first-generation of bio-based substrate needed for food production, thereby reducing land use requirements (Dessie et al., 2020). However, in some metropolitan cities, land management currently receives higher priority than value-addition of wastes because of the shortage of dumping grounds.

### 7.7.5 Carbon Neutral

Fossil fuel utilization releases numerous amounts of $CO_2$ and CO into the environment but fails to capture it in the process. Unlike fossil fuels, plant biomass naturally absorbs $CO_2$ from nature for its growth. When waste biomass is used, $CO_2$ is released by combustion, and at the same time, it is taken by the plant as the source of carbon for living; therefore, there is no net gain/loss of $CO_2$ in the environment. However, there are various advanced methods for utilizing waste biomass with minimal or nearly zero $CO_2$ emissions.

### 7.7.6 Replacement for Petrochemical Resources

Among all the sustainable resources such as wind and solar energy, waste biomass material is considered the only carbon source in the environment that replaces fossil fuel resources (Gandeepan & Ackermann, 2018). Additionally, waste biomass can be considered a potential source for chemicals and energy production by substituting not only the fossil fuel resources but also 1G biomass feedstock.

### 7.7.7 Non-competitiveness with Food

It was estimated that the world would require about 70 to 100% more food in 2050 (Bank, 2007). The rate of food consumption rapidly increases with the population, thus creating growing demand for land, energy, and freshwater (Godfray et al., 2010). Therefore, conversion of waste biomass for valuable material production minimizes the usage of crops for energy generation or other material production that can easily be economically and potentially generated from the waste biomass. Thus, waste biomass does not compete with food and feed.

### 7.7.8 Other Benefits

Conversion of waste biomass for valuable product generation adds revenue, especially with the integrated biorefinery process.Moreover, valorization of wastes presents various socioeconomic advantages such as odor minimization, avoidance of fire incidents, prevention of leachate generation, minimal land slides, and avoidance of illegal settlements. Additionally, incineration of waste biomass materials in domestic sites and landfills generates air pollution, thus causing human health risk which can potentially be controlled by waste biomass utilization into marketable chemicals, fuels, and medicines (Chen et al., 2017). Then again, most of the country still depends on fossil fuel-based fuels like petrol and diesel for transportation. Fuel supply from other states or countries may lead to increased fuel prices, which significantly affects the nation's economic system (Singhvi & Gokhale, 2019). Fortunately, unlike fossil fuels, biomass resources are evenly distributed worldwide (Hahn-Hägerdal et al., 2006 B39) and can be produced domestically, which minimizes the transportation cost of fuels and generates high-tech jobs.

## 7.8 VALUE-ADDED PRODUCTS FROM WASTE BIOMASS

Advanced bioconversion technologies provide promising opportunities for the best possible utilization of waste biomass, which is less economical than the primary material. A wide range of cost-effective marketable value-added products can be generated via the bioconversion process, especially enzymatic hydrolysis of different

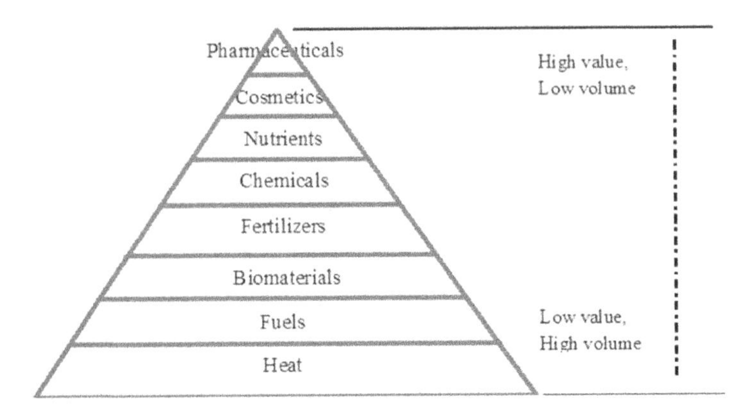


**FIGURE 7.1** Hierarchy of various value-added products according to their volume and value (Dessie et al., 2020).

lignocellulosic biomass wastes (LBW) (Mtui, 2009). These value-added products include: reducing sugars (carbohydrates), enzymes, proteins, resins, biosorbents, surfactants, fertilizers, food and feeds, biopesticides, lipids, phenols, degradable plastic composites, amino acids, activated carbon, methane, and ethanol (Demirbas & Ozturk, 2005; Mtui, 2007; Tengerdy & Szakacs, 2003; Ubalua, 2007). Figure 7.1 lists valuable bio-based products based on their value and volume.

### 7.8.1 Nutrients

Biological conversion of waste residues by cultivating mushrooms and production of single-cell proteins may contribute to protein-rich food products and simultaneously reduce the volume of waste biomass from the environment. Mushroom cultivation provides an alternative and economically viable pathway to produce food with superior taste and quality that does not need separate purification (Israilides & Philippoussis, 2003; Philippoussis et al., 2007).

### 7.8.2 Energy/Fuel

The biorefinery concept is sustainable and renewable once established and when generated from lignocellulosic waste biomass. The sustainable nature of biorefinery is shown in Figure 7.2 (Arevalo-Gallegos et al., 2017). The term "biorefinery" is defined as a sustainable approach where biotransformation of raw materials, that is, waste biomass to energies, are performed in an eco-friendly and economical way. A schematic summary of the biorefinery concept is shown in Figure 7.3 (Arevalo-Gallegos et al., 2017).

As fossil fuel-based energies are limited and becoming costlier as time passes, converting a nonrenewable energy system to a renewable energy system (biomaterial-based) is gaining a high level of interest on the industrial scale. This chapter elucidates the utilization of various types of biomass wastes into bioenergy via enzymatic hydrolysis treatment technologies. Bioenergy is produced from various

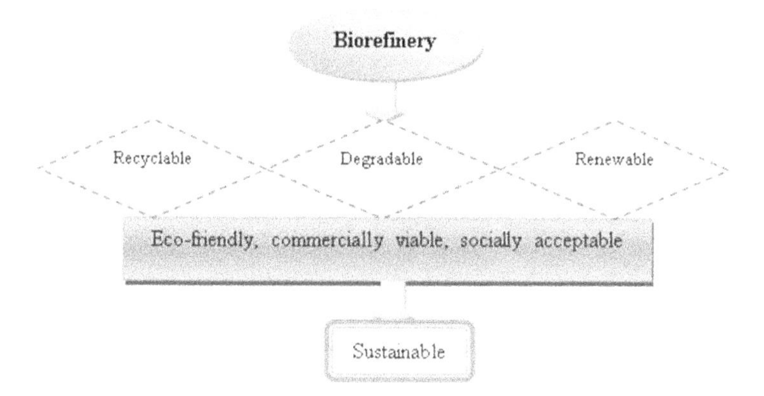


**FIGURE 7.2** Sustainability of biorefinery (Arevalo-Gallegos et al., 2017).

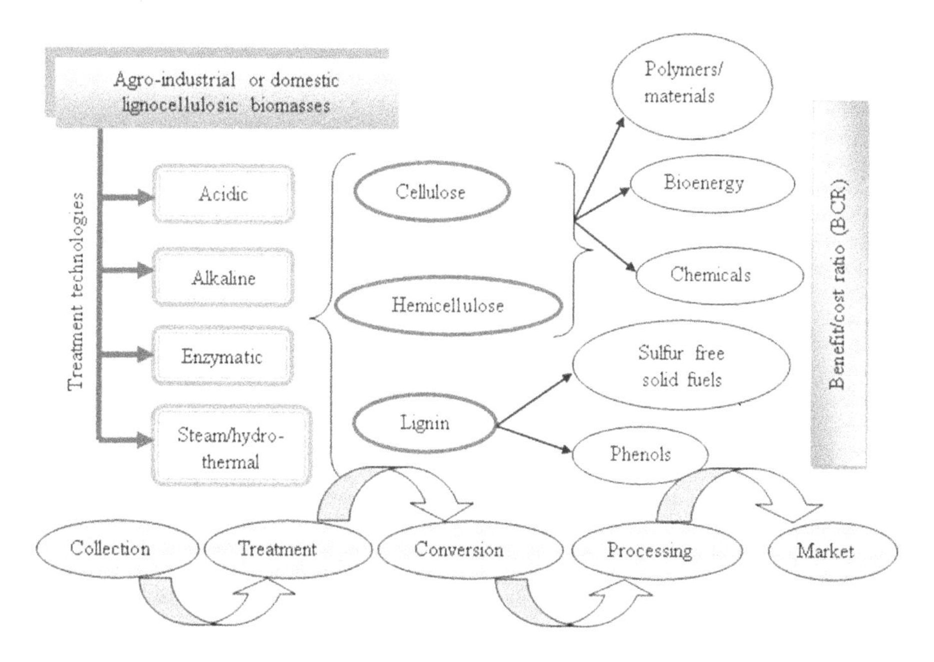


**FIGURE 7.3** Schematic overview of biorefinery concept (Arevalo-Gallegos et al., 2017).

biomass resources such as energy crops, by-products, residues, forestry, waste from food industries, and activated sludge. Table 7.4. illustrates bioenergy generation using various raw materials by enzymatic hydrolysis treatment methods. In the last few years, research has provided promising results for the improvement of conversion processes regarding the usage of ligninolytic enzymes such as manganese peroxidase (MnP) and lignin peroxidase (LiP), and laccase, their purification, and immobilization in biotechnological applications (Asgher et al., 2011; Asgher & Iqbal, 2011).

**TABLE 7.4**
**Bioenergy Generation from Different Feedstocks via Enzymatic Hydrolysis Treatment**

| Types of Biorefinery | Feedstock | Enzyme Used | Methodology Adopted | Explanation | References |
|---|---|---|---|---|---|
| Bioethanol | Corncob | - | Fermentation by yeast | 1. Delignification – liberation of cellulose, hemicellulose from lignin<br>2. Depolymeriza-tion (of carbohydrate polymer) – to produce free sugar by enzymatic or acid hydrolysis<br>3. Fermentation – (of hexose and pentose sugars) to produce ethanol | Xu et al. (2017) |
| Bioethanol | Sugarcane bagasse | Cellulase and xylanase | Separate hydrolysis and fermentation (SHF) | 1. Solid part is basically cellulose that undergoes saccharification process and easily accessible to enzyme<br>2. Fermentation by *Saccharomyces ceevisiae* | Cardona et al. (2010) |
| Bioethanol | News paper | Cellulo-lytic enzyme | Separate hydrolysis and fermentation (SHF) | 1. Enzymatic hydrolysis – by a mixture of cellulolytic enzymes.<br>2. Surfactant pretreatment improved the digestibility of enzyme by 45%.<br>3. Fermentation by *Saccharomyces ceevisiae* | Guerfali et al. (2015) |

The second-generation energy, that is, biofuel, is currently generated from LCB wastes instead of energy crops such as switchgrass, jatropha, willow, and hybrid poplar. There is a high demand for land, freshwater, and manpower for cultivating and processing energy crops. The usage of food crops such as sugarcane and corn for biofuel production has recently been strongly discouraged due to the increasing cost of food products and shortage of land for crop production. One way to minimize the food-feed-fuel conflicts is converting all biowastes into biomass economy (Mahro & Timm, 2007). Bioethanol production from organic waste biomass has been one of the main areas of research since the 1980s. The achievement of cellulose-to-ethanol from lignocellulosic waste materials is one result of pretreatment of cellulosic fiber, selection of enzyme, and processing optimum conditions (Champagne, 2007). The amount of ethanol production may differ depending on the amount of cellulosic components present in the biomass. Advanced biotechnological methods have accelerated fermentable sugar extraction from waste biomass. The application of advanced processing technologies and the use of novel enzymes have made it possible to produce bioethanol several thousand times more efficiently than previous technologies. Various factors affecting the enzymatic hydrolysis of cellulose include crystallinity of cellulosic fibers, the porosity of waste biomass, moisture content, and the amount of hemicellulose and lignin components. Simultaneous saccharification and fermentation (SSF) is an effective method for removing glucose, which inhibits cellulose activity, thereby increasing the rate and yield of the hydrolysis process.

### 7.8.3 Biopolymers

Biodegradable polymers can be defined as a class of loosely bound polymers that microbial activity can break down. These polymers can be produced in four different ways according to the desired material to be produced and the availability of feedstock biomass materials (Maraveas, 2020). The feedstock material should be cheap, that, is zero or less economic value, and available in large amounts. Among all developing and developed countries, India and China have the leading capacity to produce biopolymer from various agricultural wastes (fruits and vegetable wastes) and have high potential to fulfill global demands (Sharma et al., 2020). The production processes of such polymers are classified into (a) chemical synthesis, (b) microbial synthesis, (c) biopolymer blends, and (d) renewable resources (Satyanarayana et al., 2009). However, the primary focus of this chapter is to produce biodegradable polymers from the lignocellulosic biomass waste materials, that is, renewable resources. Polymers produced from various agricultural wastes have been recently gaining interest from researchers and industrialists because of their unique and novel properties such as high tensile strength, high aspect ratio, and elongation capacity. Yield strength and high tensile strength of such polymers are essential parameters in the field of construction, whereas the percentage of elongation determines the utility in packaging applications. It is shown from the literature that poly (glycolic acid) has the best modulus of elasticity and high tensile strength; however, it shows low elongation capacity at break (Satyanarayana et al., 2009).

Agricultural wastes are very rich in reducing sugars such as carbohydrates, which are considered the essential material for the development of plasticizers (Di Donato et al., 2020). Plasticizer enhances the material strength and elasticity of the biodegradable polymer and thus the stability and functionality of biocomposites. Some commercially available polymers are poly (lactic acid), polyhydroxyalkanoates, polycaprolactone, poly (butylene succinate-co-butylene adipate), poly (butylene succinate), poly (ethylene glycol) (Mtui, 2009). Fibers extracted from lignocellulosic biomasses are also known as natural fibers, which are considered eco-friendly, non-toxic, cost-effective, and sustainable feedstocks (Georgopoulos et al., 2005). An extensive amount of research work has been performed and published on the production of biocomposites from LCB wastes over the last 20 years. They can be made from cellulosic fibers present in the plant cell. The fibers increase the composite material's mechanical strength, that is; flexural strength, stiffness, strength, and so on. Implementation of lignocellulosic polymer into thermoplastic polymer gives the best polymer composites and is an emerging trend in the utilization of biomass wastes.

### 7.8.4 Fertilizers

Recycling waste biomass for fertilizer production using various microbes has received a massive response as both a cleaner environment and value-addition can beachieved. Various types of waste biomass, such as sewage sludge, food waste, animal manure, and municipal solid waste are used to produce organic waste as agricultural fertilizer. Organic waste can be subdivided into several categories, namely: (i) compost (source: food waste, plant source), (ii) animal-based organic waste (source: animal manure), and (iii) urban waste (source: household waste, sewage sludge) (Uysal et al., 2014). Compost is an organic fertilizer, produced by humifying organic substances present in the waste biomass material. It is highly rich in nutrients. Compost is also known as a soil conditioner, which is added to the soil to increase its physical properties, such as fertility. In vermicompost, earthworms and their enzymes play a vital role to convert solid waste into high-quality fertilizer.

### 7.8.5 Chemicals

Organic acids are chemical substances produced commercially by eco-friendly processes such as fermentation or chemical synthesis of ligninolytic compounds. Organic acids include volatile fatty acids (propionic acids, lactic acids, and butyric acids), fumaric acid, gluconic acid, and succinic acid. Fermentable sugar is the essential value-added product in the current situation where bioenergy, that is, bioethanol production, is necessary for depleting nonrenewable fossil fuels. The primary reducing sugars includes glucose, xylitol, cellobiose, xylose, pentose, arabinose, and galactose (Kim et al., 2008; Li et al., 2008; Singh et al., 2008).

### 7.8.6 Medicines

Lignocellulosic biomass wastes (LBWs) are an area of great interest for the pharmaceutical industry. They provide a suitable environment for mushroom cultivation which largely contributes to pharmaceutical products. Remarkably, some mushrooms such as *Tremella fuciformis, Lentiluna edodes*, and *Genoderma lucidum* contain bioactive compounds (antitumor, antibacterial, antivirus, and anti-inflammatory polysaccharides) (Mtui, 2009). Future prospects for research on bioactive compounds from waste biomass by enzymatic hydrolysis may have tremendous results and lead to the development of research in antiviral, anticancer, and antibacterial chemotherapies.

## 7.9 CONCLUSION AND FUTURE PROSPECTS

Shifting from a well-established petroleum-based economy to a biomass economy is not easy because of several challenges such as initial investment, large-scale production, biomass supply chain, and technological barriers. Development of optimum and effective waste biomass monitoring, available enzymes, treatment, biomass digestibility, and valorization techniques may overcome the existing limitations. Moreover, impressive research work so far has proven the production of valuable materials by utilizing the waste biomass can be performed on a large scale. The best way to commercialize the research findings is to implement academic and industrial partnerships. Therefore, multidisciplinary collaboration can provide socioeconomic and environmental benefits for waste biomass transformation into valuable, marketable products.

## REFERENCES


a Waste W. (2018). *2.0: A global snapshot of solid waste management to 2050/Silpa Kaza*, Lisa Yao, Perinaz Bhada-Tata and Frank Van Woerden. World Bank Group.

Abdel-Hamid A. M., Solbiati J. O., & Cann I. K. (2013). Insights into lignin degradation and its potential industrial applications. *Advances in Applied Microbiology*, 82, 1–28.

Agrawal K., Shankar J., & Verma P. (2020). Multicopper oxidase (MCO) laccase from *Stropharia* sp. ITCC-8422: An apparent authentication using integrated experimental and in silico analysis. *3 Biotech*, 10(9), 1–18.

Agrawal K., & Verma P. (2020a). Production optimization of yellow laccase from *Stropharia* sp. ITCC 8422 and enzyme-mediated depolymerization and hydrolysis of lignocellulosic biomass for biorefinery application. *Biomass Conversion and Biorefinery*, 1–20.

Agrawal K. & Verma P. (2020b). Laccase-mediated synthesis of bio-material using agro-residues. In: Sadhukhan P. C. & Premi S. (eds.), *Biotechnological applications in human health* (pp. 87–93). Springer, Singapore.

Ahorsu R., Medina F., & Constantí M. (2018). Significance and challenges of biomass as a suitable feedstock for bioenergy and biochemical production: A review. *Energies*, 11(12), 3366.

Akhtar N., Gupta K., Goyal D., & Goyal A. (2016). Recent advances in pretreatment technologies for efficient hydrolysis of lignocellulosic biomass. *Environmental Progress & Sustainable Energy*, 35(2), 489–511.

Alayi R., Sobhani E., & Najafi A. (2020). Analysis of environmental impacts on the characteristics of gas released from biomass. *Anthropogenic Pollution*, 4(1), 1–14.

Alma A., Agrawal K. and Verma P. (2021). Fungi and its by-products in food industry: An unexplored area. In *Microbial products for health, environment and agriculture* (pp. 103–120). Springer, Singapore.

Annepu R. K. (2012). Sustainable solid waste management in India. Columbia University, New York, 2(01), 1–189.

Arantes V., & Saddler J. N. (2010). Access to cellulose limits the efficiency of enzymatic hydrolysis: The role of amorphogenesis. *Biotechnology for Biofuels*, 3(1), 1–11.

Arevalo-Gallegos A., Ahmad Z., Asgher M., Parra-Saldivar R., & Iqbal H. M. (2017). Lignocellulose: A sustainable material to produce value-added products with a zero waste approach—A review. *International Journal of Biological Macromolecules*, 99, 308–318.

Arora R., Behera S., & Kumar S. (2015). Bioprospecting thermophilic/thermotolerant microbes for production of lignocellulosic ethanol: A future perspective. *Renewable and Sustainable Energy Reviews*, 51, 699–717.

Asgher M., Ahmed N., & Iqbal H. M. N. (2011). Hyperproductivity of extracellular enzymes from indigenous white rot fungi (*Phanerochaete chrysosporium*) by utilizing agro-wastes. *BioResources*, 6(4), 4454–4467.

Asgher M., & Iqbal H. M. (2011). Characterization of a novel manganese peroxidase purified from solid state culture of *Trametes versicolor* IBL-04. *BioResources*, 6(4), 4317–4330.

Bank W. (2007). *World development report 2008: Agriculture for development*. The World Bank.

Baruah J., Nath B. K., Sharma R., Kumar S., Deka R. C., Baruah D. C., & Kalita E. (2018). Recent trends in the pretreatment of lignocellulosic biomass for value-added products. *Frontiers in Energy Research*, 6, 141.

Bashir A., Jabeen S., Gull N., Islam A., Sultan M., Ghaffar A., … Jamil T. (2018). Co-concentration effect of silane with natural extract on biodegradable polymeric films for food packaging. *International Journal of Biological Macromolecules*, 106, 351–359.

Bhardwaj N., Agrawal K., Kumar B., & Verma P. (2021). Role of enzymes in deconstruction of waste biomass for sustainable generation of value-added products. In: *Bioprospecting of enzymes in industry, healthcare and sustainable environment* (pp. 219–250). Springer, Singapore.

Bhardwaj N., Kumar B., Agrawal K., & Verma P. (2020). Bioconversion of rice straw by synergistic effect of in-house produced ligno-hemicellulolytic enzymes for enhanced bioethanol production. *Bioresource Technology Reports*, 10, 100352.

Bhardwaj N., & Verma P. (2020). Extraction of fungal xylanase using ATPS-PEG/sulphate and its application in hydrolysis of agricultural residues. In: Sadhukhan P. C. & Premi S. (eds.), *Biotechnological applications in human health* (pp. 95–105). Springer, Singapore.

Bhatia S. K., Kim S.-H., Yoon J.-J., & Yang Y.-H. (2017). Current status and strategies for second generation biofuel production using microbial systems. *Energy Conversion and Management*, 148, 1142–1156.

Bholay A., Borkhataria B. V., Jadhav P. U., Palekar K. S., Dhalkari M. V., & Nalawade P. (2012). Bacterial lignin peroxidase: A tool for biobleaching and biodegradation of industrial effluents. *Universal Journal of Environmental Research & Technology*, 2(1), 58–64.

Biko O. D., Viljoen-Bloom M., & Van Zyl W. H. (2020). *Microbial lignin peroxidases: Applications, production challenges and future perspectives*. Enzyme and Microbial Technology, 109669.

Brinton L. T., Sloane H. S., Kester M., & Kelly K. A. (2015). Formation and role of exosomes in cancer. *Cellular and Molecular Life Sciences*, 72(4), 659–671.

Brown M. E., & Chang M. C. (2014). Exploring bacterial lignin degradation. *Current Opinion in Chemical Biology*, 19, 1–7.

Bundela P., Gautam S., Pandey A., Awasthi M., & Sarsaiya S. (2010). Municipal solid waste management in Indian cities–A review. *International Journal of Environmental Sciences*, 1(4), 591–606.

Cardoen D., Joshi P., Diels L., Sarma P. M., & Pant D. (2015). Agriculture biomass in India: Part 1. Estimation and characterization. *Resources, Conservation and Recycling*, 102, 39–48.

Cardona C., Quintero J., & Paz I. (2010). Production of bioethanol from sugarcane bagasse: Status and perspectives. *Bioresource Technology*, 101(13), 4754–4766.

Champagne P. (2007). Feasibility of producing bio-ethanol from waste residues: A Canadian perspective: Feasibility of producing bio-ethanol from waste residues in Canada. *Resources, Conservation and Recycling*, 50(3), 211–230.

Chen J., Li C., Ristovski Z., Milic A., Gu Y., Islam M. S., … He C. (2017). A review of biomass burning: Emissions and impacts on air quality, health and climate in China. *Science of the Total Environment*, 579, 1000–1034.

Chen S., Zhang X., Singh D., Yu H., & Yang X. (2010). Biological pretreatment of lignocellulosics: Potential, progress and challenges. *Biofuels*, 1(1), 177–199.

Conti J., Holtberg P., Diefenderfer J., LaRose A., Turnure J. T., & Westfall L. (2016). *International energy outlook 2016 with projections to 2040*. Retrieved from https://www.osti.gov/servlets/purl/1296780

Culler S. (2016). A bioengineering platform to industrialize biotechnology. *Chemical Engineering Progress*, 112, 42–51.

De Bhowmick G., Sarmah A. K., & Sen R. (2018). Lignocellulosic biorefinery as a model for sustainable development of biofuels and value added products. *Bioresource Technology*, 247, 1144–1154.

Demirbas A., & Ozturk T. (2005). Anaerobic digestion of agricultural solid residues. *International Journal of Green Energy*, 1(4), 483–494.

Dessie W., Luo X., Wang M., Feng L., Liao Y., Wang Z., … Qin Z. (2020). Current advances on waste biomass transformation into value-added products. *Applied Microbiology and Biotechnology*, 104(11), 4757–4770.

Dessie W., Zhu J., Xin F., Zhang W., Jiang Y., Wu H., … Jiang M. (2018). Bio-succinic acid production from coffee husk treated with thermochemical and fungal hydrolysis. *Bioprocess and Biosystems Engineering*, 41(10), 1461–1470.

Dhiman S. S., Haw J.-R., Kalyani D., Kalia V. C., Kang Y. C., & Lee J.-K. (2015). Simultaneous pretreatment and saccharification: Green technology for enhanced sugar yields from biomass using a fungal consortium. *Bioresource Technology*, 179, 50–57.

Di Donato P., Taurisano V., Poli A., d'Ayala G. G., Nicolaus B., Malinconinco M., & Santagata G. (2020). Vegetable wastes derived polysaccharides as natural eco-friendly plasticizers of sodium alginate. *Carbohydrate Polymers*, 229, 115427.

Dixit A., & Srivastava R. (2016). An estimate of contaminated land area due to industrial hazardous waste generation in India. *International Journal of Advanced Research in Education & Technology*, 2(3), 117–125.

Dos Santos B. V., Rodrigues P. O., Albuquerque C. J. B., Pasquini D., & Baffi M. A. (2019). Use of an (hemi) cellulolytic enzymatic extract produced by Aspergilli species consortium in the saccharification of biomass sorghum. *Applied Biochemistry and Biotechnology*, 189(1), 37–48.

EPA. (2019). *Waste definitions*. Retrieved from https://www.epa.sa.gov.au/files/4771336_guide_waste_definitions.pdf

Fiksel J., & Lal R. (2018). Transforming waste into resources for the Indian economy. *Environmental Development*, 26, 123–128.

Fillat Ú., Ibarra D., Eugenio M. E., Moreno A. D., Tomás-Pejó E., & Martín-Sampedro R. (2017). Laccases as a potential tool for the efficient conversion of lignocellulosic biomass: A review. *Fermentation*, 3(2), 17.

Galbe M., & Zacchi G. (2002). A review of the production of ethanol from softwood. *Applied Microbiology and Biotechnology*, 59(6), 618–628.

Gandeepan P., & Ackermann L. (2018). Transient directing groups for transformative C–H activation by synergistic metal catalysis. *Chem*, 4(2), 199–222.

Georgopoulos S. T., Tarantili P., Avgerinos E., Andreopoulos A., & Koukios E. (2005). Thermoplastic polymers reinforced with fibrous agricultural residues. *Polymer Degradation and Stability*, 90(2), 303–312.

Giacobbe S., Pezzella C., Lettera V., Sannia G., & Piscitelli A. (2018). Laccase pretreatment for agrofood wastes valorization. *Bioresource Technology*, 265, 59–65.

Global-Agriculture. (2009). Global Agriculture Towards 2050. Retrieved from http://www.fao.org/fileadmin/templates/wsfs/docs/Issues_papers/HLEF2050_Global_Agriculture.pdf

Godfray H. C. J., Beddington J. R., Crute I. R., Haddad L., Lawrence D., Muir J. F., Pretty J., Robinson S., Thomas S. M., & Toulmin C. (2010). Food security: The challenge of feeding 9 billion people. *Science*, 327(5967), 812–818.

Gopal K., & Sathiyagnanam A. (2018). Mathematical correlation of different emission characteristics analysis of DI-diesel engine fueled with lignocellulosic biomass derived n-butanol/diesel blend using response surface methodology. *International Journal for Research in Applied Science and Engineering Technology*, 6(1), 2592–2608.

Goswami R. K., Agrawal K., & Verma P. (2021). Microalgae-based biofuel-integrated biorefinery approach as sustainable feedstock for resolving energy crisis. In: Srivastava M., Srivastava N., & Singh R. (eds.), *Bioenergy research: commercial opportunities & challenges. Clean energy production technologies* (pp. 267–293). Springer, Singapore.

Goswami R. K., Agrawal K., & Verma P. (2022a). An exploration of natural synergy using microalgae for the remediation of pharmaceuticals and xenobiotics in wastewater. *Algal Research*, 64, 102703.

Goswami R. K., Agrawal K., & Verma P. (2022b). Microalgal-based remediation of wastewater: A step towards environment protection and management. *Environmental Quality Management*. https://doi.org/10.1002/tqem.21850

Goswami R. K., Mehariya S., Karthikeyan O. P. & Verma P. (2022c). Influence of carbon sources on biomass and biomolecule accumulation in *Picochlorum* sp. cultured under the mixotrophic condition. *International Journal of Environmental Research and Public Health*, 19(6), 3674.

Guerfali M., Saidi A., Gargouri A., & Belghith H. (2015). Enhanced enzymatic hydrolysis of waste paper for ethanol production using separate saccharification and fermentation. *Applied Biochemistry and Biotechnology*, 175(1), 25–42.

Hahn-Hägerdal B., Galbe M., Gorwa-Grauslund M. F., Lidén G., & Zacchi G. (2006). Bio-ethanol–the fuel of tomorrow from the residues of today. *Trends in Biotechnology*, 24(12): 549–556

Hassan S. S., Williams G. A., & Jaiswal A. K. (2018). Emerging technologies for the pretreatment of lignocellulosic biomass. *Bioresource Technology*, 262, 310–318.

Heap L., Green A., Brown D., van Dongen B., & Turner N. (2014). Role of laccase as an enzymatic pretreatment method to improve lignocellulosic saccharification. *Catalysis Science & Technology*, 4(8), 2251–2259.

Hiloidhari M., Das D., & Baruah D. (2014). Bioenergy potential from crop residue biomass in India. *Renewable and Sustainable Energy Reviews*, 32, 504–512.

Israilides C., & Philippoussis A. (2003). Bio-technologies of recycling agro-industrial wastes for the production of commercially important fungal polysaccharides and mushrooms. *Biotechnology and Genetic Engineering Reviews*, 20(1), 247–260.

Jedvert K., & Heinze T. (2017). Cellulose modification and shaping–a review. *Journal of Polymer Engineering*, 37(9), 845–860.

Jeoh T., & Agblevor F. (2001). Characterization and fermentation of steam exploded cotton gin waste. *Biomass and Bioenergy*, 21(2), 109–120.

Kadarmoidheen M., Saranraj P., & Stella D. (2012). Effect of cellulolytic fungi on the degradation of cellulosic agricultural wastes. *International Journal of Applied Microbiology Science*, 1(2), 13–23.

Kalita E., Nath B., Deb P., Agan F., Islam M. R., & Saikia K. (2015). High quality fluorescent cellulose nanofibers from endemic rice husk: Isolation and characterization. *Carbohydrate Polymers*, 122, 308–313.

Kalyani D., Dhiman S. S., Kim H., Jeya M., Kim I.-W., & Lee J.-K. (2012). Characterization of a novel laccase from the isolated Coltricia perennis and its application to detoxification of biomass. *Process Biochemistry*, 47(4), 671–678.

Kapoor R., Ghosh P., Kumar M., Sengupta S., Gupta A., Kumar S. S., Vijay V., Kumar V., Vijay V. K., & Pant D. (2020). Valorization of agricultural waste for biogas based circular economy in India: A research outlook. *Bioresource Technology*, 304, 123036.

Kaur G., Brar Y. S., & Kothari D. (2017). Potential of livestock generated biomass: Untapped energy source in India. *Energies*, 10(7), 847.

Keller F. A., Hamilton J. E., & Nguyen Q. A. (2003). Microbial pretreatment of biomass. In: Davison B. H., Lee J. W., Finkelstein M., & McMillan J. D. (eds.), *Biotechnology for fuels and chemicals. Applied biochemistry and biotechnology* (pp. 27–41). Humana Press, Totowa, NJ.

Kim Y., Hendrickson R., Mosier N. S., Ladisch M. R., Bals B., Balan V., & Dale B. E. (2008). Applied Enzyme hydrolysis and ethanol fermentation of liquid hot water and AFEX pretreated distillers' grains at high-solids loadings. *Bioresource Technology*, 99(12), 5206–5215.

Kumar B., Bhardwaj N., Agrawal K., & Verma P. (2020a). Bioethanol production: Generation-based comparative status measurements. Springer, 155–201.

Kumar B., Bhardwaj N., & Verma P. (2020b). Microwave assisted transition metal salt and orthophosphoric acid pretreatment systems: Generation of bioethanol and xylo-oligosaccharides. *Renewable Energy*, 158, 574–584.

Kumar B., & Verma P. (2020). Enzyme mediated multi-product process: A concept of bio-based refinery. *Industrial Crops and Products*, 154, 112607.

Kumar R., Sharma R. K., & Singh A. P. (2017). Cellulose based grafted biosorbents-Journey from lignocellulose biomass to toxic metal ions sorption applications-A review. *Journal of Molecular Liquids*, 232, 62–93.

Li K., Wang X., Wang J., & Zhang J. (2015). Benefits from additives and xylanase during enzymatic hydrolysis of bamboo shoot and mature bamboo. *Bioresource Technology*, 192, 424–431.

Li M., Luo N., & Lu Y. (2017). Biomass energy technological paradigm (BETP): Trends in this sector. *Sustainability*, 9(4), 567.

Li W.-Z., Xu J., Wang J., Yan Y.-J., Zhu X.-F., Chen M.-Q., & Tan Z.-C. (2008). Studies of monosaccharide production through lignocellulosic waste hydrolysis using double acids. *Energy & Fuels*, 22(3), 2015–2021.

Lin Z., Huang H., Zhang H., Zhang L., Yan L., & Chen J. (2010). Ball milling pretreatment of corn stover for enhancing the efficiency of enzymatic hydrolysis. *Applied Biochemistry and Biotechnology*, 162(7), 1872–1880.

Long H., Li X., Wang H., & Jia J. (2013). Biomass resources and their bioenergy potential estimation: A review. *Renewable and Sustainable Energy Reviews*, 26, 344–352.

Lynd L. R., Elander R. T., & Wyman C. E. (1996). Likely features and costs of mature biomass ethanol technology. Paper presented at the Seventeenth Symposium on Biotechnology for Fuels and Chemicals.

Mahro B., & Timm M. (2007). Potential of biowaste from the food industry as a biomass resource. *Engineering in Life Sciences*, 7(5), 457–468.

Mann P., Gahagan L., & Gordon M. B. (2003). Tectonic setting of the world's giant oil and gas fields. In: Halbouty M. (ed.), *Giant oil and gas fields of the decade 1990–1999*, vol. 78 (pp. 15–105). AAPG Memoir.

Maraveas C. (2020). Production of sustainable and biodegradable polymers from agricultural waste. *Polymers*, 12(5), 1127.

Meng X., Sun Q., Kosa M., Huang F., Pu Y., & Ragauskas A. J. (2016). Physicochemical structural changes of poplar and switchgrass during biomass pretreatment and enzymatic hydrolysis. *ACS Sustainable Chemistry & Engineering*, 4(9), 4563–4572.

Misra V., & Pandey S. (2005). Hazardous waste, impact on health and environment for development of better waste management strategies in future in India. *Environment International*, 31(3), 417–431.

Moreno A. D., Ibarra D., Alvira P., Tomás-Pejó E., & Ballesteros M. (2016). Exploring laccase and mediators behavior during saccharification and fermentation of steam-exploded wheat straw for bioethanol production. *Journal of Chemical Technology & Biotechnology*, 91(6), 1816–1825.

Mtui G., & Nakamura Y. (2007). Characterization of lignocellulosic enzymes from white-rot fungus *Phlebia crysocreas* isolated from a marine habitat. *Journal of Engineering and Applied Science*, 2(10), 1501–1508.

Mtui G. Y. (2007). Trends in industrial and environmental biotechnology research in Tanzania. *African Journal of Biotechnology*, 6(25).

Mtui G. Y. (2009). Recent advances in pretreatment of lignocellulosic wastes and production of value added products. *African Journal of Biotechnology*, 8(8).

Nair L. G., Agrawal K., & Verma P. (2022a). An overview of sustainable approaches for bio-energy production from agro-industrial wastes. *Energy Nexus*, 100086.

Nair L. G., Agrawal K., & Verma P. (2022b). An insight into the principles of lignocellulosic biomass-based zero-waste biorefineries: A green leap towards imperishable energy-based future. *Biotechnology and Genetic Engineering Reviews*, 1–51.

NREL. (2012). Glossary of biomass terms. (National Renewable Energy Laboratory). Retrieved from https://www.nrel.gov/biomass/glossary.html

Nunes L. A., Silva M. L., Gerber J. Z., & Kalid R. D. A. (2020). Waste green coconut shells: Diagnosis of the disposal and applications for use in other products. *Journal of Cleaner Production*, 255, 120169.

Nyashina G. S., Vershinina K. Y., Shlegel N. E., & Strizhak P. A. (2019). Effective incineration of fuel-waste slurries from several related industries. *Environmental Research*, 176, 108559.

Oliva-Taravilla A., Moreno A. D., Demuez M., Ibarra D., Tomás-Pejó E., González-Fernández C., & Ballesteros M. (2015). Unraveling the effects of laccase treatment on enzymatic hydrolysis of steam-exploded wheat straw. *Bioresource Technology*, 175, 209–215.

Ozturk M., Saba N., Altay V., Iqbal R., Hakeem K. R., Jawaid M., & Ibrahim F. H. (2017). Biomass and bioenergy: An overview of the development potential in Turkey and Malaysia. *Renewable and Sustainable Energy Reviews*, 79, 1285–1302.

Peisach J., Aisen P., & Blumberg W. E. (1966). Biochemistry of copper. Paper presented at the Symposium on Copper in Biological Systems (1965: Harriman, NY).

Philippoussis A., Diamantopoulou P., & Israilides C. (2007). Production of functional food from the sporophores of the medicinal mushroom *Lentinula edodes* through exploitation of lingocellulosic agricultural residues. *International Biodeterioration and Biodegradatio*, 59(3), 216–219.

Piscitelli A., Del Vecchio C., Faraco V., Giardina P., Macellaro G., Miele A., Pezzella C., & Sannia G. (2011). Fungal laccases: Versatile tools for lignocellulose transformation. *Comptes Rendus Biologies*, 334(11), 789–794.

Qi B., Aldrich C., Lorenzen L., & Wolfaardt G. (2005). Acidogenic fermentation of lignocellulosic substrate with activated sludge. *Chemical Engineering Communications*, 192(9), 1221–1242.

Qing Q., Guo Q., Zhou L., Gao X., Lu X., & Zhang Y. (2017). Comparison of alkaline and acid pretreatments for enzymatic hydrolysis of soybean hull and soybean straw to produce fermentable sugars. *Industrial Crops and Products*, 109, 391–397.

Reddy C. A., & D'Souza T. M. (1994). Physiology and molecular biology of the lignin peroxidases of Phanerochaete chrysosporium. *FEMS Microbiology Reviews*, 13(2–3), 137–152.

Reddy G., Babu P. R., Komaraiah P., Roy K., & Kothari I. (2003). Utilization of banana waste for the production of lignolytic and cellulolytic enzymes by solid substrate fermentation using two *Pleurotus* species (*P. ostreatus* and *P. sajor-caju*). *Process Biochemistry*, 38(10), 1457–1462.

Rico A., Rencoret J., Del Río J. C., Martínez A. T., & Gutiérrez A. (2014). Pretreatment with laccase and a phenolic mediator degrades lignin and enhances saccharification of Eucalyptus feedstock. *Biotechnology for Biofuels*, 7(1), 1–14.

Rodríguez G., Lama A., Rodríguez R., Jiménez A., Guillén R., & Fernández-Bolanos J. (2008). Olive stone an attractive source of bioactive and valuable compounds. *Bioresource Technology*, 99(13), 5261–5269.

Roig A., Cayuela M. L., & Sánchez-Monedero M. (2006). An overview on olive mill wastes and their valorization methods. *Waste Management*, 26(9), 960–969.

Santana R. F., Bonomo R. C. F., Gandolfi O. R. R., Rodrigues L. B., Santos L. S., dos Santos Pires A. C., de Oliveira C. P., Fontan R. D. C. I., & Veloso C. M. (2018). Characterization of starch-based bioplastics from jackfruit seed plasticized with glycerol. *Journal of Food Science and Technology*, 55(1), 278–286.

Saratale G. D., Kshirsagar S. D., Sampange V. T., Saratale R. G., Oh S.-E., Govindwar S. P., & Oh M.-K. (2014). Cellulolytic enzymes production by utilizing agricultural wastes under solid state fermentation and its application for biohydrogen production. *Applied Biochemistry and Biotechnology*, 174(8), 2801–2817.

Satyanarayana K. G., Arizaga G. G., & Wypych F. (2009). Biodegradable composites based on lignocellulosic fibers—An overview. *Progress in Polymer Science*, 34(9), 982–1021.

Scholl A. L., Menegol D., Pitarelo A. P., Fontana R. C., Zandoná Filho A., Ramos L. P., Dillon A. J. P., & Camassola M. (2015). Elephant grass pretreated by steam explosion for inducing secretion of cellulases and xylanases by *Penicillium echinulatum* S1M29 solid-state cultivation. *Industrial Crops and Products*, 77, 97–107.

Shamel A., Alayi R., & Abbaszadeh L. (2014). The assessing and prediction of biogas production and dissemination rate in ardebil city landfills and chemical analysis of obtained biogas. *International Journal of Engineering and Advanced Technology*, 4(1), 84–88.

Sharholy M., Ahmad K., Mahmood G., & Trivedi R. (2008). Municipal solid waste management in Indian cities–A review. *Waste Management*, 28(2), 459–467.

Sharma P., Gaur V. K., Kim S.-H., & Pandey A. (2020). Microbial strategies for bio-transforming food waste into resources. *Bioresource Technology*, 299, 122580.

Shawky B. T., Mahmoud M. G., Ghazy E. A., Asker M. M., & Ibrahim G. S. (2011). Enzymatic hydrolysis of rice straw and corn stalks for monosugars production. *Journal of Genetic Engineering and Biotechnology*, 9(1), 59–63.

Sindhu R., Binod P., & Pandey A. (2016). Biological pretreatment of lignocellulosic biomass–An overview. *Bioresource Technology*, 199, 76–82.

Singh D. P., & Dwevendi A. (2019). Production of clean energy by green ways. In: Dwevedi A. (ed.), *Solutions to environmental problems involving nanotechnology and enzyme technology* (pp. 49–90). Academic Press, Waltham, MA, USA.

Singh P., Suman A., Tiwari P., Arya N., Gaur A., & Shrivastava A. (2008). Biological pretreatment of sugarcane trash for its conversion to fermentable sugars. *World Journal of Microbiology and Biotechnology*, 24(5), 667–673.

Singhvi M. S., & Gokhale D. V. (2019). Lignocellulosic biomass: Hurdles and challenges in its valorization. *Applied Microbiology and Biotechnology*, 103(23), 9305–9320.

Sudha P., & Ravindranath N. (1999). Land availability and biomass production potential in India. *Biomass and Bioenergy*, 16(3), 207–221.

Sudha P., Somashekhar H., Rao S., & Ravindranath N. (2003). Sustainable biomass production for energy in India. *Biomass and Bioenergy*, 25(5), 501–515.

Sun S., Sun S., Cao X., & Sun R. (2016). The role of pretreatment in improving the enzymatic hydrolysis of lignocellulosic materials. *Bioresource Technology*, 199, 49–58.

Sun Y., & Cheng J. (2002). Hydrolysis of lignocellulosic materials for ethanol production: A review. *Bioresource Technology*, 83(1), 1–11.

Tengerdy R., & Szakacs G. (2003). Bioconversion of lignocellulose in solid substrate fermentation. *Biochemical Engineering Journal*, 13(2–3), 169–179.

Thrän D., Seidenberger T., Zeddies J., & Offermann R. (2010). Global biomass potentials—Resources, drivers and scenario results. *Energy for Sustainable Development*, 14(3), 200–205.

Thurston C. F. (1994). The structure and function of fungal laccases. *Microbiology*, 140(1), 19–26.

Toklu E. (2017). Biomass energy potential and utilization in Turkey. *Renewable Energy*, 107, 235–244.

Ubalua A. (2007). Cassava wastes: Treatment options and value addition alternatives. *African Journal of Biotechnology*, 6(18), 2065–2073.

UN. (2019). WPP2019_Volume-I_Comprehensive-Tables. Retrieved from https://population.un.org/wpp/Publications/Files/WPP2019_Volume-I_Comprehensive-Tables.pdf

Usmani R. A. (2020). Potential for energy and biofuel from biomass in India. *Renewable Energy*, 155, 921–930.

Uysal A., Demir S., Sayilgan E., Eraslan F., & Kucukyumuk Z. (2014). Optimization of struvite fertilizer formation from baker's yeast wastewater: Growth and nutrition of maize and tomato plants. *Environmental Science and Pollution Research*, 21(5), 3264–3274.

Van Nes E. H., Scheffer M., Brovkin V., Lenton T. M., Ye H., Deyle E., & Sugihara G. (2015). Causal feedbacks in climate change. *Nature Climate Change*, 5(5), 445–448.

Wan C., & Li Y. (2010). Microbial delignification of corn stover by Ceriporiopsis subvermispora for improving cellulose digestibility. *Enzyme and Microbial Technology*, 47(1–2), 31–36.

Wan C., & Li Y. (2012). Fungal pretreatment of lignocellulosic biomass. *Biotechnology Advances*, 30(6), 1447–1457.

Wang Z., Keshwani D. R., Redding A. P., & Cheng J. J. (2010). Sodium hydroxide pretreatment and enzymatic hydrolysis of coastal Bermuda grass. *Bioresource Technology*, 101(10), 3583–3585.

Wyman C. E., Dale B. E., Elander R. T., Holtzapple M., Ladisch M. R., & Lee Y. (2005). Coordinated development of leading biomass pretreatment technologies. *Bioresource Technology*, 96(18), 1959–1966.

Xiao X., Bian J., Li M.-F., Xu H., Xiao B., & Sun R.-C. (2014). Enhanced enzymatic hydrolysis of bamboo (*Dendrocalamus giganteus* Munro) culm by hydrothermal pretreatment. *Bioresource Technology*, 159, 41–47.

Xie Y., Niu X., Yang J., Fan R., Shi J., Ullah N., Feng X., & Chen L. (2020). Active biodegradable films based on the whole potato peel incorporated with bacterial cellulose and curcumin. *International Journal of Biological Macromolecules*, 150, 480–491.

Xu Q.-Q., Zhao M.-J., Yu Z.-Z., Yin J.-Z., Li G.-M., Zhen M.-Y., & Zhang Q.-Z. (2017). Enhancing enzymatic hydrolysis of corn cob, corn stover and sorghum stalk by dilute aqueous ammonia combined with ultrasonic pretreatment. *Industrial Crops and Products*, 109, 220–226.

Zavarzina A., Lisov A., & Leontievsky A. (2018). The role of ligninolytic enzymes laccase and a versatile peroxidase of the white-rot fungus *Lentinus tigrinus* in biotransformation of soil humic matter: Comparative in vivo study. *Journal of Geophysical Research: Biogeosciences*, 123(9), 2727–2742.

Zhang Y. H. P., Ding S. Y., Mielenz J. R., Cui J. B., Elander R. T., Laser M., Himmel M. E., McMillan J. R., & Lynd L. R. (2007). Fractionating recalcitrant lignocellulose at modest reaction conditions. *Biotechnology and Bioengineering*, 97(2), 214–223.

# 8 Valorization of Recalcitrant Feather-waste by Extreme Microbes

*Sanket K. Gaonkar*
Goa University, Goa, India
Post Graduate Department of Microbiology,
P.E.S's R.S.N College of Arts and Science, Goa, India

*Irene J. Furtado*
Goa University, Goa, India

## CONTENTS



DOI: 10.1201/9781003187684-8

## 8.1 INTRODUCTION

In the agriculture sector, poultry industries that convert feed to food (Mata-Alvarez et al. 2014) are expected to grow at a higher rate in 2050 and lead to the accumulation of feathers as the main waste (Kanani et al. 2020). Currently, the United States is the largest meat producer in the world, followed by China, Brazil, and Russia. India ranks fifth in the poultry industry, and contributes about 3.5 million tons of poultry waste. The Department of Agriculture in the United States estimated the generation of more than 4.7 million tons of feathers per every 100.5 million tons of poultry meat produced (Qiu et al. 2020).

The major insoluble recalcitrant structural protein of feathers is keratin, natural amino acid, and minerals source. Insolubility and slow natural degradation of feather keratin limit its use and yet hydrolysate, which is rich in protein has found its way as raw material for the production of various bioproducts such as feather meal, bioactive peptides, animal feed supplements, biofertilizer, and bioplastic (Bhari et al. 2021; Chaturvedi et al. 2021). Innovations are emerging to transform the keratin hydrolysate from chicken feathers into sponges, films, fibers, and blend with other natural or synthetic polymers into bioplastics.

However, presently available physical and chemical methods for valorizing keratinous waste are environmentally nonviable and lead to the destruction of products that otherwise would be valuable biological resources (Tiwary and Gupta 2012). These processes are expensive, which results in a low-quality protein product which is primarily prohibited. In 1985, Papadopoulos and team discussed that extreme temperature and pressure destroyed essential amino acids (Papadopoulos et al. 1985). Results showed that there was considerable degradation of the cysteine, lysine, and methionine content of the hydrolysate. Attempts were made by Eggum (1970) to evaluate the protein quality of feather meal treated using heat and pressure as primary physicochemical factors.

Microbial degradation of feathers under mild conditions is an eco-friendly approach for managing poultry feather-waste (Tamreihao et al. 2019). Current well-developed physicochemical and biological technologies for chicken feather-waste as bio-resource for valorization into high-value bio-products have been discussed lately in a review by Chaturvedi et al. (2021) and Bhari et al. (2021).

## 8.2 RECALCITRANT FEATHER-WASTE

### 8.2.1 Physical Characteristics of Feather-waste

Feathers consist of rachis/quill or central shaft (7 inches in length) arranged as 78% β-sheet and 18% α-helix giving it high tensile strength; barbs, (1–4.5 cm in length); and barbules (0.3–0.5 mm in size) mainly in α-helical arrangements and some β-sheet structures (Figures 8.1a and 8.2b). Feathers have about 90–95% protein, out of which 80–92% is the insoluble keratin, 5–10% fat, and 0.69% ashes (Onifade et al. 1998; Bhari et al. 2021). Extensive supercoiling is the cause of the formation of octamers and tetramers linked by disulfide bonds, hydrogen bonds, and hydrophobic interaction with the side chain of amino acids in proteins. The disulfide linkages confer maximum mechanical stability to the keratin structure, making it resistant to water, salts, acids, solvents, alkali, and even most proteases (Brandelli 2008; Korniłłowicz-Kowalska and Bohacz 2011).

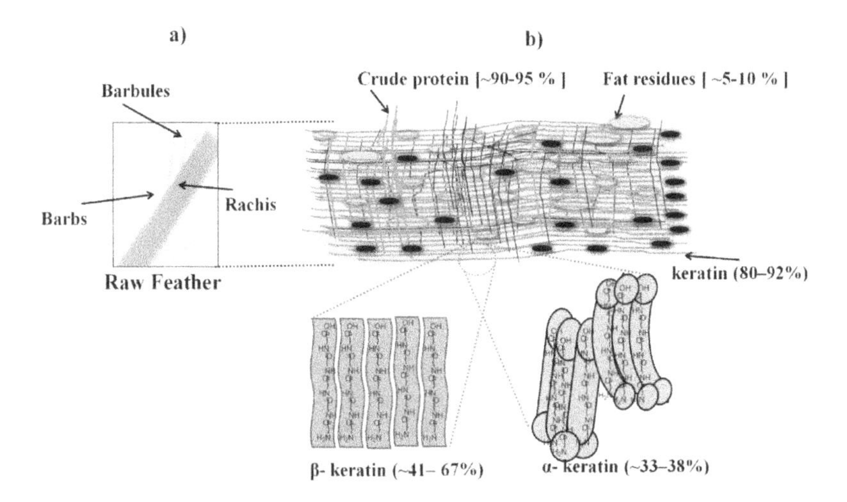


**FIGURE 8.1** Structure (a) and primary components (b) of a chicken feather.

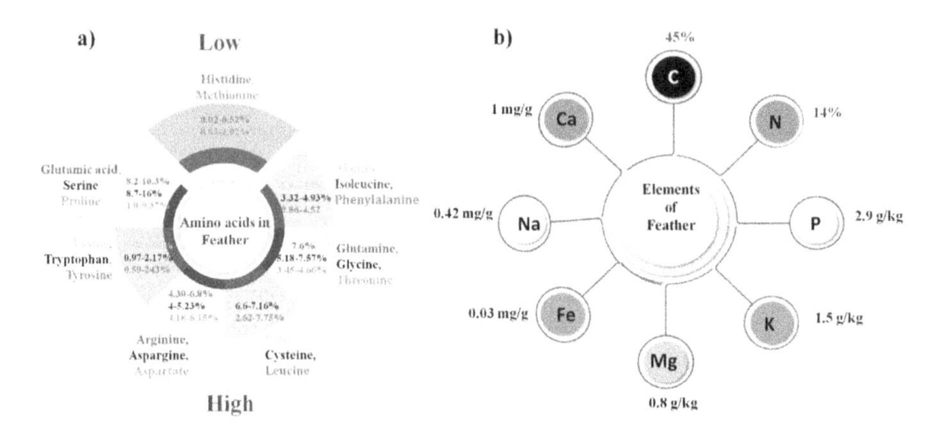


FIGURE 8.2 Amino acids (a) and elements (b) of a chicken feather.

### 8.2.2 Chemical Characteristics

Chicken feathers are a unique source of amino acids such as aspartic acid, glutamic acid, arginine, proline, glycine, phenylalanine, alanine, cysteine, valine, isoleucine, leucine, tyrosine, threonine, and serine (Bhari et al. 2021). However, the relative content of amino acids depends on feed, age, breed, and the environment, which has been widely reported in the literature, as seen in Figure 8.2a (Bhari et al. 2021). Generally, feathers contained a high proportion of glutamic acid, serine, proline, and cysteine (Tesfaye et al. 2017). Cysteine enables the formation of disulfide bonds which directly stabilizes the structural protein keratin in feathers. While they have lower amounts of essential amino acids such as histidine (0.02–0.52%) and methionine (0.03–1.02%).

A kilo of the feathers contains carbon (45%), nitrogen (14%), phosphorus (2.9 g/kg), potassium (1.5 g/kg), and magnesium (0.8 g/kg), and microelements such as calcium, iron, sodium, manganese, sulfur, copper and zinc as depicted in Figure 8.2b (Tamreihao et al. 2019; Nurdiawati et al. 2019; Nurdiawati et al. 2017).

## 8.3 HAZARDS AND MANAGEMENT OF FEATHER-WASTE

Management of recalcitrant feather-waste raises environmental concerns and economic disposal challenges (Brandelli et al. 2015). Feathers, which contain a high load of organic nitrogen, are helpful as organic fertilizer to enhace soil quality (Bolan et al. 2010; Kelleher et al. 2002). However, direct land application without treatment negatively impacts the environment via eutrophication, production of toxic by-products, emission of greenhouse gases, and spread of pathogens (Kelleher et al. 2002; Holm-Nielsen et al. 2009).

Landfill or disposal of feathers by composting, milling, and incineration can indirectly contribute to global warming due to methane pollution. Also, the accumulation of untreated feather- waste has resulted in putrefaction, leaching, and increase of nitrates and phosphorous in surface runoff, which stimulates rapid growth of microbial pathogens in lakes, waterways, and groundwater (Tamreihao et al. 2019). Cost-effective and safe disposal methods for feather-waste are the need of the day.

**TABLE 8.1**
**Comparison of Different Keratinous Waste Disposal Methods**

| Method | Operation | Advantages | Disadvantages | References |
|---|---|---|---|---|
| Incineration | Processes keratinous waste at 850°C. | Destroys infectious agents. | Expensive. Emission of greenhouse gases. Foul odors. | Ningthoujam et al. 2018; Stingone and Wing 2011; Dube, Nandan, and Dua 2014 |
| Landfill | Disposal of waste in landfills. | Less expensive. | Requires more space. Leachate, emission of greenhouse gases, growth of pathogens and retroviruses. | Kumawat et al. 2018; Ningthoujam et al. 2018; Franke-Whittle and Insam 2013; Franke-Whittle and Insam 2013 |
| Composting | Aerobic process for the biodegradation of organic material in an open or closed system. | Eco-friendly, economical recycling method. Reduces the pathogens. | Ammonia – inhibits the methanogens. | Wrońska and Cybulska 2016; Bhari et al. 2021 |

## 8.4 TRADITIONAL DISPOSAL METHODS FOR FEATHER-WASTE

Incineration, burning, milling, landfill, and composting are traditional methods for the disposal of keratin-rich waste. However, their use has been discontinued due to high energy requirements, expensive processes, and loss of valuable bio-molecules (Gupta and Ramnani 2006). European Union Regulation (EC) No. 1774/2002 strictly prohibits the use of incineration (Bhari et al. 2021). Both landfill and composting methods involve microbes producing leachates that cause eutrophication and surface and ground water contamination. However, microbes in the compost are primarily regulated through the addition of select microbial consortia. Hence, Directive 1999/31/EC of the European Communities Commission in 1999 came up with strict principles for the safe dumping of feathers to prevent contamination of soil and water bodies (Bhari et al. 2021). Table 8.1 gives a detailed comparison of keratinous waste disposal methods.

## 8.5 PRE-TREATMENT TECHNOLOGIES FOR HYDROLYSIS OF FEATHER-WASTE

Pre-treatment of recalcitrant biomass is an essential step for any type of integrated bio-refineries with a high conversion rate for the generation of bio-products of biotechnological importance (Bhardwaj et al., 2020; Kumar and Verma 2020; Kumar et al., 2020; Verma, 2022). Feathers being recalcitrant, resist complete biodegradation and cause bottlenecks in techno-conversion methods, and are thus expensive to process and environmentally nonfriendly. Therefore, there is a need for green and economic pre-treatment technologies that are efficient in the hydrolysis of the components in feathers to get rid of undesirable compounds and allow easy access,

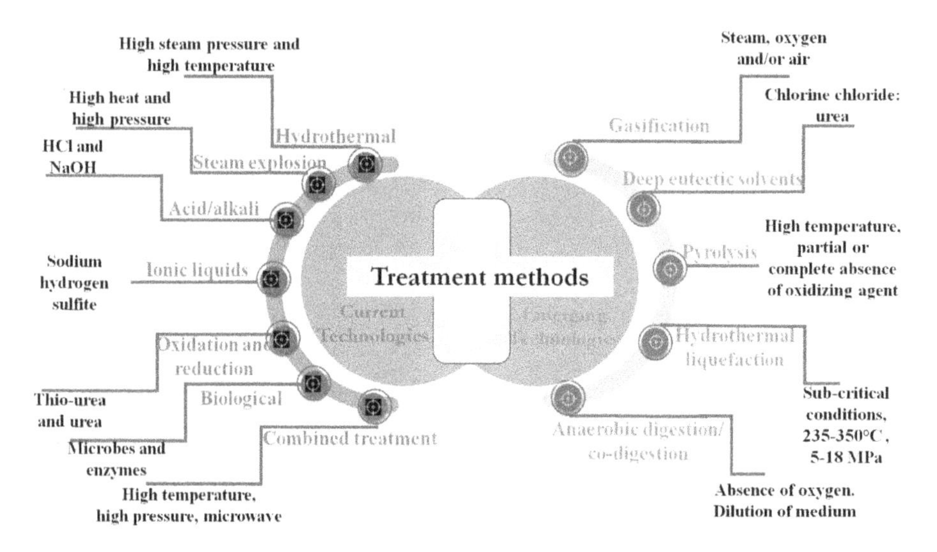


FIGURE 8.3 Current and emerging technologies for hydrolysis of feather-waste.

leading to the complete degradation of feather biomass. These methods are categorized into "current" and "emerging technologies," as in Figure 8.3, of which some are discussed in detail.

## 8.6 CURRENT METHODS

### 8.6.1 Hydrothermal

The hydrothermal method employs the use of 80–140°C temperature, 10–15 psi steam pressure with simultaneous addition of acids (HCl) or bases (NaOH) for a breakdown of peptide bonds of keratin during degradation (Ningthoujam et al. 2018; Kumawat et al. 2018). Though the process yields soluble peptides, several amino acids are destroyed, and non-nutritive amino acids like lanthionine and lysinoalanine are derived from cysteine and lysine. Papadopoulos et al. (1985) and Latshaw et al. (1994) argued that the formation of lanthionine is indicative of over-processing of feather meal. Alkali and heat treatment, on the other hand, influenced the quality of protein due to racemization. It was observed that feather meal autoclaved with NaOH showed less digestibility of amino acids than individual treatments. Overall hydrothermal processes are costly, destroy amino acids, and require a longer time (16 h). However, the ash produced in this process is rich in nutrients that have high bio-fertilizer potential (Ningthoujam et al. 2018).

### 8.6.2 Superheated Process or Thermal Hydrolysis

This involves hydrolysis of biomass with water at various temperatures and pressures to obtain oligo-peptides. In a review by Chilakamarry et al. (2021), the superheated process is described as a method for extracting keratin from feathers using a sealed cell with water at 20mg/ml, under developed pressure. The assembly is then placed in a pre-heated oven. As observed by Yin et al. (2007), the dissolution rate is governed by the time and temperature of the reaction. Tasaki (2020) described this hydrolysis as a two-step process consisting of initial denaturation of the keratin protein and breakdown of the disulfide bonds between the keratinous fibers resulting in 70% of total recovery of keratin.

### 8.6.3 Steam Explosion/Steam Pressure Cooking

This is a subtype of hydrothermal treatment wherein feathers are subjected to 275–415 kPa pressure for 30–60 minutes, and is carried out as an explosive decompression event involving short exposure and pressure release (Chilakamarry et al. 2021; Tonin et al. 2007). The feather meal generated by employing such treatment has low amino acids (Papadopoulos et al. 1985; Eggum 1970).

### 8.6.4 Acid and Alkali Treatments

The chemical hydrolysis method requires large, expensive industrial equipment (Kumawat et al. 2018), increasing the solubility of feather-waste by destroying the keratin structure (Sinkiewicz et al. 2017). Generally, acidic hydrolysis of keratinous waste is highly efficient, but hydrolysate has a low nutrient value (Chilakamarry et al. 2021). Some limitations of chemical treatments are, for instance, lack of essential amino acids, low biological activity, and poor digestibility. Al-Bahri and team stated that strong acids and alkali attack peptide bonds to release soluble peptides and amino acids (Al-Bahri et al. 2009).

### 8.6.5 Oxidation and Reduction Method

Oxidation and reduction of the disulfide linkages produce undegraded amino acids in soluble keratin (Sinkiewicz et al. 2017). Sulfitolysis is the foremost step for the reduction of keratin. Various reducing agents such as 2-mercaptoethanol, glutathione, dithiothreitol (DTT), thioglycolic acid, bisulfate, and sulfites disrupts disulfide bonds generating free cysteine thiol and cysteine-S-sulphonate (Hill et al. 2010; Schrooyen et al. 2001; Tonin et al. 2007; Vasconcelos et al. 2008; Brown et al. 2016).

Schrooyen et al. (2001) and Nakamura et al. (2002) reported the application of urea, thiourea, surfactant, metal hydroxides, and their combination to remove the majority of the trapped keratin present in the protective structures. Sierpinski et al. (2008) checked the efficiency of potassium permanganate, organic per-acids, hydrogen peroxides to convert disulfide groups to sulfonate groups and finally form cysteic acid derivatives. Sinkiewicz et al. (2017) reported 82 and 84% yield of soluble keratin after 2-hour reduction of feather keratin with sodium bisulfite and 2-mercaptoethanol, respectively. In another report by Stiborova et al. (2016), the combined treatment of 0.6% potassium hydroxide and 70°C hydrolyzed 85.9% feathers was observed to release only 12.8% of essential amino acids. Also, the effects of oxidative agents such as hydrogen peroxide and sodium hypochlorite and reductive agents like sodium dithionite were studied on chicken feathers to assess their pre-treatment efficiency (Tesfaye et al. 2018). Treatment of feathers with potassium per-sulfate and vapor of formic acid resulted in consistent ammonia and soluble feather protein release, respectively (Yang and Reddy 2013).

### 8.6.6 Ionic Liquids

Ionic liquids are safe; recyclable solvents consist of organic, inorganic anions, and bulky cations. They are durable liquids with high solvation capacity, and high melting and boiling points. They are used as catalysts, dissolution solvents for various polymers in conductive media (Chilakamarry et al. 2021). Ionic liquid containing chloride is an efficient solvent for keratin dissolution due to chloride concentration and its high nucleophilic activity breakdown of the hydrogen bonds.

For instance, Br, PF6 chloride-containing ionic liquid can produce keratin hydrolysate (Xie et al. 2005). The extracted keratin has mainly β-sheets exhibiting high thermal stability. Most of the studies have used chloride or bromide derivatives for the preparation of ionic liquid to carry out feather hydrolysis (Ji et al. 2014; Sun et al. 2009; Wang and Cao 2012). The keratin could also be easily extracted from the reaction mixture for further use.

### 8.6.7 Microbial Hydrolysis

In the last two decades, microorganisms inheriting proteolytic, lipolytic, and disulfide reductase activities have been investigated to hydrolyze the recalcitrant feather-wastes. Various mesophilic bacteria, actinomycetes, fungal isolates obtained from feather disposal sites, poultry processing plants, slaughterhouse sites, and tannery effluents are employed (Chilakamarry et al. 2021; Bhari et al. 2021). Recently, some keratinolytic extremophilic microbes growing at high temperatures have been retrieved from hypersaline, alkaline, and high-temperature environments (Bhari et al. 2021).

Generally, the savinase enzyme is used to extract keratin from feathers (Chilakamarry et al. 2021). Microbial and enzymatic hydrolysis of keratinous waste, although requiring low energy, employs specific reducing agents such as thio-urea and mercaptoethanol to break disulfide bonds holding the keratin structure (Chilakamarry et al. 2021). Complete feather degradation, consisting of ~90% protein keratin and ~2–12% of fat by a microorganism, occurs through three synergistic processes: proteolysis, sulfitolysis, and deamination. Proteolysis and sulfitolysis cleave disulfide bonds which hold the keratin chain in protein to release cysteine and S-sulfocysteine. Insoluble keratin protein then releases molecules with free amino groups through microbial deamination (Chilakamarry et al. 2021; Bhari et al. 2021). Recently, it has been confirmed that wax esterases or lipases attack fat components in feathers, resulting in their complete degradation (Barcus et al. 2017). Lipase/esterase hydrolyzes the fat by breaking ester bonds releasing free fatty acids and glycerol, which are often metabolized by the microbial cells. The crude protease and lipase produced can be purified to be used for various commercial applications, and the protein-rich hydrolysate has biotechnological potential (Figure 8.4). Thus, hydrolysis of feathers and other keratinous wastes using microbes and their enzymes offers a better technology than physicochemical treatments. The treatment is eco-friendly, cost-effective, and thus has gained popularity in various biotechnological processes (Feroz et al. 2020).

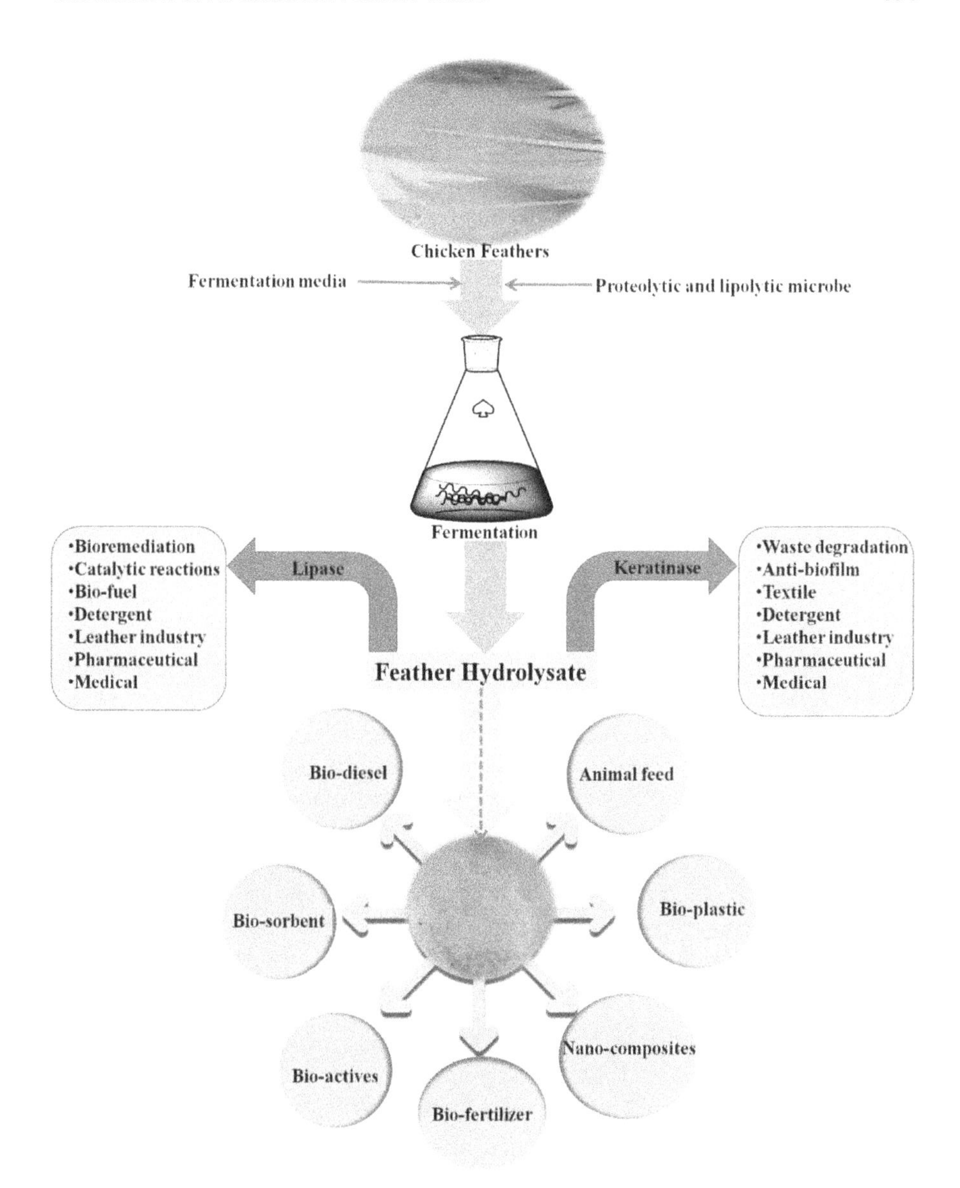


**FIGURE 8.4** Production of value-added products through fermentation of feather-waste using proteolytic and lipolytic microbial strains.

## 8.7 HYDROLYSIS USING A COMBINATION OF BIOLOGICAL AND PHYSICOCHEMICAL METHODS

Combined treatments using biological methods along with temperature/pressure/chemicals are emerging (Bhardwaj et al. 2020; Kumar et al. 2020). Hydrolysis of feathers by microorganisms and their enzymes are known to be improved by physical and/or chemical pre-treatment of feathers. Eremeev et al. (2009) worked out a process exposing feathers and water in 1:4 ratio to 170–180°C (moist heat) for one minute, followed by treatment with protease from *Acremonium chrysogenum*. They augmented the reduction efficiency by further exposing for 12 hours to 0.5% $Na_2SO_3$, 55°C, pH 10.0. Alkali treatment (NaOH) and autoclaving improved the solubilization of chicken feathers, increasing the protein digestibility of the hydrolysate (Al-Souti et al. 2018). In contrast, microwave and alkali treatment studied by Lee et al. (2016) reduced disulfide bonds and improved the production of protein hydrolysate. Feathers autoclaved in the presence of sodium sulfite (10 mM) followed by hydrolysis using *Bacillus cereus* could solubilize 86.3% feathers (Łaba and Szczekala 2013).

## 8.8 EMERGING TECHNOLOGIES FOR POULTRY WASTE HYDROLYSIS

The following methods convert poultry litter, rotten eggs, and bones, including feather-waste (Kanani et al. 2020). These ways, however, need optimization and detailed research investigation for scale-up of the operation.

### 8.8.1 Gasification

This is a thermo-chemical process of converting organic waste materials to syngas (carbon monoxide and hydrogen), methane, and carbon dioxide using steam, oxygen, and/or air and catalyst (Dai et al. 2015). The technology has programs targeted to convert organic materials of choice and selectively recover the desired/pre-programmed product (Heidenreich and Foscolo 2015). For the first time, Dudyński et al. (2012) used a fixed-bed gasifier for hydrolysis of feathers in a 3.5 MW industrial installation of 10,000 tons of feathers per year capacity. The products had promise as a source of energy for heating, power, and transportation. However, as nitrogen and sulfur oxides were released at the operating temperature and oxygen conditions, the emission levels were still an environmental eyesore (Taupe et al. 2016). Further, the generation of fossils in the reactor reduced the efficiency of gasification with time.

### 8.8.2 Pyrolysis

Pyrolysis is the thermal degradation of organic matter in the partial or complete absence of an oxidizing agent to produce bio-oil/tar, char (solid), and syngas (non-condensable gases) (Kanani et al. 2020; Goswami et al. 2021). The bio-oil (condensable gases) contains hundreds of organic compounds (Isahak et al. 2012) and syngas can be used for the production of ethanol (Mackaluso 2007) or hydrogen (Voitic et al. 2016). It is a potentially viable technology for the valorization of poultry waste due to its flexibility, maturity, and simple infrastructure. The addition of wood shavings during pyrolysis of feathers influenced the density and properties of the oils (Mante and Agblevor 2010). Another report by Brebu and Spiridon (2011) recounted the pyrolysis of feathers, sheep wool, and human hair and identified primary compounds such as phenol and 4-methyl phenol by thermal degradation. However, Simbolon et al. (2019) obtained char, bio-oil, and syngas by carrying low temperature (350 to 450°C) pyrolysis of feathers mixed chicken litter, waste feed, and manure.

### 8.8.3 Sub/Super-critical Water Hydrolysis Treatment/Hydrothermal Liquefaction

Sub/super-critical water hydrolysis can be used for liquefaction of protein-rich substrates such as distillers' grains, cakes, vinasse, fish silage, feathers, and by-products from the processing of coffee, tea, and other agricultural products. The process involves the use of two thermodynamic conditions for water. The sub-critical process is between 100 and 374°C and higher pressure than the water saturation pressure, while super-critical processes are carried out at water temperatures of > 374°C and pressures higher than 22.1 MPa. Sub-critical water acts as an excellent solvent for the recovery of ionic and polar compounds. It also allows the extraction of non-polar compounds at high temperatures and pressure in the system. Sub-critical and super-critical water processes are green alternatives that present a strong potential for extracting high value-added compounds. Also, the technologies are cost-effective, require low volumes of organic reagents, and do not produce undesirable residual by-products (Di Domenico Ziero et al. 2020). In a review by Brunner (2009), it was reported that sub-critical water is an efficient process for the hydrolysis of protein-rich feathers and hair to recover value-added amino acids.

### 8.8.4 Deep Eutectic Solvents

Deep eutectic solvents (DESs) are green solvents, generally inexpensive, and simple, which effectively extract proteins from protein-rich biomasses (Nuutinen et al. 2019; Chao et al. 2021). DESs were first reported for metal extraction as an alternative for ionic liquids (Abbott et al. 2004). Mixing two solid components at a specific molar ratio until the system reaches its lowest melting point due to hydrogen bonding between components leads to a decrease in lattice energy (Smith et al. 2014; Abbott et al. 2004). DESs are nonflammable and have a low vapor pressure. DESs are relatively easy to prepare using non-toxic, biocompatible, and biodegradable components (Smith et al. 2014; Wahlström et al. 2017). DES, choline chloride (ChCl)–urea (1: 2) with a melting point of 12°C, is notably the most cited DES prepared from biodegradable and readily available components such as urea and chlorine (Abbott et al. 2004). Recently, Chao et al. (2021) used DES consisting of choline chloride/ethylene glycol to hydrolyze feathers and the keratin into low-cost sashes used for packing instant tea in just ten minutes.

### 8.8.5 Anaerobic Digestion and Anaerobic Co-digestion

Anaerobic digestion is the conversion of organic matter into high energy methane ($CH_4$) (~65%) and carbon dioxide ($CO_2$) (~35%) constituting biogas in the absence of oxygen, leaving the digestate as a by-product (Evangelisti et al. 2014; Wu et al. 2016). Anaerobic digestion is a straightforward, reliable technique to convert wet biomass to biogas without pre-treatment. Recently, the method has been used for treating farm, food, and garden wastes (Evangelisti et al. 2014). Kanani et al. (2020) reviewed various research reports on anaerobic digestion and concluded it was a suitable technology for converting poultry litter, including feathers, for biogas generation. In recent advancements, anaerobic co-digestion technology was used for the production of biogas. Here, process efficiency is increased by adding one or more low nitrogen substrates to reduce ammonia inhibition. The process is employed at laboratory and large-scale commercial units and is more environmentally and economically feasible than single substrate anaerobic digestion (Hagos et al. 2017; Mata-Alvarez et al. 2014).

## 8.9 BIO-HYDROLYSIS OF FEATHER-WASTE BY MICROORGANISMS AND THEIR ENZYMES

Microbial/enzymatic degradation of keratin-rich poultry waste has been regarded as an eco-friendly approach in recent years (Gupta and Ramnani 2006). Also, the use of some keratinolytic microorganisms and their proteases have been commercialized and patented (Shih et al. 2003; Hansted et al. 2016). Figure 8.5 represents a schematic illustration of the advantages of microbial degradation of feather-waste over physicochemical methods.

This section presents detailed information on the biological methods for feather hydrolysis using microorganisms, the mechanism of complete hydrolysis using synergistic action of protease, lipase, and disulfide reductase, and its subsequent value addition to produce protein-rich hydrolysate from biotechnological applications.

### 8.9.1 Hydrolysis of Feathers by Mesophilic Bacteria, Actinomycetes, and Fungi

Till the 1990s, the degradation of keratin wastes was limited to the mesophilic fungi of genera *Doratomyces* and *Aspergillus*. However, these strains are pathogenic and required up to 40 days to carry out more than 50% of keratin digestion (Santos et al. 1996).

However, the demand for bacterial keratinases in the biotechnological field emerged in 1992, when the first feather-degrading *Bacillus licheniformis* strain PWD-1 was retrieved from poultry feather-waste. The research was primarily focused on the utilization of feather-meal by microorganisms and purification of keratinase to check the digestibility of feather-meal (Lin et al. 1992). Later, several authors described mesophilic bacterial strains of *Bacillus licheniformis* and *Bacillus pumilis, Bacillus subtilis*, and *Bacillus cereus*, actinobacteria such as *Streptomyces thermoviolaceus* degrading native as well as denatured keratin from feathers (Bhari et al. 2021).

While extensive studies in the last two decades have shifted to isolate non-pathogenic keratinolytic microbial strains for their detailed use in biotechnological applications, most of these microbial strains which are isolated from poultry feather-waste, slaughterhouses, soil from poultry farms, feather-dumping sites, hot springs, and so forth, have feather-degrading ability (Bhari et al. 2021). Table 8.2 depicts examples of some keratinolytic microbes for feather-waste degradation.

The research investigations of past decades have indicated that the keratinase approach is inefficient for complete hydrolysis of feather-waste and requires the prior breakdown of disulfide bonds through sulfitolysis agents (Bhari et al. 2021).

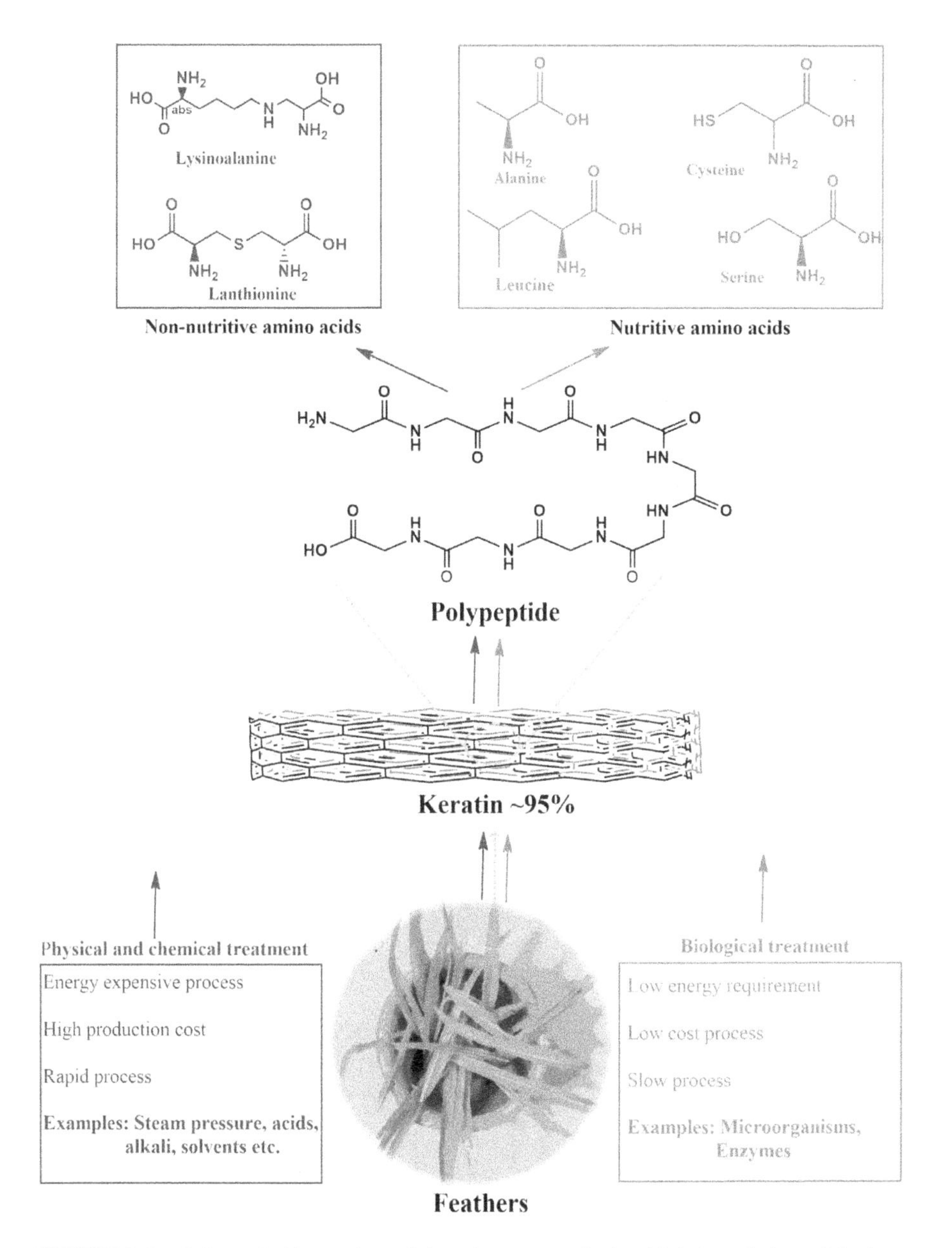


**FIGURE 8.5** Schematic illustration of the advantages of microbial degradation of feather-waste over physicochemical methods.

Trials with recombinant keratinase are found to be economically unviable (Li et al. 2020). Instead, Li et al. (2020) observed microorganisms growing on feathers to colonize the bio-material, produce and import soluble peptides, amino acids, and fatty acids into their central metabolic pathway for cell growth. Further, processes of oxygen utilization, iron uptake, spore formation, and amino acid metabolism are also prominent.

**TABLE 8.2**
**Keratinolytic Microbes Degrading Feather-waste**

| Microbe Bacteria/Fungi | Econiche | % Feather Degradation | pH | Temp | Incubation Period (Days) | Product | References |
|---|---|---|---|---|---|---|---|
| *Bacillus mycoides* (G2) | Tannery effluent | 58 | 8 | 37 | 2 | Feather compost | Beryl et al. (2021) |
| *Bacillus licheniformis*-K51 | Poultry feather-waste | - | 8 | 50 | 2 | Feed supple-ment | Dada and Wakil (2021) |
| *Bacillus amyloliquefaciens* KB1 | Chicken farm bed | 74.78 | 9 | 40 | 7 | Feather hydro-lysate | Prajapati et al. (2021) |
| *Ochrobactrum intermedium* | Soil | - | 7 | 40 | 4 | Keratin-ase | Sharma and Kango (2021) |
| *Bacillus pumilus* JYL | Poultry dump yard | - | 6.5 | 35 | 2 | Feather | Sun et al. (2021) |
| *Bacillus pumilis* | Iranian soil | - | 7 | 37 | 3 | Keratin-ase | Alahya-ribeik et al. (2020) |
| *Bacillus licheniformis* | | - | 10 | 40 | 3 | | |
| *Rhodococcus erythropolis* | | - | 7 | 37 | 5 | | |
| *Geobacillus stearothermophilus* | | - | 7-8 | 55 | 4 | | |
| *Bacillus haynesii* ALW2 | Leather waste | - | 7 | 37 | 3 | Keratin-ase | Emran, et al. (2020) |
| *Bacillus* sp. NKSP-7 | Poultry dumping sites | 88.02 | - | 37 | 1 | Keratin-ase | Haq et al. (2020) |
| *Streptomyces* sp. SCUT-3 | Feather-waste dumping site | 93.6 | - | 37 | 2 | Feed and fertilizer | Li et al. (2020) |
| *Bacillus* sp. FPF-1 | Compost-ing site | 86.0 | 5 | 25 | 5 | Keratin-ase | Nnolim, et al. (2020) |
| *Bacillus* sp. CSK2 | Dumpsite soil | - | 5 | 30 | 2 | Keratin-ase | Nnolim and Nwodo (2020) |
| *Bacillus aerius* NSMk2 | Poultry dump soil | 100 | 7 | 37 | 3 | Keratin-ase | Bhari et al. (2019) |
| *Chryseobacterium sediminis* | Feather dumping site | 100 | 7 | 30 | 2 | Hydroly-sate | Kshetri et al. (2019) |
| *Kocuria rhizophilia* | Poultry waste | 52 | 7 | 25 | 4 | Hydroly-sate | Łaba et al. 2018 |
| *Streptomyces sampsonii* | Type strain | 100 | 8 | 28 | 5 | Soil condition-er | Jain et al. (2016) |
| *Trichoderma harzianum* | Soil | 100 | 6 | 30 | 8 | Keratin-ase | Bagewadi et al. (2018) |
| *Chryseobacterium* sp. kr6 | Decompos-ing feathers | 75 | - | - | 4 | Hydrolys-ate | Maciel et al. (2017) |

| | | | | | | | |
|---|---|---|---|---|---|---|---|
| *Bacillus pumilus* | Poultry dump yard | 100 | 7 | 37 | 1 | Keratin-ase | Ramakrishna Reddy et al. (2017) |
| *Trichosporon loubieri* | Discarded leather scraps | 89 | 5.5 | 28 | 7 | Keratin-ase | de Medeiros et al. (2016) |
| *Bacillus subtilis* | Feather dumping site | 100 | 7 | 32 | 5 | Fertilizer | Bhange et al. (2016) |
| *Arthroderma multifidum* | Poultry farm | 39.2 | 7 | 30 | 1 | Keratin-ase | Kumawat et al. (2016) |
| *Pseudomonass aeuroginosa* | Poultry processing plants | 100 | 7 | 37 | 5 | - | Bishmi et al. (2015) |
| *Bacillus safensis* | Feather dumping site | 100 | 7.5 | 30 | 9 | Keratin-ase | Lateef et al. (2015) |
| *Pseudomonas aeruginosa* strain SDS3 | Detergent contamin-ated pond | 80 | 7.5 | 30 | 5 | Hydroly-sate | Chaturvedi and Verma (2014) |
| *Brevundimonas terrae* | Poultry soil | 100 | 7.5 | 37 | 3 | Keratin-ase | Kulkarni and Jadhav (2014) |
| *Bacillus sonorensis* | Soil | 100 | 7 | 25 | 7 | Keratin-ase | Mehta et al. (2014) |
| *Aspergillus fumigatus* | Poultry waste | 100 | 6 | 30 | 4 | Keratin-ase | Paul et al. (2014) |
| *Paenibacillus woosongensis* TKB2 | Feather dumping site | 86.99 | 8 | 30 | 3 | Fertilizer | Paul et al. (2013) |
| *Pseudomonas aeruginosa* C11 | Feather dumping soil | 100 | 7.5 | 30 | 4 | Keratin-ase | Han et al. (2012) |
| *Aspergillus terreus* | Chicken feather | 100 | 7.5 | 30 | 7 | Keratin-ase | Koutb et al. (2012) |
| *Bacillus altitudinis* | Waste dump yard | 100 | 9 | 37 | 2 | Feed | Kumar et al. (2011) |
| *Stenotrophomonas maltophilia* | Rhizosphere Soil | 100 | 7 | 30 | 3 | Hydroly-sate | Jeong et al. (2010) |
| *Elizabethkingia meningoseptica* | Dropped off feathers | 82.50 | 7 | 37 | 6 | Hydroly-sate | Nagal and Jain (2010) |
| *Bacillus megaterium* | Poultry waste | 100 | 7.5 | 30 | 7 | Hydroly-sate | Park and Son (2009) |
| *Bacillus pseudofirmus* | Poultry farm soil | 100 | 10.5 | 30 | 3 | Keratin-ase | Kojima et al. (2006) |
| *Microbacterium* sp. | Industrial poultry waste | 100 | 7 | 30 | 2 | Keratin-ase | Thys et al. (2004) |
| *Vibrio* sp. | Feather dumping soil | 100 | 7 | 30 | 5 | Keratin-ase | Sangali and Brandelli (2000) |
| *Kocuria rosea* | Soil | 51 | 7.5 | 40 | 3 | - | Vidal et al. (2000) |

Strains belonging to genera *Bacilli* such as *Bacillus aerius, Bacillus pumilus, Bacillus subtilis, Bacillus safensis, Bacillus sonorensis, Bacillus altitudinis, Bacillus megaterium* and *Bacillus pseudofirmis* efficiently carried out 100% feather degradation in the range of one to nine days. Other bacteria including *Chrysobacterium sediminis, Pseudomonas aeruginosa, Brevundimonas terrae, Pseudomonas aeruginosa* C11, *Xanthomonas maltophilia*, and *Vibrio* sp. degrade 100% feather-waste in two, five, three, four, three, and five days, respectively.

Actinobacteria such as type strain *Streptomyces sampsonii* and *Microbacterium* sp. retrieved from industrial poultry waste degrade 100% of feather-waste in five and two days at pH 8 and pH 7, respectively. Among the fungi reported for feather hydrolysis, *Trichoderma harzianum, Aspergillus fumigatus*, and *Aspergillus terreus* showed 100% feather degradation at 30°C in the range of four to eight days.

### 8.9.2 Microbial Cells and Mixed Culture Consortia for Large-scale Fermentation of Feather-waste

Mixed culture or co-culture is a technique in which two or more different populations of cells are grown together to improve culturing, study interactions, and enhance the efficiency of specific processes (Goers et al. 2014). Biodegradation of more than 50 g/L of chicken feathers by microorganisms is difficult because of the complex keratinous structure. One study, by Patinvoh et al. (2016), reported *Bacillus* sp. C4 degrading 75% of chicken feathers (5%, w/v) in eight days. Additionally, recombinant *Bacillus subtilis* DB 100 has wholly degraded 2% (w/v) chicken feathers in a 14 L fermenter (Bio Flo 110) (Zaghloul et al. 2011). Inoculation of keratinolytic *Bacillus licheniformis* and *Streptomyces* sp. consortia accelerated degradation of feathers more than native microflora of compost dumps (Ichida et al. 2001).

Hydrolysis of feathers using consortia of microorganisms allows more efficient conversion than individual strains and isolated enzymes. Additionally, the process is environmentally friendly, and the hydrolysate is rich in amino acids, soluble proteins, and peptides (Bhari et al. 2021). Nevertheless, detailed research is desired to optimize the degradation and conversion rate. Downstream processes are required to recover the high-value amino acids before they are utilized for microbial metabolism.

Osman et al. (2017) successfully developed a robust method to employ new mixtures of alkaliphilic microorganisms that can degrade both white and black feathers by applying a Central Composite Design (CCD).

Recently, Peng et al. (2019) reported the use of bacterial consortia consisting of keratinolytic *Bacillus licheniformis* BBE11-1 (Liu et al. 2014a) and *Stenotrophomonas maltophilia* BBE11-1 (Fang et al. 2014) for efficient degradation of feathers. Both the strains can hydrolyze feathers individually but have different growth and enzyme production conditions. They cultured the two strains separately and/or together in a 3 L fermenter (3L) with 50 g/L of the feather-waste (the upper limit for the fermenter). In this integrated and innovative co-culture fermentation, enzyme activities increased in the first 12 hours and remained high from 12 to 48 hours. An 81.8% degradation was achieved in 48 hours. A co-culture system appears to be a potential alternative to improve feather-waste degradation in minimal time.

### 8.9.3 Immobilized Whole Cells/Enzyme for Hydrolysis of Feathers

Whole-cell immobilization is a simple and effective technique and can be applied for biodegradation. It has been reported that whole immobilized cells have more productivity when compared to crude or purified microbial enzymes for bioconversion of waste. Dhiva et al. (2020) investigated the ability of free cells and immobilized *Bacillus* sp CBNRBT2 in an alginate matrix, and attained 14% feather degradation in 14 days in comparison to 38% by the whole cells. Prakash et al. (2010a) carried out immobilization of the whole cells in alginate for continuous keratinase production and subsequent feather degradation.

### 8.9.4 Hydrolysis of Feathers by Extremophilic Bacteria, Actinobacteria, Fungi, and Archaea

Many keratinases/proteases and lipases have been identified from non-extremophilic microorganisms. However, the global enzyme market for keratinases which are active at high temperatures and other physicochemical factors is very small (Reis et al. 2020). Therefore, it is imperative to isolate and identify the extremophiles with polyextremophilic keratinases, lipases, and disulfide reductases that can synergistically carry out complete feather degradation. Table 8.3. gives examples of some keratinolytic extremophilic microbes that have been reported to hydrolyze feathers in extreme conditions.

Intagun and Kanoksilapatham (2017) reviewed keratinolytic extremophiles, which degrade keratin-rich, waste-like feathers at 70°C. The first known extreme thermophilic bacterium, *Fervidobacterium pennivorans*, hydrolyzed feathers at 70°C and pH 6.5 (Friedricht and Antranikian 1996). Later, some thermophilic and halophilic keratinolytic bacteria/actinobacteria degrading 100% feather-waste such as *Fervidobacterium islandicum* AW-1 (Nam et al. 2002), *Bacillus halodurans* PPKS-2 (Prakash et al. 2010b), *Keratinibaculum paraultuenense* (Yan Huang et al. 2013), *Actinomadura viridilutea* DZ50 (Habbeche et al. 2014), *Meiothermus* sp. (Kuo et al. 2012) and *Thermoactinomyces* sp. YT06 (Wang et al. 2017) and isolated from extreme econiches such as solfatric muds, hot springs, geothermal/hydrothermal vents have been reported. Among these, strains of *Fervidobacterium* and *Keratinibaculum* produced maximal keratinase at 70°C under neutral conditions. However, biodegradation under strict anaerobic conditions creates a hindrance for scaling up the process of feather disintegration (Intagun and Kanoksilapatham 2017).

Keratinolytic fungi of dermatophytes penetrate the surface of keratin substrate with frond mycelia and through boring hyphae in non-dermatophytes. However, keratinolytic activities of dermatophytes are higher than non-dermatophytes (Sharma and Devi 2018). Recently, a fungus *Onygena corvina* with a natural ability to colonize hooves, horns, and feathers, has been reported to bring about complete keratin degradation (Huang et al. 2015).

Among domain Archaea, very few reports on the representative's cultures of thermophiles such as *Thermococcus* VC13 (Tsiroulnikov et al. 2004), *Desulfurococcus kamchatkensis* 122ln$^T$ (Kublanov et al. 2009), *Desulfurococcus* sp. (Bidzhieva et al. 2014), that can hydrolyze keratin-rich substrates have been reported. *Thermococcus*

**TABLE 8.3**
**Some Keratinolytic Extremophilic Microbes Reported to Hydrolyze Feathers in Extreme Conditions**

| Microbe Bacteria/Archaea | Econiche | % Feather Degrad-ation | Opti-mum pH | Temp. | Incubation (Days) | References |
|---|---|---|---|---|---|---|
| *Pedobacter* sp. | Antarctica | - | 6 | 20 | 8 | Bezus et al. (2021) |
| *Bacillus halodurans* SW-X | Hot spring | - | - | 47 | 2 | Kaewsalud et al. (2020) |
| *Meiothermus taiwanensis* WR-220 | Hot spring | - | - | 65 | 2 | Wu et al. (2017) |
| *Thermoactinomyces* sp. YT06 | Poultry compost | 100 | 8 | 60 | 2 | Wang et al. (2017) |
| *Caldicoprobacter algeriensis* | Hydrothermal hot springs | - | 7.4 | 50 | 1 | Bouacem et al. (2016) |
| *Actinomadura viridilutea* DZ50 | Fishing port | - | 7.4 | 45 | 5 | Elhoul et al. (2015) |
| *Desulfurococcus* sp. 3008g | Hot spring | 100 | 6.1 | 85 | 7 | Bidzhieva et al. (2014) |
| *Desulfurococcus* sp. 2355k, | | 100 | 6.1 | 92 | 7 | |
| *Actinomadura keratinilytica* Cpt 29 | Poultry compost | 100 | 8.5 | 45 | 8 | Habbeche et al. (2014) |
| *Coprinopsis* sp. strain Kucc595 | Soil | 64 | 7 | 40 | 21 | Al-Musallam, et al. (2013) |
| *Keratinibaculum paraultunense* | Grassy marshland | 100 | 7 | 70 | 1 | Huang et al. (2013) |
| *Meiothermus* sp. I40 | Hot spring | 100 | 7.5 | 55 | 1 | Kuo et al. (2012) |
| *Bacillus halodurans* PPKS-2 | Rice mill effluents | 100 | 11 | 37 | 5 | Prakash et al.(2010a) |
| *Desulfurococcus kamchatkensis of*1221n$^{T}$ | Hot spring | - | 6.5 | 85 | - | Kublanov et al. (2009) |
| *Clostridium sporogenes* | Solfataric muds | 100 | 7 | 42 | 7 | Ionata et al. (2008) |
| *Thermococcus* VC13 | Black smoker | - | - | 80 | 3 | Tsiroulnikov et al. (2004) |
| *Fervidobacterium islandicum* | Geothermal hot stream | 100 | 7 | 70 | 2 | Nam et al. (2002) |
| *Fervidobacterium pennavorans* | Hot spring | - | 6.5 | 70 | 2 | Friedricht and Antranikian (1996) |

VC13 isolated from black smoker belonging to phylum *Euryarcheota* hydrolyzed keratin at 80°C in three days. The *Desulfurococcus kamchatkensis* strain 1221n$^T$ from hot spring hydrolyzed keratin substrates at pH 6.5 and 85°C. In another study by Bidzhieva et al. (2014), *Desulfurococcus* sp. 3008g and *Desulfurococcus* sp. 2355k hydrolyzed keratin and other high molecular weight proteins molecules at pH 6.1, 85°C and pH 6.5, 92°C, respectively in seven days. They also studied the capacity of *Desulfurococcus fermentans, Desulfurococcus amylolyticus* subsp. *Amylolyticus* and *Desulfurococcus mucosus* subsp. *Mobilis* type strains for β-keratin disintegration. The ability of strains *Desulfurococcus* sp. 3008g and *Desulfurococcus* sp. 2355k to grow on poorly hydrolyzable proteins such as keratins and other proteins revealed that keratinases from hyperthermophilic archaea were not only confined to keratin, but also other protein molecules of allochthonous and autochthonous origin.

Birbir and Caglayan (2018) summarized the results of the diversity of halophilic microorganisms from salted hides to understand their biodegradation capabilities. For instance, 94% of salted hide samples from 131 cattle hides harbored extremely halophilic archaea, of which 53% exhibited protease activities. While 91% of samples of 35 salt-cured hides in France and Russia consisted of extremely halophilic archaea. Akpolat in 2015 isolated 101 strains of haloarchaea of the genera *viz. Halorubrum, Halococcus, Natrinema, Halostagnicola*, and *Haloterrigena* from salted sheepskins with 15% protease and 5% lipase activity, while salted cattle hides from Australia and England reported three and two species of *Halorubrum* and *Natrinema*, respectively. Proteolytic activity detected was 20%. Further, 186 strains of haloarchaea from eight salted hides and cured skins belonged to *Natronococcus* sp., *Natrialba* sp., *Halovivax* sp., *Halococcus* sp., *Halorubrum* sp., *Natrinema* sp., *Haloterrigena* sp., and *Halobacterium* sp. Among which 12% and 57% of the isolates showed positive activity for caseinase and protease. The presence of proteolytic and lipolytic haloarchaea in curing salt contributed to the deterioration of salted sheepskins reported by Birbir et al. (2020).

Recently, Gaonkar and Furtado (2021) reported *Haloferax* sp. strain GUBF-2 from salt pans to degrade 51.8 ± 0.71% feathers at 42°C, 30% NaCl, and pH 5 under submerged conditions. The protease and lipase activity of 82.0 ± 4.97 U/mL and 3.66 ± 0.49 U/mL, respectively, were attained after 72 hours of fermentation. The feather hydrolysate was analyzed for the presence of soluble peptides, amino acids, soluble fatty acids using various characterization techniques. Further, large-scale fermentation of feathers in 5% NaCl by strain GUBF-2 was carried out in a 5 L Carboy glass fermenter. The produced nutrient-rich hydrolysate was used as a bio-stimulant for rice (*Khorgut* var.) growth in a pot experiment under saline conditions (Gaonkar and Furtado 2022).

## 8.10 MECHANISM OF COMPLETE DEGRADATION OF FEATHERS BY THE SYNERGISTIC ACTION OF PROTEASE, DISULFIDE REDUCTASE, AND LIPASE

Biodegradation of feather-wastes by microorganisms is an attractive and potential option to utilize this keratin-rich waste by producing keratinase enzyme. However, complete degradation of a feather can be achieved by the synergistic action of disulfide reductase, keratinolytic protease (Prakash et al. 2010b; Brandelli et al. 2010), and lipase (Barcus et al. 2017; Li et al. 2020). Generally, keratinolytic microorganisms produce inducible disulfide reductase (EC.1.8.1.14) and keratinase (EC.3.4.99.11) enzymes in the presence of keratin-rich substrates (Anbu et al. 2008). There has also been an extensive discussion on the different mechanisms like proteolysis, sulfitolysis, lipolysis, and secretions of metabolites by the microbial cell that bring about complete feather degradation to produce protein-rich hydrolysate.

Based on the literature, a schematic illustration of the possible mechanism for complete feather degradation by the microbial cell is presented in Figure 8.6.

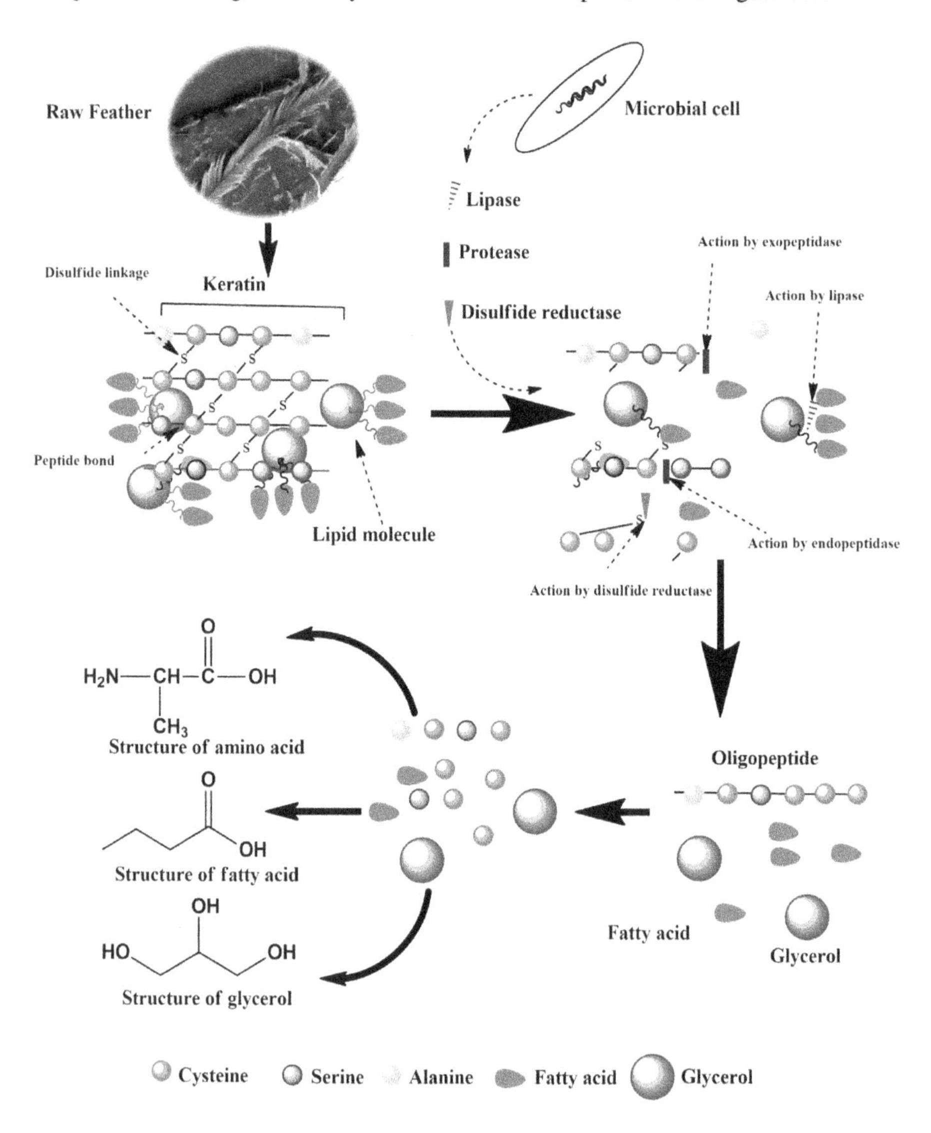


**FIGURE 8.6** Schematic illustration of the possible mechanism for complete degradation of a feather by microbes producing keratinase, disulfide reductase, and lipase.

### 8.10.1 Keratinase

The biodegradation of feather and other keratin-rich waste using enzymes will allow recycling of the materials with efficient energy conservation and overall reduction of load in the environment (Brandelli et al. 2010).

Most of the identified keratinases from microorganisms are serine proteases. However, a few acidic keratinases and metalloproteases are reported (Bhari et al. 2021). Keratinases are used in various biotechnological sectors such as fertilizer, feed, leather, detergent, textile, pharmaceutical, and biomedical applications. Keratinase improves the quality of the leather and biodegradation of components in tannery effluents (Gupta and Ramnani 2006). This has resulted in the large-scale production of keratinase on an industrial scale.

A recombinant thermostable keratinolytic protease from *Geobacillus* sp. AD-11 efficiently degraded keratin waste without any loss of nutritional elements (Gegeckas et al. 2015). Similarly, keratinase from *Thermoanerobacter keratinophilus* (Riessen and Antranikian 2001) and *Brevibacillus* sp. AS-S10-II (Rai and Mukherjee 2011) were reported to degrade alpha and beta keratin. An extracellular alkaline serine keratinase from *Actinomadura viridilutea* DZ50 is highly active at alkaline pH and high temperature (Elhoul et al. 2016).

### 8.10.2 Disulfide Reductase

The involvement of disulfide reductase in the breakdown of disulfide bonds in keratin degradation has been relatively understudied. The process of microbial keratinolysis for complete hydrolysis of keratin is composed of two main steps: 1) reduction of disulfide bonds (sulfitolysis), and 2) breakdown of the peptide bond (Proteolysis) (Kunert 1976).

In a complex pathway of proteolysis, the reduction of strong disulfide linkages between the two keratin residues by either living cell or disulfide reductase or chemical reducing agents (mercaptoethanol, glutathione, dithiothreitol, cysteine, thioglycolate, sodium sulfite) is essential for complete hydrolysis of keratin (Onifade et al. 1998; Ramnani et al. 2005).

The process of sulfitolysis is initiated by the breakdown of disulfide linkages polypeptide keratin chains liberating thiol groups (Kunert 1976). This modifies the keratin configuration exposing more active sites for further degradation by proteases (Monod 2008).

Biologically, the intracellular fraction contributes mainly to disulfide or thiosulfate, or sulfite reductases that synergistically aid keratinase to cleave disulfide bonds, thereby degrading the tough protein-keratin-yielding hydrolysate, rich in amino acids (Ramnani et al. 2005; Yamamura et al. 2002).

According to Rahayu (2012), keratinase activity on the natural keratin-rich substrates such as wool and feathers was significantly increased due to the initial action of disulfide reductase. Verma et al. (2016) stated that the proper hydrolytic mechanism of fungal keratinase and reductases is observed only on complex keratinous structures which are not utilized by other microorganisms producing proteases. Filamentous actinomycetes and fungi initiated mycelial growth on keratin substrates and secreted sulfite for the breakdown of disulfide bonds, thereby releasing oligopeptides for the action of proteases.

Yamamura et al. (2002) reported disulfide reductase-like protein from *Stenotrophomonas* sp. having the ability to degrade keratin with the cooperation of protease D-1. While, keratinase I (disulfide reductase-like protein) and keratinase II (protease) have been purified from *Bacillus haloduran* PPKS-2 (Prakash et al. 2010b). It has been reported that keratinase and disulfide reductase alone, without live cells, could not carry out feather degradation. Complete feather degradation was initiated by the colonization of cells and initiation of various redox mechanisms such as the production of sulfite and reductases (Ramnani et al. 2005; Ghosh et al. 2008; Prakash et al. 2010b).

### 8.10.3 Lipase

Lipase is the enzyme that catalyzes the hydrolysis of long-chain triacylglycerides. It can increase the strength and keratinase compatibility for the complete hydrolysis of feathers with zero-waste generation in poultry waste management.

Chicken feathers consist of 2-12% of fat, contributing to the recalcitrant property of the feather. It has been reported that 38.2% and 12.7% of wax esters and free fatty acids contribute to the total content of the feather lipids (Barcus 2017). Kondamudi et al. (2009) estimated 11% of fat in the chicken feather that can be used to produce biodiesel (Kondamudi et al. 2009). It should be noted that chickens secrete lipids from the uropygial gland onto their plumage as protection against preening (Barcus 2017).

Most of the studies reported are centered on the isolation of keratinolytic microorganisms and keratinase falls short in the complete hydrolysis of keratin-rich feathers (Lange et al. 2016). A lipolysis step for hydrolysis of lipid is important to aid keratinolytic and other reducing enzymes to access their respective substrates. Also, as free fatty acids are the product of lipid hydrolysis, it is vital to remove the fatty acids to increase keratinase activity.

The academic research on lipids' characterization in feathers and identifying the biological methods to remove the lipid barriers have been extensively studied by Barcus (2017). The research mainly focused on identifying lipolytic enzymes from *Streptomyces fradia* var. k11 capable of hydrolyzing the waxy lipids in the feathers. They identified two lipases and characterized one as wax ester hydrolases of 29 kDa that hydrolyzed various waxy lipids present on the chicken feather. The study highlighted *Streptomyces fradiae* var. k11 as a potential candidate for application in industrial feather hydrolysis. Also, many keratinases and reductases in *S Streptomyces fradiae* have been identified (Gupta and Ramnani 2006).

It has been reported by Li et al. (2020) that genes involved in lipid digestion and catabolism are up-regulated in *Streptomyces* SCUT-3 grown on feather medium, including secretion of extracellular esterase, cell wall carboxylesterase, fatty acid CoA ligases, fatty acid β-oxidation, acetyl CoA synthetase, steroid, and cholesterol catabolism. This concluded the efficiency of SCUT-3 to catabolize amino acids and lipids to provide energy and growth metabolites in feather medium. Table 8.4 lists examples of isolated protease (keratinase), disulfide reductase and/or lipase from microorganisms for efficient feather degradation.

**TABLE 8.4**
**Efficient Degradation of Feathers by Microbial Protease (Keratinase), Disulfide Reductase, and/or Lipase**

| Microbial Source | Characteristics | References |
|---|---|---|
| *Streptomyces* SCUT-3 | • Culture initiates feather degradation by activation of sulfite and free cysteinyl groups which reduced disulfide bonds<br>• Protease/keratinase break down peptide bonds<br>• Cell bound esterase hydrolyze feather lipids.<br>• Culture efficiently degrades feathers to high-amino-acid-containing products of nutritional use to animals and microorganisms. | Li et al. (2020) |
| *Bacillus amyloliquifaciens* 3-2 | • Produced four crude compound enzymes (protease, oligopeptidase, lipase and disulfide reductase)<br>• Hydrolysis of feather-waste produces 20–65 times more pro-solubility and 10-27 times more digestibility than fresh feather meal. | Zhou et al. (2020) |
| *Streptomyces fradia* VarK11 | • Wax ester hydrolase of 29 kDa was purified, showing activity against para-nitro phenyl palmitate.<br>• Hydrolysis of wax lipids, suggesting a potential application in feather disintegration. | Barcus et al. (2017) |
| *Bacillus subtilis* CH-1 | • Protease, γ- glutamyl transpeptidase and glyoxal methylglyoxal reductase were purified, showing optimum activities at 45°C/pH 7, 40°C/pH8 and 50°C/pH 6, respectively. | Liu et al. (2014b) |
| *Bacillus* sp. MTS | • Purified 13 and 35 kDa bands of disulfide reductase protein and 16, 32 and 50 kDa protein bands of keratinase separated on SDS-PAGE.<br>• The mutual action of keratinase and disulfide reductase on the feather and wool is higher than with the individual enzyme. | Rahayu et al. (2012) |
| *Stenotrophomonas maltophilia* strain R13 | • Used feathers as a sole source of nitrogen and carbon.<br>• Possessed keratinolytic and disulfide reductase activity. | Jeong et al. (2010) |
| *Bacillus halodurans* PPKS-2 | • Purified 30 kDa keratinase I disulfide reductase-like protein and 66 kDa keratinase II are stable at 70°C for 3 hours.<br>• Both the enzymes showed optimal activity at pH 11 and 60-70°C | Prakash et al. (2010b) |
| *Stenotrophomonas* sp. strain D1 | • Keratinase of 40 kDa showed optimal activity at pH 8-10 and 40°C.<br>• A reductase-like protein of 15 kDa showed maximum activity at pH 7 and 30°C.<br>• 50% fold increase in keratinolytic activity attained using a mixture of two enzymes. | Yamamura et al. (2002) |

## 8.11 ADVANTAGES OF BIOCHEMICAL PROPERTIES OF EXTREMOZYMES OVER NON-EXTREMOZYMES

The primary factor limiting the wide usage of mesophilic keratinolytic microorganisms is the contamination of the process, which reduces keratinase production. On the other hand, extremely halophilic microorganisms and their keratinase are functional under extreme pH, temperature, salinity which determine the growth of mesophilic microbial contaminants. These characteristics make extremophilic keratinase potential candidates for reaction requiring harsh physicochemical conditions (Intagun and Kanoksilapatham 2017).

Further, the decomposition of recalcitrant proteins at >50°C allows the protein to regain high plasticity resulting in efficient degradation by producing extreme protease (Suzuki et al. 2006). Further, non-specific enzymes from extremophiles viz. thermophiles are advantageous for decontaminating infectious prion proteins because of high stability at high temperatures and denaturing conditions (Vieille et al. 2001). The suitability of extremophilic enzymes over non-extremophiles is detailed in Figure 8.7.

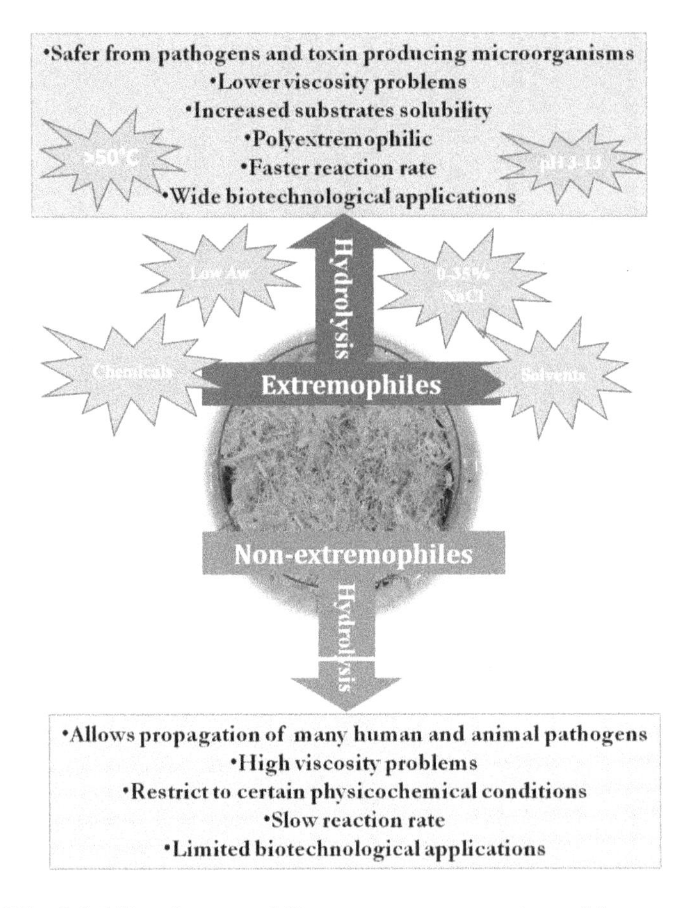


**FIGURE 8.7** Suitability of extremophilic enzymes over non-extremophiles.

Haloarchaea are extremophiles that form the largest culturable group of the phylum *Euryarcheota* and consist of six families, 60 genera, and 267 described species. Recent technologies using high-throughput Illumina sequencing of 16S rRNA gene aided detailed characterization of the haloarchaea (Mani et al. 2020; Zhu et al. 2021). The studies revealed an abundance of sequences of members belonging to Euryarcheota with predominant genera of *Halorubrum, Halobacterium, Haloarcula, Halorhabdus, Halopenitus, Haloplanus, Methanothrix, Halomicrobium*, and *Natronomonas*. A wide variety of biotechnological products such as polyhydroxyalkanoate, nanoparticles, carotenoids, exopolysaccharides, and extremozymes have been reported from Haloarchaea active under harsh physicochemical conditions of pH, temperature, salinity, and low Mw (Amoozegar et al. 2017; Oren 2010).

Haloarchaea have been screened for extremozymes keratinases and collagenases. However, some efforts have been made recently on the isolation and screening of haloarchaea from salt-treated/conserved raw hides and skins for various hydrolytic enzymes such as keratinase, caseinase, esterase, lipase, amylase, cellulase, and DNase. A study showed that the haloarchaeon *Natrinema sp.* exhibited the highest caseinase, keratinase, collagenase, and lipase enzyme activities. Further quantitative measurements of the keratinase and collagenase using 0.4% keratin-azure substrate at 37°C, pH 7.5 in the presence of 2M NaCl resulted in 0.6 U/ml and 1.7 U/ml activity, respectively (Caglayan et al. 2015; Birbir and Caglayan 2018; Birbir et al. 2020).

Among hyperthermophiles, a novel, strictly anaerobic, *Desulfurococcus kamchatkensis* strain 1221n$^T$ produced 120 kDa proteinases that hydrolyzed a-keratin and 40 kDa acting on gelatin and albumin proteins (Kublanov et al. 2009). Another strain *Thermococcus* VC13 synthesized several extracellular endopeptidases ranging from 45 to 100 kDa, active optimally at pH 7.2 and 80°C against alpha and beta keratin. Moreover, the cell-free supernatant could hydrolyze proteins from wool (Tsiroulnikov et al. 2004). Keratinases from *Fervidobacterium pennivorans* (Friedricht and Antranikian 1996) and *Fervidobacterium islandicum* AW-1 (Nam et al. 2002) exhibited optimal activities between 80 and 100°C and hydrolyzed feathers to produce soluble peptides and amino acids. Table 8.5 details the characteristics and applications of isolated keratinase from extremophiles.

## 8.12 BIOREFINERY CIRCULAR ECONOMY PLATFORM CONCEPT FOR CONTINUOUS CLEAN TECHNOLOGY FOR FEATHER HYDROLYSIS

A bio-refinery is an integrative and multifunctional concept for converting raw biomass using different methods and technologies into sustainable intermediates and bio-products such as bio-energy; biosynthesis of chemicals, materials, and fuels. The idea leads to the fullest possible use of raw materials and ensures minimum process wastes. Conceptually the bio-refinery chain includes the pre-treatment and hydrolysis of organic biomass, the primary separation of components, subsequent secondary conversion, and final product separations. Bio-refineries for ligno-cellulosic biomass are commonly reported in the literature (Fernando et al. 2006). The concept for feather-waste bio-refinery is of potential interest and prospective recognition soon

**TABLE 8.5**
**Characteristics and Applications of isolated keratinase from Extremophiles**

| Extremophilic microbe | Features and applications | References |
|---|---|---|
| *Actinomadura keratinilytica* strain cpt20 | • Purified keratinase of 71 and 19 kDa was optimally active at 50°C, pH 8 and 40°C, pH 7, respectively.<br>• High catalytic efficiency with dehairing potential and feather degradation application. | Kerouaz et al. (2021) |
| *Deinococcus geothermalis* | • Recombinant keratinase showed maximum activity at 70°C and pH 9.<br>• Enzyme degrades feathers at 70°C in 60 mins. | Tang et al. (2021) |
| *Thermoactinomyces vulgaris* strain CDF | • Recombinant subtilis-like protease displayed optimal activity at 60–70°C and pH 10.<br>• Efficiently hydrolyzed soluble keratin released from chicken feathers.<br>• Notably exhibited 100% activity at 3 M NaCl and enhanced thermostability with increase in NaCl concentration. | Ding et al. (2020) |
| *Bacillus* sp. CSK2 | • The keratinase showed optimal catalytic efficiency between 60 and 80°C and pH 8.<br>• Detergent stability. | Nnolim and Nwodo (2020) |
| *Bacillus aereus* NSMK2 | • Thermostable halotolerant 9 kDa keratinase highly active between 20 and 100°C and pH 4–11.<br>• The keratinase retained more than 95% activity in the presence of 20% NaCl.<br>• Detergent and dehairing applications. | Bhari et al. (2019) |
| *Thermoactinomyces* sp. YT06 | • The 35 kDa keratinase maintained optimum activity between 65 and 75°C and pH 8–11.<br>• Great potential in degradation of chicken feathers into high-value products. | Wang et al. (2017) |
| *Caldicoprobacter algeriensis* | • Keratinase is active at 30°C and pH 7.<br>• Displayed higher level of feather hydrolysis and used for dehairing application. | Bouacem et al. (2016) |
| *Actinomadura viridilatea* DZ50 | • The 195 kDa keratinase is active at 80°C and pH 11.<br>• Displayed a higher level of feather-meal hydrolysis. | Elhoul et al. (2015) |
| *Desulfurococcus* 2355K<br>*Desulfurococcus* 3008 | • Both endo-peptidase of >200 kDa active at 85°C and pH 8.8 against gelatin and casein.<br>• Showed enhanced hydrolytic activity of the keratinase on α- keratin and β- keratin-rich feathers. | Bidzhieva et al. (2014) |

| | | |
|---|---|---|
| *Actinomadura keratinilytica* cpt29 | • The 292 kDa keratinase is active at 70°C and pH 10.<br>• Stable in the presence of detergents and carried 100% feather degradation. | Habbeche et al. (2014) |
| *Keratinibaculum paraultunense* strain KD-1 | • Serine-type protease possessed optimal activity at 70°C and pH 8.5. | Huang et al. (2013) |
| *Bacillus* sp. JB99 | • The 29 kDa keratinase showed optimal activity at 70°C and pH 11.<br>• Feather disintegration and dehairing of buffalo and goat hide. | Shrinivas and Naik (2011) |
| *Bacillus halodurans* PPKS-2 | • Keratinase displayed maximum activity at 60°C, pH 11 and 0-16% NaCl.<br>• The enzyme effectively hydrolyzed feather and was used for dehairing application. | Prakash et al. (2010b) |
| *Desulfurococcus kamchatkensis* 1221nT | • Cell bound proteinase of 120 kDa active on α- keratin and 40 kDa on albumin and peptone proteins. | Kublanov et al. (2009) |
| *Thermococcus* VC13 | • Endopeptidase of 45–100 kDa optimally active at 80°C and pH 7.2 against α- keratin and β- keratin.<br>• Partial dissolution of protein meat, proteins of wool, blood and mixed meal.<br>• Protease hydrolyzed keratin of feathers and hairs. | Tsiroulnikov et al. (2004) |
| *Fervidobacterium islandicum* | • The homo-multimeric keratinase of >200 kDa and 97 kDa was optimally active at 100°C and pH 9.<br>• Higher specificity towards casein and soluble keratin. | Nam et al. (2002) |
| *Fervidobacterium pennavorans* | • The 130 kDa keratinase is active at 80°C and pH 8.<br>• The enzyme system converts feather meal to amino acids and peptides. | Friedricht and Antranikian (1996) |

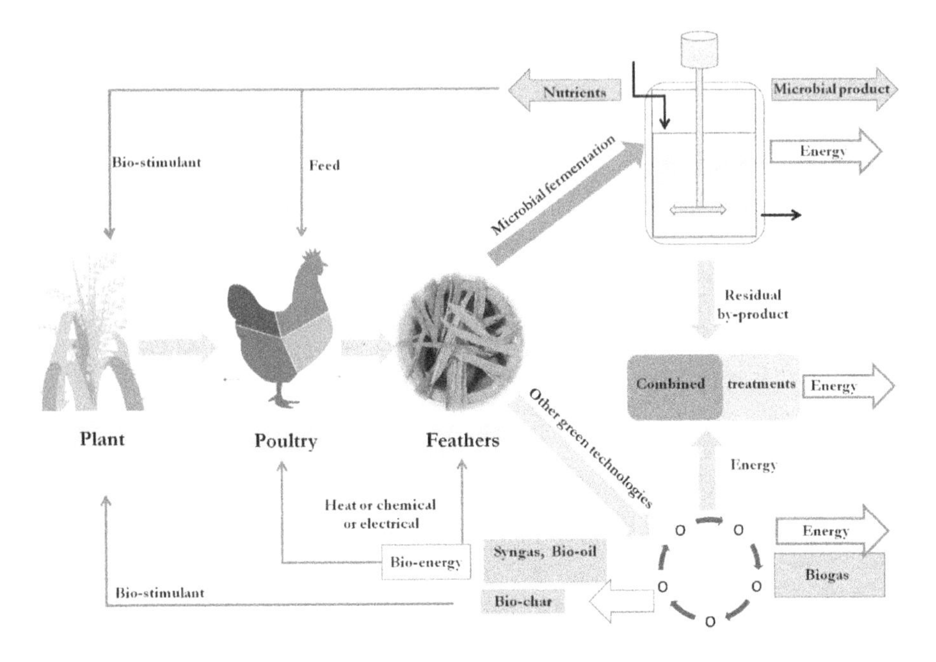


**FIGURE 8.8** Bio-refinery circular economy concept to convert chicken feathers from the poultry industry into value-added products.

for its contribution to the bio-based economy toward cleaner production and circular economy (Ramirez et al. 2021). The following section outlines a modeled feather-waste bio-refinery to produce high-value-added compounds that would be recycled in the supply chain for resource sustainability. The possible use of a combination of well-developed current and emerging technologies for feather degradation and hydrolysis is predicted.

Figure 8.8 describes a green bio-refinery-circular economy model to convert chicken feathers into value-added products. Protein is present at about 90% in feathers. It is a rich and renewable source of keratin which is highly complex and inaccessible. Hydrolyzed keratin has many biotechnological applications, including fertilizer, cosmetics, animal feed, leather tanning, bio-plastics, bio-composites, and bio-actives. For instance, cosmetic grade keratin hydrolysate currently costs between 50 and 130 USD/kg (Poultry Hub 2020). Therefore, efficient conversion of feathers into protein-rich hydrolysate can add value to the waste and mitigate the disposal problem.

The currently well-developed and emerging technologies can expedite economic, social, and environmental impact. Previous treatment methods to add the value derived from feathers, such as hydrothermal treatment, and ionic liquids to produce protein-rich hydrolysate, have enabled the poultry industry to raise their revenues and make high-quality protein despite increased costs and market instability.

Currently, Dr. Netsanet and her team are investigating combining ultrasonic and biological treatments to recover high-quality keratin-rich protein hydrolysate from feather-waste. The application of ultrasound can enhance enzyme kinetics, reducing enzyme dose and reaction time. To date, they have successfully managed to develop product hydrolysate with 80% purity with excellent amino acids profile and content, solubility, and foaming and emulsification properties comparable to whey protein. The hydrolysate also exhibited high antioxidant activity compared to strawberries (Poultry Hub 2020).

Recovery and recycling of feather-waste can lead to significant sustainability if processes are performed within the poultry waste generation chain. Organic nutrient-rich hydrolysate will decrease greenhouse gas emissions from raw waste and enhance the soil's fertility index. Similarly, biofuel (biogas, syngas, biochar, bio-oil) produced from feather-waste by composting or anaerobic digestion will reduce the consumption of fossil fuels. Concerning the circular economy, bio-compost and biochar will increase agricultural productivity, and biogas can generate heat and electricity (Kanani et al. 2020; Ramirez et al. 2021).

The success of advanced technology for sustainable feather-waste bio-refinery embedded in the circular economy will depend on the robustness and benefits to society by minimizing the impact on the environment and maximizing the product's value.

## 8.13 ECONOMIC ASSESSMENT OF THE FEATHER HYDROLYSIS PROCESS AND FEASIBILITY OF THE PROPOSED MODEL

The processing of poultry feathers is a difficult and relatively expensive task. There are few published technologies suitable for processing large quantities of feathers to guarantee sustainable development. Cost evaluation of any product can be ascertained as total cost for production, which consists of production, sales, and business overheads, or direct production costs. Economic assessment of the raw material and hydrolysis process plays an essential role. It also considers the cost of the method used for disposal of waste, energy input, and type of product formed (Kumar and Verma 2021a; 2021b).

In the present case, chicken feathers are considered as waste for which the poultry farmer or industry takes the cost of feather-waste management care. Therefore, environmentally sick hydrolysis processes that cut down the cost associated with disposal are engaged (Solcova et al. 2021).

On the other hand, when the farmer or industrialist is aware and has technologies targeted to produce protein and other products rich in hydrolysate, the feather-waste will be eyed through a commercial lens. It will no longer be considered as waste but necessary raw material. The product has added value and the farmer/industrialist as the manufacturer will have a share in the economic benefits of the product. Figure 8.9 shows the proposed model of using proteolytic, disulfide reductase, and lipolytic haloarchaeal strain to support feather-waste bio-refinery for value addition considering the economic evaluation of the process.

The future bio-refinery model for feather hydrolysis is conceptualized using haloarchaeal extremophilic microbes and their extremozymes. We presumed that the conversion rate would match other reported green technologies expected to be cost-effective and reduce the energy input, either as heat/chemical or other means. The proposed model using haloarchaea with detailed usage cost for the laboratory

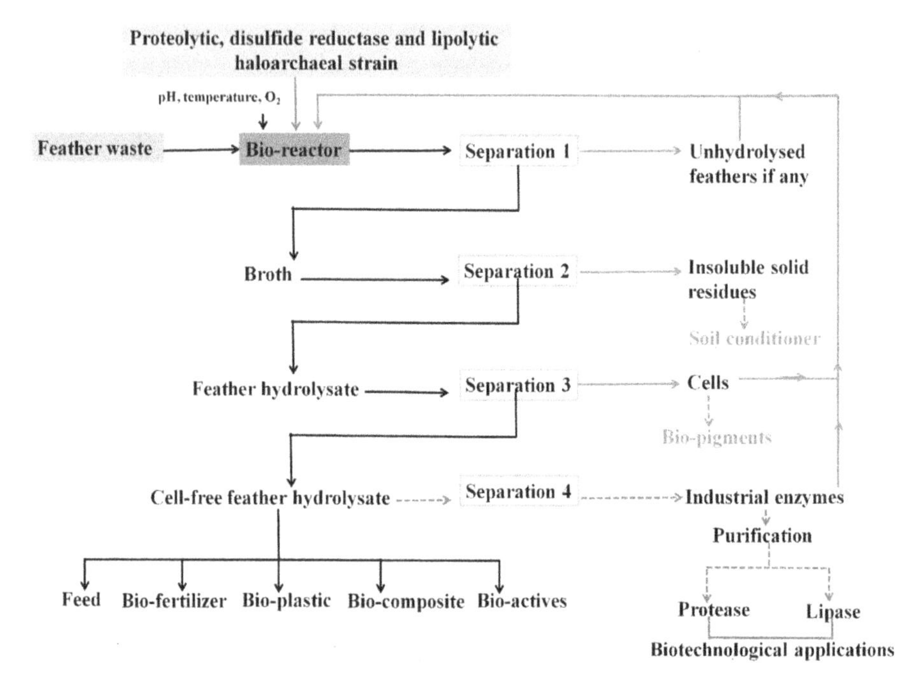


**FIGURE 8.9** A simplified flow chart of the future bio-refinery proposed model for feather-waste valorization using haloarchaeal strain.

scale bioreactor setup is one unit of glass carboy reactor (5 L capacity) assembled with an overhead stirrer (60 rpm) and tanks for retention, collection, and filtration of hydrolysate utilized from the laboratory infrastructure. However, the overall cost of the apparatus would be INR 10,000. At the same time, assembly tools such as knobs, taps, silicone tubing, and charcoal for filtration of hydrolysate would cost about INR 500. Thus, for the laboratory scale aerobic bioreactor setup used in this work, the average cost is about INR 10,500 (Gaonkar and Furtado 2022). Solcova et al. (2021) devised a green technology model for processing feathers to hydrolysate through the addition of malic acid. The process was scaled up to 8000 L with 340 kg of feathers per batch. It reduced the calculated cost significantly per batch due to the use of malic acid, and the input of heat/electricity was around 280–294 EUR in addition to the cost of the labor. Policies relevant to recovery, waste management renewable energy, sustainability, safety, and feather-waste-derived products must be in place even as the concept grows into full-fledged technology.

## REFERENCES

Abbott, A. P., D. Boothby, G. Capper, D. L. Davies, and R. K. Rasheed. 2004. "Deep Eutectic Solvents Formed between Choline Chloride and Carboxylic Acids: Versatile Alternatives to Ionic Liquids." *Journal of the American Chemical Society* 126 (29): 9142–47. https://doi.org/10.1021/ja048266j.

Al-Bahri, M. B., S. A. Al-Naimi, and S. H. Ahammed. 2009. "Study the Effect of Hydrolysis Variables on the Production of Soya Proteins Hydrolysis." *Al-Khwarizmi Engineering Journal* 5 (4): 25–38.

Al-Musallam, A. A., D. H. Al-Gharabally, and N. Vadakkancheril. 2013. "Biodegradation of Keratin in Mineral-Based Feather Medium by Thermophilic Strains of a New *Coprinopsis* Sp." *International Biodeterioration and Biodegradation* 79: 42–48. https://doi.org/10.1016/j.ibiod.2012.11.011.

Al-Souti, A., W. Gallardo, M. Claereboudt, and O. Mahgoub. 2018. "Effects of Autoclaving, Addition of Sodium Hydroxide and Their Combination on Protein Content and in Vitro Digestibility of Chicken Feathers." *International Journal of Poultry Science* 17 (8): 356–61. https://doi.org/10.3923/ijps.2018.356.361.

Alahyaribeik, Samira, S. D. Sharifi, F. Tabandeh, S. Honarbakhsh, and S. Ghazanfari. 2020. "Bioconversion of Chicken Feather Wastes by Keratinolytic Bacteria." *Process Safety and Environmental Protection* 135: 171–78. https://doi.org/10.1016/j.psep.2020.01.014.

Amoozegar, M. A., M. Siroosi, S. Atashgahi, H. Smidt, and A. Ventosa. 2017. "Systematics of Haloarchaea and Biotechnological Potential of Their Hydrolytic Enzymes." *Microbiology (United Kingdom)* 163 (5): 623–45. https://doi.org/10.1099/mic.0.000463.

Anbu, P., Hilda, A., Sur, H. W., Hur, B. K., and Jayanthi, S. 2008. "Extracellular keratinase from *Trichophyton* sp. HA-2 isolated from feather dumping soil." *International Biodeterioration & Biodegradation* 62 (3): 287–292.

Bagewadi, Z. K., S. I. Mulla, and H. Z. Ninnekar. 2018. Response Surface Methodology Based Optimization of Keratinase Production from *Trichoderma Harzianum* Isolate HZN12 Using Chicken Feather Waste and Its Application in Dehairing of Hide. *Biochemical Pharmacology*. Elsevier B.V. https://doi.org/10.1016/j.jece.2018.07.007.

Barcus, M. 2017. "Kinetics and Structural Characterization of Lipolytic Enzymes for Hydrolyzing Poultry Feathers." Doctoral dissertation, Cornell University, https://doi.org/10.7298/X4PC30JK

Barcus, M., D. Mizrachi, and X. G. Lei. 2017. "Identification and Kinetics Characterization of a Wax Ester Hydrolase from a Feather-Degrading Actinomycete." *BioRxiv*, 178673. https://doi.org/10.1101/178673.

Beryl, G. P., B. Thazeem, M. Umesh, K. Senthilkumar, M. N. Kumar, and K. Preethi. 2021. "Bioconversion of Feather Composts Using Proteolytic *Bacillus mycoides* for Their Possible Application as Biofertilizer in Agriculture." *Waste and Biomass Valorization*, no. 0123456789. https://doi.org/10.1007/s12649-021-01472-4.

Bezus, B., F. Ruscasso, G. Garmendia, S. Vero, I. Cavello, and S. Cavalitto. 2021. "Revalorization of Chicken Feather Waste into a High Antioxidant Activity Feather Protein Hydrolysate Using a Novel Psychrotolerant Bacterium." *Biocatalysis and Agricultural Biotechnology* 32. https://doi.org/10.1016/j.bcab.2021.101925.

Bhange, K., V. Chaturvedi, and R. Bhatt. 2016. "Ameliorating Effects of Chicken Feathers in Plant Growth Promotion Activity by a Keratinolytic Strain of *Bacillus Subtilis* PF1." *Bioresources and Bioprocessing* 3 (1). https://doi.org/10.1186/s40643-016-0091-y.

Bhardwaj, N., B. Kumar, and P. Verma. 2020. "Microwave-assisted Pretreatment Using Alkali Metal Salt in Combination with Orthophosphoric Acid for Generation of Enhanced Sugar and Bioethanol." *Biomass Conversion and Biorefinery*: 1–8.

Bhari, R., M. Kaur, and R. S. Singh. 2019. "Thermostable and Halotolerant Keratinase from *Bacillus aerius* NSMk2 with Remarkable Dehairing and Laundary Applications." *Journal of Basic Microbiology* 59 (6): 555–68. https://doi.org/10.1002/jobm.201900001.

Bhari, R., M. Kaur, and R. S. Singh. 2021. "Chicken Feather Waste Hydrolysate as a Superior Biofertilizer in Agroindustry." *Current Microbiology* 78 (6): 2212–30. https://doi.org/10.1007/s00284-021-02491-z.

Bidzhieva, S., K. Kh, S. Derbikova, I. V. Kublanov, and E. A. Bonch-Osmolovskaya. 2014. "Capacity of Hyperthermophilic Crenarchaeota for Decomposition of Refractory Proteins (α- and β-Keratins)." *Microbiology (Russian Federation)* 83 (6): 880–87. https://doi.org/10.1134/S0026261714060034.

Birbir, M., and P. Caglayan. 2018. "A Review on Catabolic Activity of Microorganisms in Leather Industry." In *International Conference on Advanced Materials and Systems*, 301–6. https://doi.org/10.24264/icams-2018.vi.3.

Birbir, M., P. Caglayan, and Y. Birbir. 2020. "The Destructive Effects of Extremely Halophilic Archaeal Strains on Sheepskins, and Proposals for Remedial Curing Processes Use of Sterile Brine or Direct Electric Current to Prevent Red Heat Damage on Salted Sheepskins." *Johnson Matthey Technology Review* 64 (4): 489–503. https://doi.org/10.1595/205651320X15943793010464.

Bishmi, A., J. Thatheyus, and D. Ramya. 2015. "Biodegradation of Poultry Feathers Using a Novel Bacterial Isolate *Pseudomonas aeruginosa*." *International Journal of Research Studies in Microbiology and Biotechnology* 1 (1): 25–30.

Bolan, N. S., A. A. Szogi, T. Chuasavathi, B. Seshadri, M. J. Rothrock, and P. Panneerselvam. 2010. "Uses and Management of Poultry Litter." *World's Poultry Science Journal* 66 (4): 673–98. https://doi.org/10.1017/S0043933910000656.

Bouacem, K., A. Bouanane-Darenfed, N. Z. Jaouadi, M. Joseph, H. Hacene, B. Ollivier, M. L. Fardeau, S. Bejar, and B. Jaouadi. 2016. "Novel Serine Keratinase from *Caldicoprobacter algeriensis* Exhibiting Outstanding Hide Dehairing Abilities." *International Journal of Biological Macromolecules* 86: 321–28. https://doi.org/10.1016/j.ijbiomac.2016.01.074.

Brandelli, A. 2008. "Bacterial Keratinases: Useful Enzymes for Bioprocessing Agroindustrial Wastes and Beyond." *Food and Bioprocess Technology* 1 (2): 105–16. https://doi.org/10.1007/s11947-007-0025-y.

Brandelli, A., D. J. Daroit, and A. Riffel. 2010. "Biochemical Features of Microbial Keratinases and Their Production and Applications." *Applied Microbiology and Biotechnology* 85: 1735–50. https://doi.org/10.1007/s00253-009-2398-5.

Brandelli, A., L. Sala, and S. J. Kalil. 2015. "Microbial Enzymes for Bioconversion of Poultry Waste into Added-Value Products." *Food Research International* 73: 3–12. https://doi.org/10.1016/j.foodres.2015.01.015.

Brebu, M., and I. Spiridon. 2011. "Thermal Degradation of Keratin Waste." *Journal of Analytical and Applied Pyrolysis* 91 (2): 288–95. https://doi.org/10.1016/j.jaap.2011.03.003.

Brown, E. M., K. Pandya, M. M. Taylor, and C.-K. Liu. 2016. "Comparison of Methods for Extraction of Keratin from Waste Wool." *Agricultural Sciences* 07 (10): 670–79. https://doi.org/10.4236/as.2016.710063.

Brunner, G. 2009. "Near Critical and Supercritical Water. Part I. Hydrolytic and Hydrothermal Processes." *Journal of Supercritical Fluids* 47 (3): 373–81. https://doi.org/10.1016/j.supflu.2008.09.002.

Caglayan, P., C. Sánchez-Porro, A. Ventosa, and M. Birbir. 2015. "Characterization of Moderately Halophilic Bacteria from Salt-Pack-Cured Hides." *Journal of the Society of Leather Technologists and Chemists* 99 (5): 250–54.

Chao, S. J., K. H. Chung, Y. F. Lai, Y. K. Lai, and S. H. Chang. 2021. "Keratin Particles Generated from Rapid Hydrolysis of Waste Feathers with Green DES/KOH: Efficient Adsorption of Fluoroquinolone Antibiotic and Its Reuse." *International Journal of Biological Macromolecules* 173: 211–18. https://doi.org/10.1016/j.ijbiomac.2021.01.126.

Chaturvedi, V., K. Agrawal, and P. Verma. 2021. "Chicken Feathers: A Treasure Cove of Useful Metabolites and Value-Added Products." *Environmental Sustainability* 4(2), 231–243. https://doi.org/10.1007/s42398-021-00160-2.

Chaturvedi, V., and P. Verma. 2014. "Metabolism of Chicken Feathers and Concomitant Electricity Generation by *Pseudomonas aeruginosa* by Employing Microbial Fuel Cell (MFC)." *Journal of Waste Management* 2014: 1–9. https://doi.org/10.1155/2014/928618.

Chilakamarry, C. R., S. Mahmood, S. N. B. M. Saffe, M. A. B. Arifin, A. Gupta, M. Y. Sikkandar, S. Sabarunisha Begum, and B. Narasaiah. 2021. "Extraction and Application of Keratin from Natural Resources: A Review." *3 Biotech* 11 (5): 1–12. https://doi.org/10.1007/s13205-021-02734-7.

Dada, M., and S. Wakil. 2021. "Conversion of Feather to Potential Feed Supplement Using Keratinase from *Bacillus licheniformis* -K51." *Journal of Applied Sciences & Environmental Sustainability* 13 (7): 10–31.

Dai, J., J. Saayman, J. R. Grace, and N. Ellis. 2015. "Gasification of Woody Biomass." *Annual Review of Chemical and Biomolecular Engineering* 6: 77–99. https://doi.org/10.1146/annurev-chembioeng-061114-123310.

Dhiva, S., R. Sreelakshmi, S. Sruthi, U. Biji, G. Narendrakumar, P. J. Jane Cypriyana, and P. Raji. 2020. "Optimization of Biomass of Keratinase Producing *Bacillus* sp CBNRBT2 to Utilize in Whole-Cell Immobilization for Feather Degradation." *Letters in Applied NanoBioScience* 9 (3): 1339–47. https://doi.org/10.33263/lianbs93.13391347.

Ding, Y., Y. Yang, Y. Ren, J. Xia, F. Liu, Y. Li, X. F. Tang, and B. Tang. 2020. "Extracellular Production, Characterization, and Engineering of a Polyextremotolerant Subtilisin-Like Protease From Feather-Degrading *Thermoactinomyces vulgaris* Strain CDF." *Frontiers in Microbiology* 11 (December): 1–13. https://doi.org/10.3389/fmicb.2020.605771.

Dube, R., V. Nandan, and S. Dua. 2014. "Waste Incineration for Urban India: Valuable Contribution to Sustainable MSWM or Inappropriate High-Tech Solution Affecting Livelihoods and Public Health?" *International Journal of Environmental Technology and Management* 17 (2–4): 199–214. https://doi.org/10.1504/IJETM.2014.061792.

Dudyński, M., K. Kwiatkowski, and K. Bajer. 2012. "From Feathers to Syngas - Technologies and Devices." *Waste Management* 32 (4): 685–91. https://doi.org/10.1016/j.wasman.2011.11.017.

Eggum, B. O. 1970. "Evaluation of Protein Quality of Feather Meal Under Different Treatments." *Acta Agriculturae Scandinavica* 20 (4): 230–34. https://doi.org/10.1080/00015127009433412.

Elhoul, M. B., N. Z. Jaouadi, H. Rekik, W. Bejar, S. B. Touioui, M. Hmidi, A. Badis, S. Bejar, and B. Jaouadi. 2015. "A Novel Detergent-Stable Solvent-Tolerant Serine Thiol Alkaline Protease from *Streptomyces koyangensis* TN650." *International Journal of Biological Macromolecules* 79: 871–82. https://doi.org/10.1016/j.ijbiomac.2015.06.006.

Elhoul, M. B., N. Z. Jaouadi, H. Rekik, M. O. Benmrad, S. Mechri, E. Moujehed, S. Kourdali, et al. 2016. "Biochemical and Molecular Characterization of New Keratinoytic Protease from *Actinomadura viridilutea* DZ50." *International Journal of Biological Macromolecules* 92: 299–315. https://doi.org/10.1016/j.ijbiomac.2016.07.009.

Emran, M. A., S. A. Ismail, and A. M. Abdel-Fattah. 2020. "Valorization of Feather via the Microbial Production of Multi-Applicable Keratinolytic Enzyme." *Biocatalysis and Agricultural Biotechnology* 27 (June): 101674. https://doi.org/10.1016/j.bcab.2020.101674.

Eremeev, N. L., I. V. Nikolaev, I. D. Keruchen'ko, E. V. Stepanova, A. D. Satrutdinov, S. V. Zinov'ev, D. Y. Ismailova, et al. 2009. "Enzymatic Hydrolysis of Keratin-Containing Stock for Obtaining Protein Hydrolysates." *Applied Biochemistry and Microbiology* 45 (6): 648–55. https://doi.org/10.1134/S0003683809060131.

Evangelisti, S., P. Lettieri, D. Borello, and R. Clift. 2014. "Life Cycle Assessment of Energy from Waste via Anaerobic Digestion: A UK Case Study." *Waste Management* 34 (1): 226–37. https://doi.org/10.1016/j.wasman.2013.09.013.

Fang, Z., J. Zhang, B. Liu, L. Jiang, G. Du, and J. Chen. 2014. "Cloning, Heterologous Expression and Characterization of Two Keratinases from *Stenotrophomonas maltophilia* BBE11-1." *Process Biochemistry* 49 (4): 647–54. https://doi.org/10.1016/j.procbio.2014.01.009.

Fernando, S., S. Adhikari, C. Chandrapal, and N. Murali. 2006. "Biorefineries: Current Status, Challenges, and Future Direction." *Energy and Fuels* 20 (4): 1727–37. https://doi.org/10.1021/ef060097w.

Feroz, S., N. Muhammad, J. Ranayake, and G. Dias. 2020. "Keratin - Based Materials for Biomedical Applications." *Bioactive Materials* 5 (3): 496–509. https://doi.org/10.1016/j.bioactmat.2020.04.007.

Franke-Whittle, I. H., and H. Insam. 2013. "Treatment Alternatives of Slaughterhouse Wastes, and Their Effect on the Inactivation of Different Pathogens: A Review." *Critical Reviews in Microbiology* 39 (2): 139–51. https://doi.org/10.3109/1040841X.2012.694410.

Friedricht, A. B., and G. Antranikian. 1996. "Keratin Degradation by *Feividobacterium pennavomns*, a Novel Thermophilic Anaerobic Species of the Order Thermotogales." *Applied and Environmental Microbiology* 62 (8): 2875–82. https://doi.org/10.1128/aem.62.8.2875-2882.1996.

Gaonkar, S. K., and I. J. Furtado. 2021. "Valorization of Low-Cost Agro-Wastes Residues for the Maximum Production of Protease and Lipase Haloextremozymes by *Haloferax lucentensis* GUBF-2 MG076878." *Process Biochemistry* 101: 72–88. https://doi.org/10.1016/j.procbio.2020.10.019.

Gaonkar, S. K., and I. J. Furtado. 2022. "Biorefinery-Fermentation of Agro-Wastes by Haloferax lucentensis GUBF-2 MG076878 to Haloextremozymes for use as Biofertilizer and Biosynthesizer of AgNPs." *Waste and Biomass Valorization* 13: 1117–33. https://doi.org/10.1007/s12649-021-01556-1

Gegeckas, A., R. Gudiukaite, J. Debski, and D. Citavicius. 2015. "Keratinous Waste Decomposition and Peptide Production by Keratinase from *Geobacillus Stearothermophilus* AD-11." *International Journal of Biological Macromolecules*: 1–8. https://doi.org/10.1016/j.ijbiomac.2015.01.031.

Ghosh, A., K. Chakrabarti, and D. Chattopadhyay. 2008. "Degradation of Raw Feather by a Novel High Molecular Weight Extracellular Protease from Newly Isolated *Bacillus cereus* DCUW." *Journal of Industrial Microbiology and Biotechnology* 35 (8): 825–34. https://doi.org/10.1007/s10295-008-0354-5.

Goers, L., P. Freemont, and K. M. Polizzi. 2014. "Co-Culture Systems and Technologies: Taking Synthetic Biology to the next Level." *Journal of the Royal Society Interface* 11 (96). https://doi.org/10.1098/rsif.2014.0065.

Goswami, R. K., K. Agrawal, P. Verma 2021. Microalgae-based Biofuel-integrated Biorefinery Approach as Sustainable Feedstock for Resolving Energy Crisis. In: *Bioenergy Research: Commercial Opportunities & Challenges* (pp. 267–93). Springer, Singapore.

Gupta, R., and P. Ramnani. 2006. "Microbial Keratinases and Their Prospective Applications: An Overview." *Applied Microbiology and Biotechnology* 70 (1): 21–33. https://doi.org/10.1007/s00253-005-0239-8.

Habbeche, A., B. Saoudi, B. Jaouadi, S. Haberra, B. Kerouaz, M. Boudelaa, A. Badis, and A. Ladjama. 2014. "Purification and Biochemical Characterization of a Detergent-Stable Keratinase from a Newly Thermophilic Actinomycete *Actinomadura keratinilytica* Strain Cpt29 Isolated from Poultry Compost." *Journal of Bioscience and Bioengineering* 117 (4): 413–21. https://doi.org/10.1016/j.jbiosc.2013.09.006.

Hagos, K., J. Zong, D. Li, C. Liu, and X. Lu. 2017. "Anaerobic Co-Digestion Process for Biogas Production: Progress, Challenges and Perspectives." *Renewable and Sustainable Energy Reviews* 76 (March): 1485–96. https://doi.org/10.1016/j.rser.2016.11.184.

Han, M., W. Luo, Q. Gu, and X. Yu. 2012. "Isolation and Characterization of a Keratinolytic Protease from a Feather-Degrading Bacterium *Pseudomonas aeruginosa* C11." *African Journal of Microbiology Research* 6 (9): 2211–21. https://doi.org/10.5897/ajmr11.921.

Hansted, J. G., K. M. Kragh, S. Shipovskov, and S. Yu. 2016. Process for the hydrolysis of keratin employing proteases. W O 2016/110523 Al, issued 2016.

Haq, I. ul, F. Akram, and Z. Jabbar. 2020. "Keratinolytic Enzyme-Mediated Biodegradation of Recalcitrant Poultry Feathers Waste by Newly Isolated *Bacillus* sp. NKSP-7 under Submerged Fermentation." *Folia Microbiologica* 65 (5): 823–34. https://doi.org/10.1007/s12223-020-00793-6.

Heidenreich, S., and P. U. Foscolo. 2015. "New Concepts in Biomass Gasification." *Progress in Energy and Combustion Science* 46: 72–95. https://doi.org/10.1016/j.pecs.2014.06.002.

Hill, P., H. Brantley, and M. Van Dyke. 2010. "Some Properties of Keratin Biomaterials: Kerateines." *Biomaterials* 31 (4): 585–93. https://doi.org/10.1016/j.biomaterials.2009.09.076.

Holm-Nielsen, J. B., T. Al Seadi, and P. Oleskowicz-Popiel. 2009. "The Future of Anaerobic Digestion and Biogas Utilization." *Bioresource Technology* 100 (22): 5478–84. https://doi.org/10.1016/j.biortech.2008.12.046.

Huang, Y., Y. Sun, S. Ma, L. Chen, H. Zhang, and Y. Deng. 2013. "Isolation and Characterization of *Keratinibaculum paraultunense* Gen. Nov., Sp. Nov., a Novel Thermophilic, Anaerobic Bacterium with Keratinolytic Activity." *FEMS Microbiology Letters* 345 (1): 56–63. https://doi.org/10.1111/1574-6968.12184.

Huang, Y., P. K. Busk, F. A. Herbst, and L. Lange. 2015. "Genome and Secretome Analyses Provide Insights into Keratin Decomposition by Novel Proteases from the Non-Pathogenic Fungus *Onygena corvina*." *Applied Microbiology and Biotechnology* 99 (22): 9635–49. https://doi.org/10.1007/s00253-015-6805-9.

Ichida, J. M., L. Krizova, C. A. LeFevre, H. M. Keener, D. L. Elwell, and E. H. Burtt. 2001. "Bacterial Inoculum Enhances Keratin Degradation and Biofilm Formation in Poultry Compost." *Journal of Microbiological Methods* 47 (2): 199–208. https://doi.org/10.1016/S0167-7012(01)00302-5.

Ionata, E., F. Canganella, G. Bianconi, Y. Benno, M. Sakamoto, A. Capasso, M. Rossi, and F. La Cara. 2008. "A Novel Keratinase from *Clostridium sporogenes* Bv. Pennavorans Bv. Nov., a Thermotolerant Organism Isolated from Solfataric Muds." *Microbiological Research* 163 (1): 105–12. https://doi.org/10.1016/j.micres.2006.08.001.

Intagun, W., and W. Kanoksilapatham. 2017. "A Review: Biodegradation and Applications of Keratin Degrading Microorganisms and Keratinolytic Enzymes, Focusing on Thermophiles and Thermostable Serine Proteases." *American Journal of Applied Sciences* 14 (11): 1016–23. https://doi.org/10.3844/ajassp.2017.1016.1023.

Isahak, W. N. R. W., M. W. M. Hisham, M. A. Yarmo, and T. Y. Y. Hin. 2012. "A Review on Bio-Oil Production from Biomass by Using Pyrolysis Method." *Renewable and Sustainable Energy Reviews* 16 (8): 5910–23. https://doi.org/10.1016/j.rser.2012.05.039.

Jain, R., A. Jain, N. Rawat, M. Nair, and R. Gumashta. 2016. "Feather Hydrolysate from Streptomyces Sampsonii GS 1322: A Potential Low Cost Soil Amendment." *Journal of Bioscience and Bioengineering* 121 (6): 672–77. https://doi.org/10.1016/j.jbiosc.2015.11.003.

Jeong, J. H., O. M. Lee, Y. D. Jeon, J. D. Kim, N. R. Lee, C. Y. Lee, and H. J. Son. 2010. "Production of Keratinolytic Enzyme by a Newly Isolated Feather-Degrading *Stenotrophomonas maltophilia* That Produces Plant Growth-Promoting Activity." *Process Biochemistry* 45 (10): 1738–45. https://doi.org/10.1016/j.procbio.2010.07.020.

Ji, Y., J. Chen, J. Lv, Z. Li, L. Xing, and S. Ding. 2014. "Extraction of Keratin with Ionic Liquids from Poultry Feather." *Separation and Purification Technology* 132: 577–83. https://doi.org/10.1016/j.seppur.2014.05.049.

Kaewsalud, T., K. Yakul, K. Jantanasakulwong, W. Tapingkae, M. Watanabe, and T. Chaiyaso. 2020. "Biochemical Characterization and Application of Thermostable-Alkaline Keratinase From *Bacillus halodurans* SW-X to Valorize Chicken Feather Wastes." *Waste and Biomass Valorization*, no. 0123456789. https://doi.org/10.1007/s12649-020-01287-9.

Kanani, F., M. D. Heidari, B. H. Gilroyed, and N. Pelletier. 2020. "Waste Valorization Technology Options for the Egg and Broiler Industries: A Review and Recommendations." *Journal of Cleaner Production* 262: 121129. https://doi.org/10.1016/j.jclepro.2020.121129.

Kelleher, B. P., J. J. Leahy, A. M. Henihan, T. F. O'Dwyer, D. Sutton, and M. J. Leahy. 2002. "Advances in Poultry Litter Disposal Technology – A Review." *Bioresource Technology* 16 (2): 27–36. https://doi.org/10.1007/s11668-016-0088-z.

Kerouaz, B., B. Jaouadi, A. Brans, B. Saoudi, and A. Habbeche. 2021. "Purification and Biochemical Characterization of Two Novel Extracellular Keratinases with Feather-Degradation and Hide-Dehairing Potential." *Process Biochemistry* 106 (April): 137–48. https://doi.org/10.1016/j.procbio.2021.04.009.

Kojima, M., M. Kanai, M. Tominaga, S. Kitazume, A. Inoue, and K. Horikoshi. 2006. "Isolation and Characterization of a Feather-Degrading Enzyme from *Bacillus sseudofirmus* FA30-01." *Extremophiles* 10 (3): 229–35. https://doi.org/10.1007/s00792-005-0491-y.

Kondamudi, N., J. Strull, M. Misra, and S. K. Mohapatra. 2009. "A Green Process for Producing Biodiesel from Feather Meal A Green Process for Producing Biodiesel from Feather Meal." *Journal of Agricultural and Food Chemistry* 57: 6163–66. https://doi.org/10.1021/jf900140e.

Korniłłowicz-Kowalska, T., and J. Bohacz. 2011. "Biodegradation of Keratin Waste: Theory and Practical Aspects." *Waste Management* 31 (8): 1689–701. https://doi.org/10.1016/j.wasman.2011.03.024.

Koutb, M., F. M. Morsy, M. M. K. Bagy, and E. A. Hassan. 2012. "Optimization of Extracellular Keratinase Production by *Aspergillus terreus* Isolated from Chicken's Litter." *Journal of Advanced Laboratory Research in Biology* 3 (3): 210–16. https://e-journal.sospublication.co.in/index.php/jalrb/article/view/128%0Ainternal-pdf://0.0.6.87/128.html.

Kshetri, P., S. S. Roy, S. K. Sharma, T. S. Singh, M. A. Ansari, N. Prakash, and S. V. Ngachan. 2019. "Transforming Chicken Feather Waste into Feather Protein Hydrolysate Using a Newly Isolated Multifaceted Keratinolytic Bacterium *Chryseobacterium sediminis* RCM-SSR-7." *Waste and Biomass Valorization* 10 (1): 1–11. https://doi.org/10.1007/s12649-017-0037-4.

Kublanov, I. V., S. K. Bidjieva, A. V. Mardanov, and E. A. Bonch-Osmolovskaya. 2009. "*Desulfurococcus kamchatkensis* Sp. Nov., a Novel Hyperthermophilic Protein-Degrading Archaeon Isolated from a Kamchatka Hot Spring." *International Journal of Systematic and Evolutionary Microbiology* 59 (7): 1743–47. https://doi.org/10.1099/ijs.0.006726-0.

Kulkarni, S. A., and A. R. Jadhav. 2014. "Isolation and Characterization of Keratinolytic Bacteria from Poultry Farm Soils." *International Research Journal of Biological Sciences* 3 (7): 2278–3202.

Kumar, B., N. Bhardwaj and P. Verma. 2020. "Microwave Assisted Transition Metal Salt and Orthophosphoric Acid Pretreatment Systems: Generation of Bioethanol and Xylo-Oligosaccharides." *Renewable Energy* 158: 574–84.

Kumar, B., and P. Verma. 2020. Application of Hydrolytic Enzymes in Biorefinery and its Future Prospects. In: *Microbial Strategies for Techno-economic Biofuel Production* (pp. 59–83). Springer, Singapore.

Kumar, B., and P. Verma. 2021a. "Life Cycle Assessment: Blazing a Trail for Bioresources Management." *Energy Conversion and Management X* 10100063: 1–17. https://doi.org/10.1016/j.ecmx.2020.100063.

Kumar, B., P. Verma. 2021b. Techno-Economic Assessment of Biomass-Based Integrated Biorefinery for Energy and Value-Added Product. In: Verma, P. (ed.) *Biorefineries: A Step Towards Renewable and Clean Energy. Clean Energy Production Technologies.* pp. 581–616. Springer, Singapore. https://doi.org/10.1007/978-981-15-9593-6_23

Kumawat, T. K., A. Sharma, V. Sharma, and S. Chandra. 2018. *Keratin Waste: The Biodegradable Polymers*. Intech. https://www.intechopen.com/books/advanced-biometric-technologies/liveness-detection-in-biometrics.

Kumawat, T. K., A. Sharma, and S. Bhadauria. 2016. "Biodegradation of Keratinous Waste Substrates by *Arthroderma multifidum*." *Asian Journal of Applied Sciences* 9 (3): 106–12. https://doi.org/10.3923/ajaps.2016.106.112.

Kunert, J. 1976. "Keratin Decomposition by Dermatophytes II. Presence of S-Sulfocysteine and Cysteic Acid in Soluble Decomposition Products." *Zeitschrift Fur Allg. Mikrobiologie* 2: 97–105.

Kuo, J. M., J. I. Yang, W. M. Chen, M. H. Pan, M. L. Tsai, Y. J. Lai, A. Hwang, B. S. Pan, and C. Y. Lin. 2012. "Purification and Characterization of a Thermostable Keratinase from *Meiothermus* sp. I40." *International Biodeterioration and Biodegradation* 70: 111–16. https://doi.org/10.1016/j.ibiod.2012.02.006.

Łaba, W., and K. B. Szczekala. 2013. "Keratinolytic Proteases in Biodegradation of Pretreated Feathers." *Polish Journal of Environmental Studies* 22 (4): 1101–9.

Łaba, W., B. Zarowska, D. Chorążyk, A. Pudło, M. Piegza, A. Kancelista, and W. Kopeć. 2018. "New Keratinolytic Bacteria in Valorization of Chicken Feather Waste." *AMB Express* 8 (1). https://doi.org/10.1186/s13568-018-0538-y.

Lange, L., Y. Huang, and P. K. Busk. 2016. "Microbial Decomposition of Keratin in Nature — A New Hypothesis of Industrial Relevance." *Applied Microbiology and Biotechnology*: 2083–96. https://doi.org/10.1007/s00253-015-7262-1.

Lateef, A., I. A. Adelere, T. B. Asafa, and L. S. Beukes. 2015. "Green Synthesis of Silver Nanoparticles Using Keratinase Obtained from a Strain of *Bacillus safensis* LAU 13." *International Nano Letters*: 29–35. https://doi.org/10.1007/s40089-014-0133-4.

Latshaw, J. D., N. Musharaf, and R. Retrum. 1994. "Processing of Feather Meal to Maximize Its Nutritional Value for Poultry." *Animal Feed Science and Technology* 47 (3–4): 179–88. https://doi.org/10.1016/0377-8401(94)90122-8.

Lee, Y. S., L. Y. Phang, S. A. Ahmad, and P. T. Ooi. 2016. "Microwave-Alkali Treatment of Chicken Feathers for Protein Hydrolysate Production." *Waste and Biomass Valorization* 7 (5): 1147–57. https://doi.org/10.1007/s12649-016-9483-7.

Li, Z. W., S. Liang, Y. Ke, J. J. Deng, M. S. Zhang, D. L. Lu, J. Z. Li, and X. C. Luo. 2020. "The Feather Degradation Mechanisms of a New *Streptomyces* sp. Isolate SCUT-3." *Communications Biology* 3 (1). https://doi.org/10.1038/s42003-020-0918-0.

Lin, X., C. G. Lee, E. S. Casale, and J. C. Shih. 1992. "Purification and Characterization of a Keratinase from a Feather-Degrading *Bacillus licheniformis* strain". *Applied and Environmental Microbiology* 58(10): 3271–75.

Liu, B., J. Zhang, Z. Fang, G. Du, J. Chen, and X. Liao. 2014a. "Functional Analysis of the C-Terminal Propeptide of Keratinase from *Bacillus licheniformis* BBE11-1 and Its Effect on the Production of Keratinase in Bacillus Subtilis." *Process Biochemistry* 49 (9): 1538–42. https://doi.org/10.1016/j.procbio.2014.04.021.

Liu, Q., T. Zhang, N. Song, Q. Li, Z. Wang, X. Zhang, X. Lu, J. Fang, and J. Chen. 2014b. "Purification and Characterization of Four Key Enzymes from a Feather-Degrading *Bacillus subtilis* from the Gut of Tarantula *Chilobrachys guangxiensis*." *International Biodeterioration and Biodegradation* 96: 26–32. https://doi.org/10.1016/j.ibiod.2014.08.008.

Maciel, J. L., P. O. Werlang, D. J. Daroit, and A. Brandelli. 2017. "Characterization of Protein-Rich Hydrolysates Produced Through Microbial Conversion of Waste Feathers." *Waste and Biomass Valorization* 8 (4): 1177–86. https://doi.org/10.1007/s12649-016-9694-y.

Mackaluso, J. 2007. "The Use of Syngas Derived from Biomass and Waste Products to Produce Ethanol and Hydrogen." *Microbiology and Molecular Genetics - 445 Basic Biotechnology EJournal* 3 (1): 98–103. www.msu.edu/course/mmg/445/.

Mani, K., N. Taib, M. Hugoni, G. Bronner, J. M. Bragança, and D. Debroas. 2020. "Transient Dynamics of Archaea and Bacteria in Sediments and Brine across a Salinity Gradient in a Solar Saltern of Goa, India." *Frontiers in Microbiology* 11 (August): 1–19. https://doi.org/10.3389/fmicb.2020.01891.

Mante, O. D., and F. A. Agblevor. 2010. "Influence of Pine Wood Shavings on the Pyrolysis of Poultry Litter." *Waste Management* 30 (12): 2537–47. https://doi.org/10.1016/j.wasman.2010.07.007.

Mata-Alvarez, J., J. Dosta, M. S. Romero-Güiza, X. Fonoll, M. Peces, and S. Astals. 2014. "A Critical Review on Anaerobic Co-Digestion Achievements between 2010 and 2013." *Renewable and Sustainable Energy Reviews* 36: 412–27. https://doi.org/10.1016/j.rser.2014.04.039.

de Medeiros, I. P., S. Rozental, A. S. Costa, A. Macrae, A. N. Hagler, J. R. Ribeiro, and A. B. Vermelho. 2016. "Biodegradation of Keratin by *Trichosporum loubieri* RC-S6 Isolated from Tannery/Leather Waste." *International Biodeterioration and Biodegradation* 115: 199–204. https://doi.org/10.1016/j.ibiod.2016.08.006.

Mehta, R. S., R. J. Jholapara, and C. S. Sawant. 2014. "Isolation of a Novel Feather-Degrading Bacterium and Optimization of Its Cultural Conditions for Enzyme Production." *International Journal of Pharmacy and Pharmaceutical Sciences* 6 (1): 194–201.

Monod, M. 2008. "Secreted Proteases from Dermatophytes." *Mycopathologia* 166 (January): 285–94. https://doi.org/10.1007/s11046-008-9105-4.

Nagal, S., and P. C. Jain. 2010. "Production of Feather Hydrolysate by *Elizabethkingia meningoseptica* KB042 (MTCC 8360) in Submerged Fermentation." *Indian Journal of Microbiology* 50 (1 SUPPL): 41–45. https://doi.org/10.1007/s12088-010-0014-0.

Nakamura, A. N., M. A. Rimoto, K. T. Akeuchi, and T. F. Ujii. 2002. "The Protein Content in a Keratinized Structure." *Biological and Pharmaceutical Bulletin* 25 (May): 569–72.

Nam, G. W., D. W. Lee, H. S. Lee, N. J. Lee, B. C. Kim, E. A. Choe, J. K. Hwang, M. T. Suhartono, and Y. R. Pyun. 2002. "Native-Feather Degradation by *Fervidobacterium islandicum* AW-1, a Newly Isolated Keratinase-Producing Thermophilic Anaerobe." *Archives of Microbiology* 178 (6): 538–47. https://doi.org/10.1007/s00203-002-0489-0.

Ningthoujam, D. S., K. Tamreihao, S. Mukherjee, R. Khunjamayum, L. J. Devi, and R. S. Asem. 2018. *Keratinaceous Wastes and Their Valorization through Keratinolytic Microorganisms*. Intech. https://www.intechopen.com/books/advanced-biometric-technologies/liveness-detection-in-biometrics.

Nnolim, N. E., and U. U. Nwodo. 2020. "*Bacillus* sp. CSK2 Produced Thermostable Alkaline Keratinase Using Agro-Wastes: Keratinolytic Enzyme Characterization." *BMC Biotechnology* 20 (1): 1–14. https://doi.org/10.1186/s12896-020-00659-2.

Nnolim, N. E., A. I. Okoh, and U. U. Nwodo. 2020. "*Bacillus* sp. FPF-1 Produced Keratinase with High Potential for Chicken Feather Degradation." *Molecules* 25 (7): 1–16. https://doi.org/10.3390/molecules25071505.

Nurdiawati, A., B. Nakhshiniev, I. N. Zaini, N. Saidov, F. Takahashi, and K. Yoshikawa. 2017. "Effect of Hydrothermal Carbonization Reaction Parameters On." *Environmental Progress & Sustainable Energy* 33 (3): 676–80. https://doi.org/10.1002/ep.

Nurdiawati, A., C. Suherman, Y. Maxiselly, M. A. Akbar, B. A. Purwoko, P. Prawisudha, and K. Yoshikawa. 2019. "Liquid Feather Protein Hydrolysate as a Potential Fertilizer to Increase Growth and Yield of Patchouli (*Pogostemon cablin benth*) and Mung Bean (*Vigna radiata*)." *International Journal of Recycling of Organic Waste in Agriculture* 8 (3): 221–32. https://doi.org/10.1007/s40093-019-0245-y.

Nuutinen, E. M., P. Willberg-Keyriläinen, T. Virtanen, A. Mija, L. Kuutti, R. Lantto, and A. S. Jääskeläinen. 2019. "Green Process to Regenerate Keratin from Feathers with an Aqueous Deep Eutectic Solvent." *RSC Advances* 9 (34): 19720–28. https://doi.org/10.1039/c9ra03305j.

Onifade, A. A., N. A. Al-Sane, A. A. Al-Musallam, and S. Al-Zarban. 1998. "A Review: Potentials for Biotechnological Applications of Keratin-Degrading Microorganisms and Their Enzymes for Nutritional Improvement of Feathers and Other Keratins as Livestock Feed Resources." *Bioresource Technology* 66: 1–11.

Oren, A. 2010. "Industrial and Environmental Applications of Halophilic Microorganisms." *Environmental Technology* 31 (8–9): 825–34. https://doi.org/10.1080/09593330903370026.

Osman, Y., A. Elsayed, A. M. Mowafy, A. Abdelrazak, and M. Fawzy. 2017. "Bioprocess Enhancement of Feather Degradation Using Alkaliphilic Microbial Mixture." *British Poultry Science* 58 (3): 319–28. https://doi.org/10.1080/00071668.2017.1278627.

Papadopoulos, M. C., A. R. El-Boushy, and A. E. Roodbeen. 1985. "The Effect of Varying Autoclaving Conditions and Added Sodium Hydroxide on Amino Acid Content and Nitrogen Characteristics of Feather Meal." *Journal of the Science of Food and Agriculture* 36 (12): 1219–26. https://doi.org/10.1002/jsfa.2740361204.

Park, G. T., and H. J. Son. 2009. "Keratinolytic Activity of Bacillus Megaterium F7-1, a Feather-Degrading Mesophilic Bacterium." *Microbiological Research* 164 (4): 478–85. https://doi.org/10.1016/j.micres.2007.02.004.

Patinvoh, R. J., E. Feuk-Lagerstedt, M. Lundin, I. S. Horváth, and M. J. Taherzadeh. 2016. "Biological Pretreatment of Chicken Feather and Biogas Production from Total Broth." *Applied Biochemistry and Biotechnology* 180 (7): 1401–15. https://doi.org/10.1007/s12010-016-2175-8.

Paul, T., A. Das, A. Mandal, S. K. Halder, P. K. DasMohapatra, B. R. Pati, and K. C. Mondal. 2014. "Production and Purification of Keratinase Using Chicken Feather Bioconversion by a Newly Isolated *Aspergillus fumigatus* TKF1: Detection of Valuable Metabolites." *Biomass Conversion and Biorefinery* 4 (2): 137–48. https://doi.org/10.1007/s13399-013-0090-6.

Paul, T., S. K. Halder, A. Das, S. Bera, C. Maity, A. Mandal, P. S. Das, P. K. Das Mohapatra, B. R. Pati, and K. C. Mondal. 2013. "Exploitation of Chicken Feather Waste as a Plant Growth Promoting Agent Using Keratinase Producing Novel Isolate *Paenibacillus woosongensis* TKB2." *Biocatalysis and Agricultural Biotechnology* 2 (1): 50–57. https://doi.org/10.1016/j.bcab.2012.10.001.

Peng, Z., X. Mao, J. Zhang, G. Du, and J. Chen. 2019. "Effective Biodegradation of Chicken Feather Waste by Co-Cultivation of Keratinase Producing Strains." *Microbial Cell Factories* 18 (1): 1–11. https://doi.org/10.1186/s12934-019-1134-9.

PoultryHub. 2020. "Reclaiming Feather Waste." Available at http://www.Poultryhub.Org/Reclaiming-Feather-Waste/ [Verified December 2020].

Prajapati, S., S. Koirala, and A. K. Anal. 2021. "Bioutilization of Chicken Feather Waste by Newly Isolated Keratinolytic Bacteria and Conversion into Protein Hydrolysates with Improved Functionalities." *Applied Biochemistry and Biotechnology*. https://doi.org/10.1007/s12010-021-03554-4.

Prakash, P., S. K. Jayalakshmi, and K. Sreeramulu. 2010a. "Production of Keratinase by Free and Immobilized Cells of *Bacillus halodurans* strain PPKS-2: Partial Characterization and Its Application in Feather Degradation and Dehairing of the Goat Skin." *Applied Biochemistry and Biotechnology* 160 (7): 1909–20. https://doi.org/10.1007/s12010-009-8702-0.

Prakash, P., S. K. Jayalakshmi, and K. Sreeramulu 2010b. "Purification and Characterization of Extreme Alkaline, Thermostable Keratinase, and Keratin Disulfide Reductase Produced by *Bacillus halodurans* PPKS-2." *Applied Microbiology and Biotechnology* 87 (2): 625–33. https://doi.org/10.1007/s00253-010-2499-1.

Qiu, J., C. Wilkens, K. Barrett, and A. S. Meyer. 2020. "Microbial Enzymes Catalyzing Keratin Degradation: Classification, Structure, Function." *Biotechnology Advances* 44 (January). https://doi.org/10.1016/j.biotechadv.2020.107607.

Rahayu, S., D. Syah, and M. T. Suhartono. 2012. "Degradation of Keratin by Keratinase and Disulfide Reductase from *Bacillus* sp. MTS of Indonesian Origin." *Biocatalysis and Agricultural Biotechnology* 1 (2): 152–58. https://doi.org/10.1016/j.bcab.2012.02.001.

Rai, S. K., and A. K. Mukherjee. 2011. "Optimization of Production of an Oxidant and Detergent-Stable AlkalineKeratinase from *Brevibacillus* sp. strain AS-S10-II: Application of Enzyme in Laundry Detergent Formulations and in Leather Industry." *Biochemical Engineering Journal* 54 (1): 47–56. https://doi.org/10.1016/j.bej.2011.01.007.

Ramakrishna Reddy, M., K. Sathi Reddy, Y. Ranjita Chouhan, H. Bee, and G. Reddy. 2017. "Effective Feather Degradation and Keratinase Production by *Bacillus pumilus* GRK for Its Application as Bio-Detergent Additive." *Bioresource Technology* 243: 254–63. https://doi.org/10.1016/j.biortech.2017.06.067.

Ramirez, J., B. McCabe, P. D. Jensen, R. Speight, M. Harrison, L. Van Den Berg, and I. O'Hara. 2021. "Wastes to Profit: A Circular Economy Approach to Value-Addition in Livestock Industries." *Animal Production Science* 61 (6): 541–50. https://doi.org/10.1071/AN20400.

Ramnani, P., R. Singh, and R. Gupta. 2005. "Keratinolytic Potential of *Bacillus licheniformis* RG1: Structural and Biochemical Mechanism of Feather Degradation." *Canadian Journal of Microbiology* 51 (3): 191–96. https://doi.org/10.1139/w04-123.

Reis, S. V. dos, W. O. Beys-da-Silva, L. Tirloni, L. Santi, A. Seixas, C. Termignoni, M. V. da Silva, and A. J. Macedo. 2020. "The Extremophile *Anoxybacillus* sp. PC2 Isolated from Brazilian Semiarid Region (Caatinga) Produces a Thermostable Keratinase." *Journal of Basic Microbiology* 60 (9): 809–15. https://doi.org/10.1002/jobm.202000186.

Riessen, S., and G. Antranikian. 2001. "Isolation of *Thermoanaerobacter Keratinophilus* sp. nov., a Novel Thermophilic, Anaerobic Bacterium with Keratinolytic Activity." *Extremophiles* 5 (6): 399–408.

Sangali, S., and A. Brandelli. 2000. "Feather Keratin Hydrolysis by a *Vibrio* sp. strain Kr2." *Journal of Applied Microbiology* 89 (5): 735–43. https://doi.org/10.1046/j.1365-2672.2000.01173.x.

Santos, R. M. D. B., A. A. P. Firmino, C. M. De Sá, and C. R. Felix. 1996. "Keratinolytic Activity of *Aspergillus fumigatus* fresenius." *Current Microbiology* 33 (6): 364–70. https://doi.org/10.1007/s002849900129.

Schrooyen, P. M. M., P. J. Dijkstra, R. C. Oberthür, A. Bantjes, and J. Feijen. 2001. "Stabilization of Solutions of Feather Keratins by Sodium Dodecyl Sulfate." *Journal of Colloid and Interface Science* 240 (1): 30–39. https://doi.org/10.1006/jcis.2001.7673.

Sharma, I., and N. Kango. 2021. "Production and Characterization of Keratinase by Ochrobactrum Intermedium for Feather Keratin Utilization." *International Journal of Biological Macromolecules* 166: 1046–56. https://doi.org/10.1016/j.ijbiomac.2020.10.260.

Sharma, R., and S. Devi. 2018. "Versatility and Commercial Status of Microbial Keratinases: A Review." *Reviews in Environmental Science and Biotechnology* 17 (1): 19–45. https://doi.org/10.1007/s11157-017-9454-x.

Shih, J., J.-J. Wang, and H. Swaisgood. 2003. Immobilization of Keratinase for Proteolysis and Keratinolysis. US 2003/0108991A1. *Optics Express*, issued 2003.

Shrinivas, D., and G. R. Naik. 2011. "Characterization of Alkaline Thermostable Keratinolytic Protease from Thermoalkalophilic *Bacillus halodurans* JB 99 Exhibiting Dehairing Activity." *International Biodeterioration and Biodegradation* 65 (1): 29–35. https://doi.org/10.1016/j.ibiod.2010.04.013.

Sierpinski, P., J. Garrett, J. Ma, P. Apel, D. Klorig, T. Smith, L. A. Koman, A. Atala, and M. Van Dyke. 2008. "The Use of Keratin Biomaterials Derived from Human Hair for the Promotion of Rapid Regeneration of Peripheral Nerves." *Biomaterials* 29 (1): 118–28. https://doi.org/10.1016/j.biomaterials.2007.08.023.

Simbolon, L. M., D. S. Pandey, A. Horvat, M. Kwapinska, J. J. Leahy, and S. A. Tassou. 2019. "Investigation of Chicken Litter Conversion into Useful Energy Resources by Using Low Temperature Pyrolysis." *Energy Procedia* 161: 47–56. https://doi.org/10.1016/j.egypro.2019.02.057.

Sinkiewicz, I., A. Śliwińska, H. Staroszczyk, and I. Kołodziejska. 2017. "Alternative Methods of Preparation of Soluble Keratin from Chicken Feathers." *Waste and Biomass Valorization* 8 (4): 1043–48. https://doi.org/10.1007/s12649-016-9678-y.

Smith, E. L., A. P. Abbott, and K. S. Ryder. 2014. "Deep Eutectic Solvents (DESs) and Their Applications." *Chemical Reviews* 114 (21): 11060–82. https://doi.org/10.1021/cr300162p.

Solcova, O., J. Knapek, L. Wimmerova, K. Vavrova, T. Kralik, M. Rouskova, S. Sabata, and J. Hanika. 2021. "Environmental Aspects and Economic Evaluation of New Green Hydrolysis Method for Waste Feather Processing." *Clean Technologies and Environmental Policy*, no. 0123456789. https://doi.org/10.1007/s10098-021-02072-5.

Stiborova, H., B. Branska, T. Vesela, P. Lovecka, M. Stranska, J. Hajslova, M. Jiru, P. Patakova, and K. Demnerova. 2016. "Transformation of Raw Feather Waste into Digestible Peptides and Amino Acids." *Journal of Chemical Technology and Biotechnology* 91 (6): 1629–37. https://doi.org/10.1002/jctb.4912.

Stingone, J. A., and S. Wing. 2011. "Poultry Litter Incineration as a Source of Energy: Reviewing the Potential for Impacts on Environmental Health and Justice." *New Solutions* 21 (1): 27–42. https://doi.org/10.2190/NS.21.1.g.

Sun, P., Z. T. Liu, and Z. W. Liu. 2009. "Particles from Bird Feather: A Novel Application of an Ionic Liquid and Waste Resource." *Journal of Hazardous Materials* 170 (2–3): 786–90. https://doi.org/10.1016/j.jhazmat.2009.05.034.

Sun, Z., X. Li, K. Liu, X. Chi, and L. Liu. 2021. "Optimization for Production of a Plant Growth Promoting Agent from the Degradation of Chicken Feather Using Keratinase Producing Novel Isolate *Bacillus pumilus* JYL." *Waste and Biomass Valorization* 12 (4): 1943–54. https://doi.org/10.1007/s12649-020-01138-7.

Suzuki, Y., Y. Tsujimoto, H. Matsui, and K. Watanabe. 2006. "Decomposition of Extremely Hard-to-Degrade Animal Proteins by Thermophilic Bacteria." *Journal of Bioscience and Bioengineering* 102 (2): 73–81. https://doi.org/10.1263/jbb.102.73.

Tamreihao, K., S. Mukherjee, R. Khunjamayum, L. J. Devi, R. S. Asem, and D. S. Ningthoujam. 2019. "Feather Degradation by Keratinolytic Bacteria and Biofertilizing Potential for Sustainable Agricultural Production." *Journal of Basic Microbiology*. https://doi.org/10.1002/jobm.201800434.

Tang, Y., L. Guo, M. Zhao, Y. Gui, J. Han, W. Lu, Q. Dai, et al. 2021. "A Novel Thermostable Keratinase from *Deinococcus geothermalis* with Potential Application in Feather Degradation." *Applied Sciences (Switzerland)* 11 (7). https://doi.org/10.3390/app11073136.

Tasaki, K. 2020. "A Novel Thermal Hydrolysis Process for Extraction of Keratin from Hog Hair for Commercial Applications." *Waste Management* 104: 33–41. https://doi.org/10.1016/j.wasman.2019.12.042.

Taupe, N. C., D. Lynch, R. Wnetrzak, M. Kwapinska, W. Kwapinski, and J. J. Leahy. 2016. "Updraft Gasification of Poultry Litter at Farm-Scale - A Case Study." *Waste Management* 50: 324–33. https://doi.org/10.1016/j.wasman.2016.02.036.

Tesfaye, T., B. Sithole, and D. Ramjugernath. 2017. "Valorisation of Chicken Feathers: A Review on Recycling and Recovery Route—Current Status and Future Prospects." *Clean Technologies and Environmental Policy* 19 (10): 2363–78. https://doi.org/10.1007/s10098-017-1443-9.

Tesfaye, T., B. Sithole, and D. Ramjugernath. 2018. "Valorisation of Waste Chicken Feathers: Optimisation of Decontamination and Pre-Treatment with Bleaching Agents Using Response Surface Methodology." *Sustainable Chemistry and Pharmacy* 8 (November 2017): 21–37. https://doi.org/10.1016/j.scp.2018.02.003.

Thys, R. C. S., F. S. Lucas, A. Riffel, P. Heeb, and A. Brandelli. 2004. "Characterization of a Protease of a Feather-Degrading *Microbacterium* Species." *Letters in Applied Microbiology* 39 (2): 181–86. https://doi.org/10.1111/j.1472-765X.2004.01558.x.

Tiwary, E., and R. Gupta. 2012. "Rapid Conversion of Chicken Feather to Feather Meal Using Dimeric Keratinase from *Bacillus licheniformis* ER-15." *Journal of Bioprocessing & Biotechniques* 02 (04): 0–4. https://doi.org/10.4172/2155-9821.1000123.

Tonin, C., A. Aluigi, C. Vineis, A. Varesano, A. Montarsolo, and F. Ferrero. 2007. "Thermal and Structural Characterization of Poly(Ethylene-Oxide)/Keratin Blend Films." *Journal of Thermal Analysis and Calorimetry* 89 (2): 601–8. https://doi.org/10.1007/s10973-006-7557-7.

Tsiroulnikov, K., H. Rezai, E. Bonch-Osmolovskaya, P. Nedkov, A. Gousterova, V. Cueff, A. Godfroy, et al. 2004. "Hydrolysis of the Amyloid Prion Protein and Nonpathogenic Meat and Bone Meal by Anaerobic Thermophilic Prokaryotes and *Streptomyces* Subspecies." *Journal of Agricultural and Food Chemistry* 52 (20): 6353–60. https://doi.org/10.1021/jf0493324.

Vasconcelos, A., G. Freddi, and A. Cavaco-Paulo. 2008. "Biodegradable Materials Based on Silk Fibroin and Keratin." *Biomacromolecules* 9 (4): 1299–1305. https://doi.org/10.1021/bm7012789.

Verma, A., H. Singh, S. Anwar, A. Chattopadhyay, K. Tiwari, S. Kaur, and G. S. Dhilon. 2016. "Critical Reviews in Biotechnology Microbial Keratinases: Industrial Enzymes with Waste Management Potential." *Critical Reviews in Biotechnology* 8551 (June). https://doi.org/10.1080/07388551.2016.1185388.

Verma, P. (2022) *Industrial Microbiology and Biotechnology*, Springer, Singapore.

Vidal, L., P. Christen, and M. N. Coello. 2000. "Feather Degradation by Kocuria Rosea in Submerged Culture." *World Journal of Microbiology and Biotechnology* 16 (6): 551–54. https://doi.org/10.1023/A:1008976802181.

Vieille, C., G. J. Zeikus, and C. Vieille. 2001. "Hyperthermophilic Enzymes: Sources, Uses, and Molecular Mechanisms for Thermostability." *Microbiology and Molecular Biology Reviews* 65 (1). https://doi.org/10.1128/MMBR.65.1.1.

Vijay Kumar, E., M. Srijana, K. Chaitanya, Y. Reddy, and G. Reddy. 2011. "Biodegradation of Poultry Feathers by a Novel Bacterial Isolate *Bacillus altitudinis* GVC11." *Indian Journal of Biotechnology* 10 (4): 502–7.

Voitic, G., S. Nestl, K. Malli, J. Wagner, B. Bitschnau, F. A. Mautner, and V. Hacker. 2016. "High Purity Pressurised Hydrogen Production from Syngas by the Steam-Iron Process." *RSC Advances* 6 (58): 53533–41. https://doi.org/10.1039/c6ra06134f.

Wahlström, R., K. Rommi, P. Willberg-Keyriläinen, D. Ercili-Cura, U. Holopainen-Mantila, J. Hiltunen, O. Mäkinen, H. Nygren, A. Mikkelson, and L. Kuutti. 2017. "High Yield Protein Extraction from Brewer's Spent Grain with Novel Carboxylate Salt - Urea Aqueous Deep Eutectic Solvents." *Chemistry Select* 2 (29): 9355–63. https://doi.org/10.1002/slct.201701492.

Wang, L., Y. Qian, Y. Cao, Y. Huang, Z. Chang, and H. Huang. 2017. "Production and Characterization of Keratinolytic Proteases by a Chicken Feather-Degrading Thermophilic Strain, *Thermoactinomyces* sp. YT06." *Journal of Microbiology and Biotechnology* 27 (12): 2190–98. https://doi.org/10.4014/jmb.1705.05082.

Wang, Y. X., and X. J. Cao. 2012. "Extracting Keratin from Chicken Feathers by Using a Hydrophobic Ionic Liquid." *Process Biochemistry* 47 (5): 896–99. https://doi.org/10.1016/j.procbio.2012.02.013.

Wrońska, I., and K. Cybulska. 2016. "Presence of Microorganisms at Various Stages of Poultry Wastes Management Part I. Keratinolytic Microorganisms." *Journal of Ecological Engineering* 17 (5): 43–48. https://doi.org/10.12911/22998993/64216.

Wu, S., P. Ni, J. Li, H. Sun, Y. Wang, H. Luo, J. Dach, and R. Dong. 2016. "Integrated Approach to Sustain Biogas Production in Anaerobic Digestion of Chicken Manure under Recycled Utilization of Liquid Digestate: Dynamics of Ammonium Accumulation and Mitigation Control." *Bioresource Technology* 205: 75–81. https://doi.org/10.1016/j.biortech.2016.01.021.

Wu, W. L., M. Y. Chen, I. F. Tu, Y. C. Lin, N. Eswarkumar, M. Y. Chen, M. C. Ho, and S. H. Wu. 2017. "The Discovery of Novel Heat-Stable Keratinases from *Meiothermus taiwanensis* WR-220 and Other Extremophiles." *Scientific Reports* 7 (1): 1–12. https://doi.org/10.1038/s41598-017-04723-4.

Xie, H., S. Li, and S. Zhang. 2005. "Ionic Liquids as Novel Solvents for the Dissolution and Blending of Wool Keratin Fibers." *Green Chemistry* 7 (8): 606–8. https://doi.org/10.1039/b502547h.

Yamamura, S., Y. Morita, Q. Hasan, K. Yokoyama, and E. Tamiya. 2002. "Erratum to 'Keratin Degradation: A Cooperative Action of Two Enzymes from *Stenotrophomonas* sp.' [Biochem. Biophys. Res. Commun. 294 (2002) 1138–1143]." *Biochemical and Biophysical Research Communications* 295 (4): 1034. https://doi.org/10.1016/s0006-291x(02)00733-7.

Yang, Y., and N. Reddy. 2013. "Potential of Using Plant Proteins and Chicken Feathers for Cotton Warp Sizing." *Cellulose* 20 (4): 2163–74. https://doi.org/10.1007/s10570-013-9956-9.

Yin, J., S. Rastogi, A. E. Terry, and C. Popescu. 2007. "Self-Organization of Oligopeptides Obtained on Dissolution of Feather Keratins in Superheated Water." *Biomacromolecules* 8 (3): 800–6. https://doi.org/10.1021/bm060811g.

Zaghloul, T. I., A. M. Embaby, and A. R. Elmahdy. 2011. "Biodegradation of Chicken Feathers Waste Directed by *Bacillus subtilis* Recombinant Cells: Scaling up in a Laboratory Scale Fermentor." *Bioresource Technology* 102 (3): 2387–93. https://doi.org/10.1016/j.biortech.2010.10.106.

Zhou, L., X. Xie, T. Wu, M. Chen, Q. Yao, H. Zhu, and W. Zou. 2020. "Compound Enzymatic Hydrolysis of Feather Waste to Improve the Nutritional Value." *Biomass Conversion and Biorefinery*. https://doi.org/10.1007/s13399-020-00643-y.

Zhu, D., G. Shen, Z. Wang, R. Han, Q. Long, X. Gao, J. Xing, Y. Li, and R. Wang. 2021. "Distinctive Distributions of Halophilic Archaea across Hypersaline Environments within the Qaidam Basin of China." *Archives of Microbiology*, no. Oren 2019. https://doi.org/10.1007/s00203-020-02181-7.

Ziero, H. D. D., L. S. Buller, A. Mudhoo, L. C. Ampese, S. I. Mussatto, and T. F. Carneiro. 2020. "An Overview of Subcritical and Supercritical Water Treatment of Different Biomasses for Protein and Amino Acids Production and Recovery." *Journal of Environmental Chemical Engineering* 8 (5). https://doi.org/10.1016/j.jece.2020.104406.

# 9 The Role of Halophilic Enzymes in Bioremediation of Waste in Saline Systems

*Sushama U. Dessai*
Governement College of Arts, Science and Commerce, Khandola, Marcela-Goa, Goa, India

*Irene J. Furtado*
Goa University, Goa, India

**CONTENTS**



## 9.1 HALOPHILES

Halophiles or halophilic organisms are salt-loving organisms, widely distributed around the world in saline habitats, such as Salt Lake (Amoozegar et al., 2009), Dead Sea (Arahal et al., 1999), salt mines (Zhang et al., 2013), marine solar salterns (Cui & Qiu, 2014), salt marshes (Munson et al., 1997), and fumaroles (Ellis et al.,

DOI: 10.1201/9781003187684-9

2008). The halophiles are divided into three different categories: (1) slight halophiles (1–3%), (2) moderate halophiles (3–15%), and (3) extreme halophiles (15–30%) based on the concentration of NaCl required for their growth (Oren, 2002; Corral et al., 2020). On the contrary, non-halophiles grow optimally at (1%) NaCl concentrations. Halotolerant microorganisms are not obligate halophiles; they can grow in the presence and absence of high salt concentrations. Phylogenetically, halophiles are distributed in all three domains of life such as archaea, bacteria, and eukaryote (Edbeib et al., 2016).

The halophilic Archaea are included in the phylum Euryarchaeota and the family Halobacteriaceae (Gibbons, 1974). However, some halophilic methanogenic species have been described from hypersaline environments. Taxonomically, the methanogenic Archaea are grouped within five orders, but only the order Methanosarcinales includes halophilic species. They are placed within the family Methanosarcinaceae (Boone et al., 2001).

Moderately or extremely halophilic species of the domain Bacteria are subdivided into the following phyla: *Proteobacteria, Firmicutes, Actinobacteria, Spirochaetes, Bacteroidetes, Thermotogae, Cyanobacteria*, and *Tenericutes* (de la Haba et al., 2011).

There are multiple reviews available on the potential applications of halophiles in biotechnological and environmental processes (Lee, 2013; Edbeib et al., 2016; Mohamedin et al., 2018; Amoozegar et al., 2019). Oren (2010) divided these potential applications of halophilic microorganisms into five categories. (1) Halophiles in manufacturing solar salt from seawater and the production of traditional fermented foods. Tapingkae et al. (2010) studied the role of 156 extreme haloarchaea to improve the quality and safety of fish sauce under high salt concentration. This was achieved by histamine degrading activity of haloarchaeon *Natrinema* sp. HDS3-1 mediated by histamine dehydrogenase. (2) Use of the salt tolerance potential of halophilic microorganisms and their biomolecules like enzymes produced by them (Table 9.1) to catalyze processes in high salt environments. (3) Use of specific properties of biomolecules produced by halophiles that contribute to their salt tolerance, such as osmotic solutes glycerol, glutamate, glutamine, glycine betaine, ectoine, proline, trehalose, and sulotrahalose (Figure 9.1). For instance, one of the most common osmotic solutes in the domain Bacteria is ectoine (1,4,5,6-tetrahydro-2- methyl-4-pyrimidinecarboxylic acid), which is produced by a great variety of halophilic and halotolerant bacteria, often together with its 5-hydroxy derivative. Due to their counteractive effects of ultraviolet UV-A-induced and accelerated skin aging, they find application in the cosmetics industry to prepare dermatological cosmetic preparations and moisturizers to care for aged, dry, or irritable skin. Among them, ectoine is, commercially produced by extracting the compound from halophilic bacteria (Kunte et al., 2014). (4) Use of biomolecules unique to halophiles, for example, bacteriorhodopsin for biophotonic device applications such as Fourier transform holographic associative processors, three-dimensional optical memories, biosensors, photovoltaic cells, and protein-based retinal prostheses (Wagner et al., 2013). (5) Use of halophilic counterparts of mesophilic biomolecules of industrial importance,for example, β-carotene, poly-β-hydroxyalkanoate, and exopolysaccharides. In conclusion, halophilic microorganisms are good sources of biomolecules like carotenoid pigments, retinal proteins, enzymes, and compatible solutes which function optimally at elevated salinities.

**TABLE 9.1**
**Halophilic Enzymes Produced by Halophilic Bacteria and Halophilic Archaea**

| Enzyme | Halophile | Extreme Conditions Tolerated by Enzyme | References |
|---|---|---|---|
| α - Amylase | *Haloarcula sp.* HS | 60°C, 25% NaCl | Gómez-Villegas et al. (2021) |
| | *Haloterrigena turkmenica* | 55°C, pH 8.5 and 2M NaCl | Santorelli et al. (2016) |
| | *Haloarcula hispanica* | - | Hutcheon et al. (2005) |
| | *Haloferax mediterranei* | 2 to 4 M NaCl, pH- 7 to 8, 50°C to 60°C | Pérez-Pomares et al. (2003) |
| Lipase | *Halococcus agarilyticus* GUGFAWS-3 | 25% NaCl | Gaonkar and Furtado (2018) |
| | *Haloarcula* sp. G41 | 70°C, pH 8.0, and 15 % NaCl | Li and Yu (2014) |
| | *Chromohalobacter canadensis* | 2.5–3.5 M NaCl, 55 °C and pH 8.5 | Ai et al. (2018) |
| Cellulase | *Haloarcula* sp. LLSG7 | 50 C, pH 8.0, and 20 % NaCl | Li and Yu (2013) |
| | *Alkalilimnicola* sp. NM-DCM1 | 55°C, pH 8.8, and 2.5 M NaCl | Mesbah and Wiegel (2017) |
| | *Halorhabdus utahensis* | 5 M NaCl, pH 11.5, 20% w/w) of ionic liquids | Zhang et al. (2011) |
| | *Haloferax sulfurifontis* GUMFAZ2 | 3.5 M NaCl | Malik and Furtado (2019) |
| Xylanase | *Thermoanaerobacterium saccharolyticum* NTOU1 | 12.5% | Hung et al. (2011) |
| | *Bacillus subtilis* cho40 | 0.5 M | Khandeparker et al. (2011) |
| | *Halorhabdus utahensis* | 5% to 15% | Wainø and Ingvorsen (2003) |
| Protease | *Halococcus agarilyticus* GUGFAWS-3 | 25% NaCl | Gaonkar and Furtado (2018) |
| | *Haloferax lucentensis* VKMM 007 | 0.85–5.13 M | Manikandan et al. (2009) |

Halophiles play a crucial role in industries such as cosmetics, bioplastics, fermented food products, preservatives, photoelectric devices, and biosensors (Corral et al., 2020) and in recent times have been investigated in the hope of harnessing them as efficient cleanup microbes for industrial effluents containing high levels of salt.

## 9.2 HALOPHILES IN BIOREMEDIATION

A plethora of reports are available on the bioremediation of contaminated sites using microorganisms. Most of these reports are limited to mesophilic microorganisms. Many saline and hypersaline environments, including salt pans, saline lakes, saline industrial effluents, salt marshes, and oil fields having considerable economic, ecological, and scientific value, are often contaminated with high levels of pollutants (Abou Khalil et al., 2021). Hypersaline environments are affected mainly by pollution due to various anthropogenic activities such as mining, uncontrolled discarding

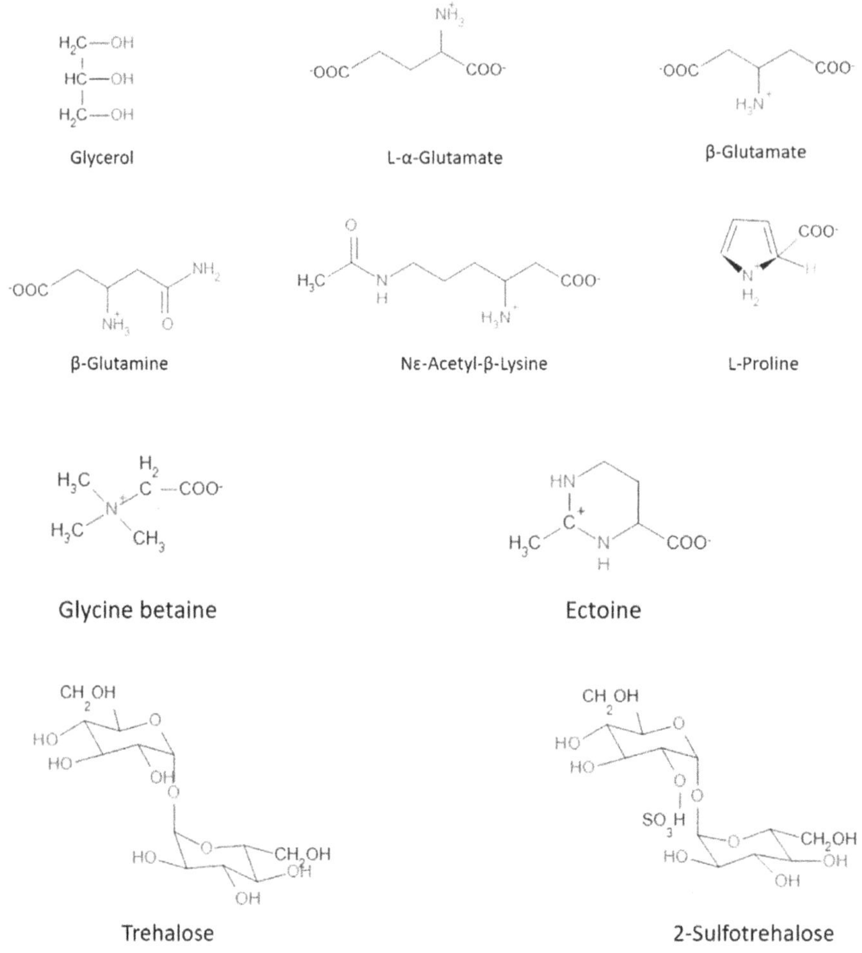

Glycerol L-α-Glutamate β-Glutamate

β-Glutamine Nε-Acetyl-β-Lysine L-Proline

Glycine betaine Ectoine

Trehalose 2-Sulfotrehalose

**FIGURE 9.1** Osmotic solutes produced by halophilic microorganisms.

of solid materials, agricultural waste and runoff of pesticides and pest control agents, and toxic pollutants from industrial processes (Paul & Mormile, 2017). Using bioremediation technology, microorganisms can be exploited to degrade toxic pollutants into harmless products such as carbon dioxide and water, and transform inorganic compounds like metal into nontoxic or less toxic forms as these processes are environmentally safe and cost-efficient (Dzionek et al., 2016). However, the application of mesophilic microorganisms for treating contaminated sites with moderate to high salinity is restricted. This is because of the lethal effects of salt on microbial life, including disruption of the cell membrane, denaturation of enzymes, less available oxygen, desiccation, and accumulation of pollutant hydrocarbons due to low solubility of hydrocarbons in this milieu. Thus, halophiles that thrive in high salt and highly salt-tolerant microorganisms hold promise for bioremediation of hypersaline environments. Processes employing these microbial cultures are expected to bioremediate saline environments without the requirement of costly dilution of salt-laden soil and water. Most studies related to this end are in their infancy, trying to isolate highly halotolerant and extremely halophilic microbes, some of them are being evaluated in vitro singly (Mnif et al., 2009) or as cocultures or as consortia to degrade a diverse variety of organic and inorganic pollutants (Patowary et al., 2016; Al-Shaikh & Jamal, 2020).

## 9.3 BIODEGRADATION OF ORGANIC POLLUTANTS IN SALINE WASTEWATER

Large amounts of wastewater are generated during crude oil and natural gas drilling and production. Around ten barrels of brackish or saline water are generated for every barrel of crude oil siphoned, containing salt concentrations ranging between 1 and 250 g $L^{-1}$. These waters also carry significant quantities of toxic metals like iron, copper, manganese, cadmium, lead, nickel, and chromium (Erakhrumen, 2015; Goswami et al., 2021) and recalcitrant petroleum hydrocarbons, and therefore are of environmental concern, as these cause hazards by contamination of soil surface, open water bodies, ground, and marine waters, if discharged directly without prior treatment, deliberately or by accident. Further, many hypersaline environments are often contaminated with high levels of petroleum hydrocarbons. Conventional microbiological treatment processes are unsuccessful as these do not function well at high salt concentrations.

Different strains of aerobic bacteria, archaea, and eukaryotes have been isolated to degrade hydrocarbons over a wide range of salinities (Table 9.2). There has been impressive progress in hydrocarbon degradation in hypersaline environments, as some of these also stand stressful pH and temperature regimes. Although many microorganisms are reported to degrade hydrocarbons in hypersaline environments, the reports on the degradation of hydrocarbons by haloarchaea are limited. These include *Halomonas* sp., *Haloferax* sp., *Halobacterium* sp., and *Haloarcula* sp. (Al-Mailem et al., 2010).

Recent studies (Abou Khalil et al., 2021) suggest that both bacteria and archaea can metabolize n-alkanes in the presence of high salt concentrations. Additionally, the biodegradation of nonoxygenated hydrocarbons in moderate to high salinity environments has been evaluated and recorded (Park & Park, 2018); however, only a few reports exist on the biodegradation of BTEX (Benzene, Toluene, Ethylbenzene, and Xylene) compounds. Among archaea, *Haloferax* (Bonfá et al., 2011), *Haloarcula* (Taran, 2011), and *Halobacterium* (Al-Mailem et al., 2010) play an essential role in the degradation of hydrocarbons at high salt concentration. Fathepure (2014) argued that pigment-mediated ATP synthesis helps these haloarchaea to degrade hydrocarbons in hypersaline environments deficient in oxygen. Also, information on genes, enzymes, and the molecular mechanism of hydrocarbon degradation in high salinity environments are unknown or sporadic and, if known then with large lacunae in understanding regarding metabolic pathways, intermediate products, and end products. Focused, dedicated research studies are wanted as these would give a jump start to prototypes/models/mock processes, directly using the polluted or simulated waters/soils.

**TABLE 9.2**
**Halophiles Involved in Hydrocarbon Degradation**

| Sr. No. | Halophile | Hydrocarbon | Salinity | References |
|---|---|---|---|---|
| ***Halophilic Bacteria*** | | | | |
| 1 | *Martelella* sp. AD-3 | Phenanthrene | 1–15% | Feng et al. (2012) |
| 2 | *Ochrobactrum* sp. VA1 | Polycyclic aromatic hydrocarbons (PAHs): anthracene, phenanthrene, naphthalene, fluorene, pyrene, benzo(k) fluoranthene and benzo(e) pyrene | 30 g/L | Arulazhagan and Vasudevan (2011) |
| 3 | *Halomonas* sp. | Diesel fuel and lubricating oil | 100 g/L | Mnif et al. (2009) |
| 4 | *Staphylococcus haemoliticus* strain 10SBZ1A | Benzo[a]pyrene (BaP) | 10% | Nzila et al. (2021) |
| 5 | *Halomonas campisalis* | Benzoate and salicylate | 50 and 100 g/L | Oie et al. (2007) |
| 6 | *Alcanivorax sp.* strain Est-02 | Long chain alkanes: tetracosane and octacosane | 10% | SadrAzodi et al. (2019) |
| 7 | *Pseudomonas* strains | BTEX: Benzene, toluene, ethylbenzene, and xylene | 7% | Hassan and Aly (2018) |
| ***Halophilic Archaea*** | | | | |
| 1 | *Haloferax* sp., *Halobacterium* sp., and *Halococcus* sp. | Crude oil, n-octadecane, and phenanthrene | 4 M | Al-Mailem et al. (2010) |
| 2 | *Haloarcula*, sp. *Haloferax* sp. | Heptadecane | 225 g/L | Tapilatu et al. (2010) |
| 3 | *Halobacterium piscisalsi, Halorubrum ezzemoulense, Halobacterium salinarium, Haloarcula hispanica, Haloferax* sp., *Halorubrum* sp., *and Haloarcula* sp | p-hydroxybenzoic acid, naphthalene, phenanthrene, and pyrene | 20% | Erdoğmuş et al. (2013) |
| 4 | *Haloferax* sp. | Naphthalene, anthracene, phenanthrene, pyrene and benzo[a]anthracene | 20% | Bonfá et al. (2011) |
| 5 | *Haloarcula* sp. IRU1 | Crude Oil | 25% | Taran (2011) |
| 6 | *Halobacterium* strain $R_1$ | Crude Oil | 25% | Raghavan and Furtado (2000) |

## 9.4 HEAVY METAL BIOREMEDIATION BY HALOPHILES

Saline environments have often become sinks for industrial discharge and thereby accumulate a range of heavy metals and hydrocarbons due to urbanization and industrialization, which includes mining, agriculture, and waste disposal (Lefebvre & Moletta, 2006; Zgórska et al., 2016). At high concentrations, metal ions are complex with other inorganic or organic substances, and influence adversely the

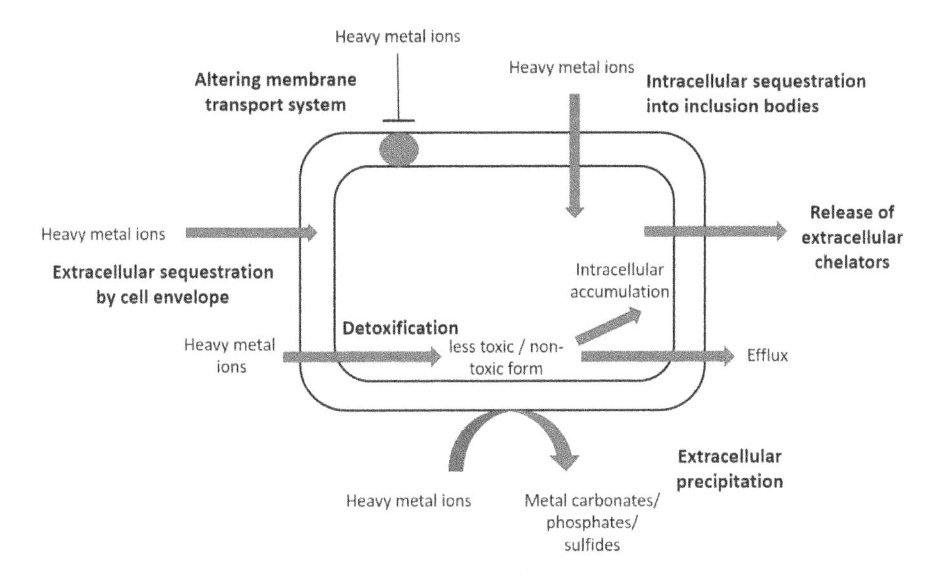


**FIGURE 9.2** Different strategies of metal resistance and detoxification in microorganisms.

biochemical activity, growth, and morphology of microbes (Tchounwou et al., 2012). Microorganisms use various metal response strategies: complex formation, extracellular precipitation, sequestration, metal reduction, metal oxidation, and methylation for reducing the transport of metal across the cell membrane and even impermeability. As portrayed in Figure 9.2 microorganisms interact with heavy metals by: (1) altering the membrane transport system involved in metal accumulation; (2) intracellular sequestration of metal using internal exclusion bodies such as polyphosphate granules (volutin); (3) extracellular sequestration using cell envelope; (4) extracellular precipitation as phosphate, sulfide, or oxalate; (4) detoxification by enzymatic oxidation or reduction; and (6) release of extracellular chelators to fix metal in the cell surrounding (Haferburg & Kothe, 2007). In some studies (Bosecker, 1997; Goswami et al., 2022), microorganisms have been utilized to remove metal contamination from wastewaters, to separate metals from sediments and soil or to foster metal solubilization or "bioleaching" for extraction.

## 9.5 HALOPHILIC ENZYMES IN BIOREMEDIATION AND MECHANISM OF ENZYME ACTION

### 9.5.1 Microbial Oxidoreductases

Oxidoreductases are the groups of enzymes involved in the detoxification of toxic organic compounds in various living forms during their metabolic activities (Alneyadi et al., 2018). In the process of energy production, electrons are transferred from reduced donor to oxidized acceptor, wherein the contaminants are finally oxidized to harmless compounds (Vidal et al., 2018). The oxidoreductases participate in the oxidation of various phenolic substances produced from the decomposition of lignin in a soil environment. In the same way, oxidoreductases can also detoxify toxic xenobiotics, such as phenolic or anilinic compounds such as 2-chlorophenol, indole, sertraline, estrone, among others (Alneyadi et al., 2018). Many bacteria reduce the heavy metals through redox reactions from an oxidized soluble form to a reduced insoluble form and subsequently precipitate it extracellularly (White et al., 1997).

### 9.5.2 Microbial Oxygenases

The major degradation pathways for petroleum hydrocarbons involve oxygenase: monooxygenase/dioxygenase, and molecular oxygen, indicating the importance of oxygen for oil degradation microbes. The substrate n-alkane, an important group in crude oil, is oxidized first to the corresponding alcohol by substrate-specific terminal monooxygenases/hydroxylases. Further, the alcohol is oxidized to the corresponding aldehyde which is then lastly converted into fatty acid. Fatty acids are converted into fatty acyl CoA and subsequently processed by β – oxidation to generate acetyl-CoA. Both terminal and subterminal oxidation has been described for short and long-chain alkanes (Abou Khalil et al., 2021).

Oxygenases are oxidoreductases that carry out the oxidation of substrates by the addition of oxygen into the substrate using either $FADH_2$ or NADH or NADPH. Based on the number of oxygen atoms used for oxygenation, they are classified into two classes: monooxygenases and dioxygenases. Oxygenase adds $O_2$ atoms into the organic molecule to promote the cleavage of aromatic rings and also carry out dehalogenation reactions of halogenated alkanes (Arora et al., 2010).

### 9.5.3 Monooxygenases

Monooxygenases carry out oxidation by the addition of one atom of the oxygen molecule into the substrate. Monooxygenases are divided into two classes based on the presence of cofactor: P450 monooxygenases and flavin-dependent monooxygenases. As the name says, flavin-dependent monooxygenases require flavin for the activity and use NADP or NADPH as a coenzyme (Huijbers et al., 2014), whereas P450 monooxygenases contain heme cofactor (Hrycay & Bandiera, 2015). Though most monooxygenases studied previously have cofactors, certain monooxygenases function independently of a cofactor.

Methane monooxygenase is involved in hydrocarbon degradation, such as alkanes, and aromatic and heterocyclic hydrocarbons. They catalyze oxidative dehalogenation reactions under oxygen-rich conditions. On the other hand, they perform reductive dichlorination under low oxygen levels (Semrau, 2011).

### 9.5.4 Dioxygenases

Dioxygenases are multicomponent enzyme systems that introduce two atoms of molecular oxygen into their substrate. Figure 9.3 shows the degradation of naphthalene by 1,2-naphthalene dioxygenase by incorporation of two oxygens atoms in naphthalene. Moreno et al. (2011) studied genes in *Halomonas organivorans* responsible for the degradation of phenol and benzoate. They reported catA gene coding

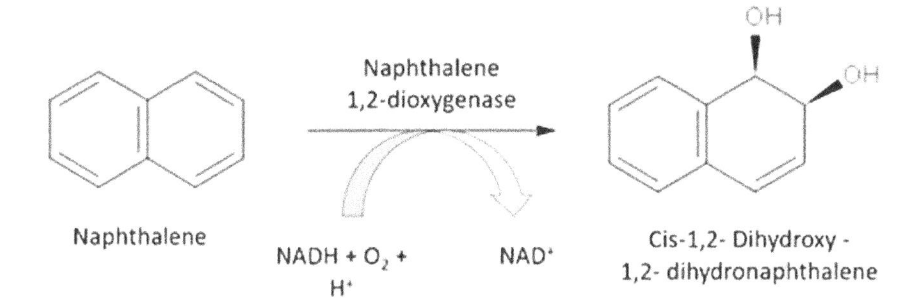


**FIGURE 9.3** Reaction catalysed by 1,2-napthalene dioxygenase.

for catechol 1,2 dioxygenases (1,2-CAT) on the gene cluster catRBCA. They also found the similarity in gene organization between *H. organivorans* and other non-halophilic eubacteria.

A study carried out by Erdoğmuş et al. (2013) described the degradation of various organic substrates by haloarchaea via ortho cleavage pathway. Nine different isolates of haloarchaea: *Halobacterium piscisalsi, Halorubrum ezzemoulense, Halobacterium salinarium, Haloarcula hispanica, Haloferax* sp., *Halorubrum* sp., *and Haloarcula* sp. were reported to degrade p-hydroxybenzoate, naphthalene, phenanthrene, and pyrene in the presence of 20% NaCl. In addition, this study showed that the isolates possessed catechol 1,2 dioxygenase and protocatechuate 3,4 dioxygenase enzyme (Erdoğmuş et al., 2013). In 2008, Kim and coworkers isolated a halophilic strain, belonging to the genus *Chromohalobacter*. This strain could grow in the presence of benzoate- and p-hydroxybenzoate. Further, the catabolic reactions of the degradation pathway of these compounds when studied using molecular and biochemical tools revealed the involvement of three enzymes, namely: 1,2-dioxygenase (1,2-CAT), p-hydroxybenzoate hydroxylase (pobA), and 3,4-PCA.

The role of genes and enzymes in the degradation of benzene was elucidated in *Arhodomonas* sp., extremely halophilic bacteria. This strain used benzene and toluene as sole carbon sources for growth. Based on the draft genome sequence, they hypothesized that benzene is first converted to phenol and catechol by phenol hydroxylase. The catechol is converted into 2-hydroxymuconic semialdehyde via meta cleavage pathway by the action of 2,3-dioxygenase (2,3-CAT). The resulting 2-hydroxymuconic semialdehyde enters the tricarboxylic acid cycle to generate energy (Dalvi et al., 2012).

### 9.5.5 Laccase

Laccase (E.C. 1.10.3.2) is p-diphenol: dioxygen oxidoreductase. It comprises a family of multicopper oxidases widely distributed in nature, produced by plants, fungi, insects, and bacteria (Singh & Gupta, 2020; Agrawal & Verma, 2020a). Multiple isoenzymes are reported for laccases and each of them are coded by distinct genes. In some instances, the genes have been expressed differently depending upon the nature of the inducer. Laccases are produced by microorganisms either intracellularly or extracellularly. Uthandi et al. (2010) reported the purification and characterization of a laccase (LccA) from the halophilic archaeon *Haloferax volcanii*. The purified enzyme was capable of oxidizing different organic substrates, including bilirubin, syringaldazine, 2,2,-azino-bis-(3-ethylbenzothiazoline-6-sulfonic acid), and dimethoxyphenol.

Apart from oxidation of phenolic compounds, laccases are known to carry out delignification of lignocellulosic materials with the formation of phenols (Karigar & Rao, 2011). Lignocellulosic biomasses like agriculture residues (e.g., corn stover), perennial grasses (e.g., miscanthus and switchgrass), and woody biomass (e.g., poplar and pine) are promising biowastes with a high potential as natural and renewable resources (Singh et al., 2017; Bhardwaj et al., 2020; Nair et al., 2022a). Valorizing significant components of lignocellulose – cellulose, hemicelluloses, and lignin – is critical for the sustainability of next-generation biorefineries (Jafari et al., 2017; Kumar et al., 2020; Nair et al., 2022b). Lignocellulose bioconversion is a laborious process requiring the use of surfactants and organic solvents. Consequently, incorporating laccases in this bioconversion requires the bioprospecting of enzymes that can remain stable under extreme conditions (Rezaie et al., 2017).

Siroosi et al. (2018) reported the laccase activity in spores of a halotolerant bacterium, *Bacillus safensis* sp. strain S31, an isolate from a chromite mine soil. The enzyme decolorized malachite green, toluidine blue, and reactive black 5 at acidic pH values in free form and reactive black 5 in immobilized form. In another report, Jafari and coworkers evaluated laccase activity from the halophilic bacterium *Chromohalobacter salexigens* for its potential application for delignification of an almond shell (Jafari et al., 2017). In another report, an extracellular laccase was produced by the halophilic isolate *Aquisalibacillus elongatus*. The purified enzyme delignified a peanut shell substrate by 45% at a pH of 8.0 and a temperature of 35°C and detoxified the toxic phenols found in peanut shell waste (Rezaie et al., 2017; Kumar & Verma, 2020; Agrawal & Verma, 2020b).

### 9.5.6 Azoreductases

Azo dyes and related compounds are known as xenobiotics. They are widely used in paint, textile, cosmetics, printing, and pharmaceutical industries for dyeing purposes, which release a high amount of untreated wastewater carrying these azo dyes that contain toxic and mutagenic nitro and amine groups (Misal & Gawai, 2018). Reduction of azo bond is displayed in Figure 9.4. Azoreductases are commonly present amongst microorganisms involved in the biotransformation and detoxification of azo dyes, nitro-aromatic, and azoic drugs. Azoreductases are flavin-containing or flavin-free enzymes, utilizing the nicotinamide adenine dinucleotide (NAD) or nicotinamide adenine dinucleotide phosphate (NADP) as a reducing equivalent. Azoreductases from anaerobic microorganisms are highly oxygen-sensitive, while their counterparts in aerobic microorganisms are usually oxygen insensitive. They have variable pH, temperature stability, and broad substrate specificity. Azo dyes, nitro-aromatic compounds, and quinones are the known substrates of azoreductase (Misal & Gawai, 2018). In the study carried out by Eslami et al. (2016), a halophilic bacterium, *Halomonas elongata*, was used to decolorize different mono- (methyl

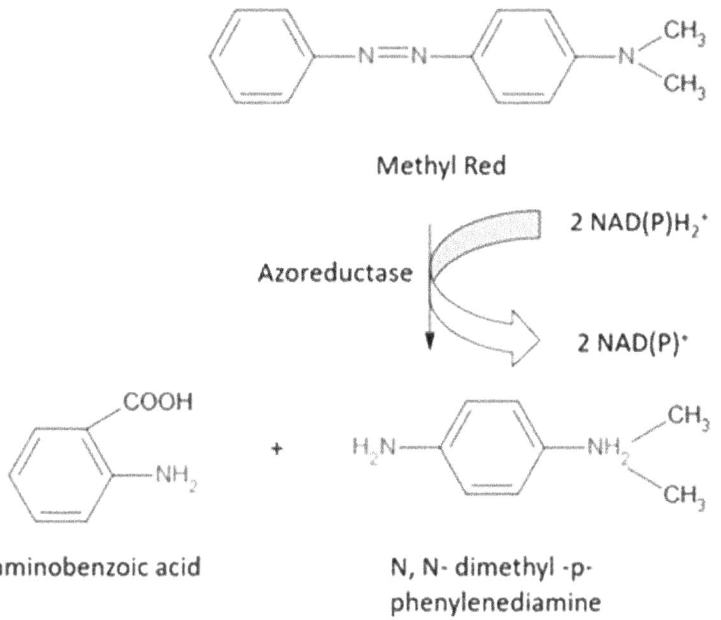

Methyl Red

Azoreductase

2 NAD(P)H₂⁺

2 NAD(P)⁺

2-aminobenzoic acid

N, N- dimethyl -p- phenylenediamine

**FIGURE 9.4** Degradation of methyl red by azoreductase.

red) and di-azo dyes (remazol black B) in anoxic conditions and at NaCl concentrations up to 15% (w/v). Also, the azoreductase gene of *H. elongata*, (acp) were cloned in *Escherichia coli*, purified and characterized (Eslami et al., 2016).

### 9.5.7 Alkane Hydroxylases

Alkane hydroxylases (AHs) (*EC 1.14.15.3*) are the crucial enzymes in the aerobic degradation of alkanes by bacteria. They hydroxylate alkanes to alcohols (Figure 9.5), which are then oxidized to fatty acids and catabolized via the bacterial β-oxidation pathway. Thus, they play an important role in the bioremediation of petroleum-contaminated environments and microbial-enhanced oil recovery. Bacteria commonly produce two AHs: integral-membrane alkane monooxygenase (AlkB)-related AHs, and cytochrome P450 CYP153 family (van Beilen et al., 2006). AlkB-related AHs are the most commonly found AHs distributed in bacteria. They use rubredoxin and rubredoxin reductase as electron transfer components for alkane hydroxylation (van Beilen et al., 2002). AlkB generally acts on n-alkanes ranging from C10 to C16 (Nie et al., 2014). Dastgheib et al. (2011) reported the presence of alkB genes in halotolerant *Alcanivorax* sp. This strain was isolated from oil-contaminated saline soils and was able to degrade tetracosane, crude oil, and diesel fuel, and utilize various pure aliphatic hydrocarbon substrates (from C12 to C34) at a wide range of NaCl concentrations. The second type of AHs is the cytochrome P450 CYP153 family involved in the degradation of n-alkanes with short and medium-chain length, commonly found in alkane-degrading bacteria lacking AlkB. Several bacteria have multiple AHs,

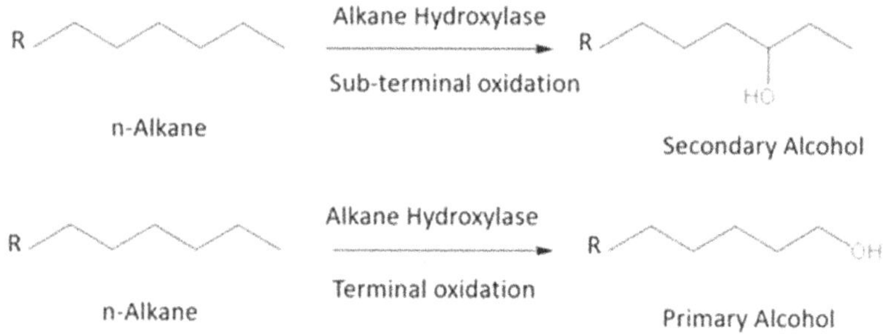

**FIGURE 9.5** Mechanism of alkane hydroxylase.

which allow the bacteria to grow on a wide range of n-alkane (Nie et al., 2014). For instance, the moderate halophile *Amycolicicoccus subflavus* DQS3-9A1T, which was isolated from oil mud precipitated from oil-produced water reported to contain four types of alkane hydroxylases, including the integral-membrane non-heme iron monooxygenase (AlkB) and cytochrome P450 CYP153, a long-chain alkane monooxygenase (LadA) and propane monooxygenase (Nie et al., 2013).

### 9.5.8 Microbial Lipases

Lipases (EC 3.1) are present in a variety of microorganisms, animals, and plants that degrade lipids by hydrolysing ester bonds between a fatty acid and glycerol or phosphate (Schreck & Grunden, 2014). Lipase activity is reported for the degradation of organic pollutants such as hydrocarbon present in the contaminated soil and bioremediation of oil spills. Mohamedin et al. (2018) reported petroleum and used engine oil degradation by moderate halophilic bacteria *Piscibacillus halophilus* and *Halomonas* sp.

### 9.5.9 Arsenite Methytransferase

Arsenic is a toxic metal that causes lesions, skin color changes, black foot disease and cancers of lungs, skin, and bladder on exposure (Guha Mazumder, 2015). Therefore, Arsenic contamination of drinking water is considered a major global health concern. Arsenic enters the surface and groundwater system by both naturally and anthropogenically derived sources. Arsenic is found naturally in rocks and soil in the environment, predominantly as inorganic arsenite As (III) and arsenate As (V). It is anthropogenically released into the environment due to coal burning, mining, and automobile emissions (Roy et al., 2020).

Many bacteria, archaea, fungi, and animals produce arsenite methytransferase that can methylate arsenite into dimethyarsinate. The enzyme has 283 amino acid residues and is coded by the arsM gene (Zeng et al., 2018). One of the resistance mechanisms in *Hobacterium* sp. strain NRC-1 is confirmed to be due to arsenite(III)-methyltransferase by studying knockouts of arsM gene on the extrachromosomal replicon pNRC100 (Wang et al., 2004).

### 9.5.10 $P_{1B}$-TYPE ATPASES

$P_{1B}$-type ATPases are a subgroup of P ATPases. They are involved in the transport of cations, including heavy metals ($Cu^{+}$, $Cu^{2+}$, $Zn^{2+}$, $Co^{2+}$, $Cd^{2+}$, $Pb^{2+}$) across membranes and thus, are considered as a key element for metal homeostasis (Argüello et al., 2007). As seen in Figure 9.6 P-type ATPase is composed of three domains: TM, ATPBD, and AD. The transmembrane helix (TM) is made up of 6–8 transmembrane segments carrying signature sequences in segments flanking the large ATP binding cytoplasmic loop. These sequences make possible the differentiation of at least four $P_{1B}$-ATPase subgroups with distinct metal selectivity: $P_{1B}$–1: $Cu^{+}$, $P_{1B}$–2: $Zn^{2+}$, $P_{1B}$–3: $Cu^{2+}$, $P_{1B}$–4: $Co^{2+}$ (Coombs & Barkay, 2005). The ATP-binding domain (ATPBD) is made up of nucleotide-binding domain (N-domain) and phosphorylation domain (P-domain). The third domain is the actuator domain (AD). It transmits changes in the ATPBD to the transmembrane region. The AD and ATPBD are hydrophilic whereas the TM domains are hydrophobic and are embedded in the cell membrane (Smith et al., 2014).

$P_{1B}$-type ATPases are widely distributed throughout the bacterial and archaeal lineages. The As(III)/Sb(III) transporting P1B-type ATPase, is present in almost all haloarchaea sequenced to date, including *Halobacterium* sp. strain NRC1, *Halalkalicoccus jeotgali, Haloarcula hispanica, Natrialba magadii, Haloarcula marismortui, Haloquadratum walsbyi*, and *Natronomonas pharaonis* (Srivastava & Kowshik, 2013). The genome of *Halobacterium salinarum* NRC-1 carried two distinct phylogenetic clusters, CopA1 (Cu(II) influx) and CopA2 (Cu(II) influx and efflux. Complete sequence analysis of *Halobacterium* sp. strain NRC-1 genome revealed the presence of Cd(II)-efflux ATPase (Ng et al., 2000).

Investigations dedicated to unveiling the capacity of halophilic microbes to degrade various aliphatic and aromatic hydrocarbons, and to transform heavy metal pollutants in varying salinities, together with our growing understanding of mechanisms of operation of halophilic enzymes will, in years to come, help in developing "Clean Bioremediation Technologies".

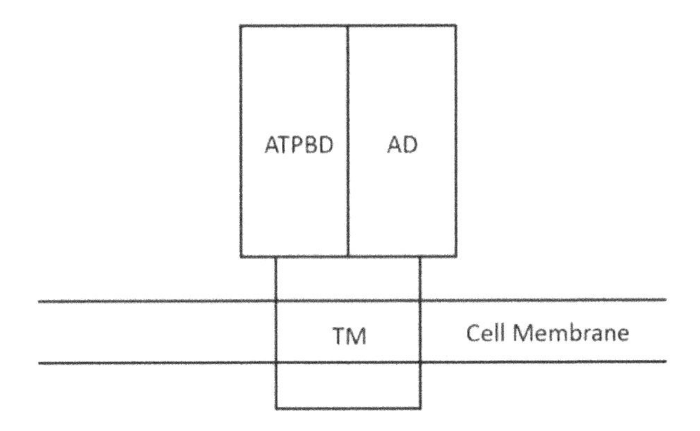


**FIGURE 9.6** Structure of $P_{1B}$-type ATPases.

## REFERENCES


Abou Khalil, C., Prince, V. L., Prince, R. C., Greer, C. W., Lee, K., Zhang, B., & Boufadel, M. C. (2021). Occurrence and biodegradation of hydrocarbons at high salinities. *Science of the Total Environment, 762*. Elsevier B.V. https://doi.org/10.1016/j.scitotenv.2020.143165

Agrawal, K., & Verma, P. (2020a). Laccase-Mediated Synthesis of Bio-material Using Agro-residues. In: *Biotechnological applications in human health* (pp. 87–93). Springer, Singapore.

Agrawal, K., & Verma, P. (2020b). Production optimization of yellow laccase from *Stropharia* sp. ITCC 8422 and enzyme-mediated depolymerization and hydrolysis of lignocellulosic biomass for biorefinery application. *Biomass Conversion and Biorefinery*, 1–20.

Ai, L., Huang, Y., & Wang, C. (2018). Purification and characterization of halophilic lipase of *Chromohalobacter* sp. from ancient salt well. *Journal of Basic Microbiology*, *58*(8), 647–657. https://doi.org/10.1002/jobm.201800116

Al-Mailem, D. M., Sorkhoh, N. A., Al-Awadhi, H., Eliyas, M., & Radwan, S. S. (2010). Biodegradation of crude oil and pure hydrocarbons by extreme halophilic archaea from hypersaline coasts of the Arabian Gulf. *Extremophiles*, *14*(3), 321–328. https://doi.org/10.1007/s00792-010-0312-9

Al-Shaikh, A. Z., & Jamal, M. T. (2020). Bioaugmentation of halophilic consortia for the degradation of petroleum hydrocarbons and petroleum wastewater treatment. *International Journal of Advanced Research in Biological Science*, 7(9), 97–112. https://doi.org/10.22192/ijarbs

Alneyadi, A. H., Rauf, M. A., & Ashraf, S. S. (2018). Oxidoreductases for the remediation of organic pollutants in water–a critical review. *Critical Reviews in Biotechnology*, *38*(7), 971–988. Taylor and Francis Ltd. https://doi.org/10.1080/07388551.2017.1423275

Amoozegar, M. A., Safarpour, A., Noghabi, K. A., Bakhtiary, T., & Ventosa, A. (2019). Halophiles and their vast potential in biofuel production. *Frontiers in Microbiology*, *10*(8). Frontiers Media S.A. https://doi.org/10.3389/fmicb.2019.01895

Amoozegar, M. A., Sánchez-Porro, C., Rohban, R., Hajighasemi, M., & Ventosa, A. (2009). Piscibacillus halophilus sp. nov., a moderately halophilic bacterium from a hypersaline Iranian lake. *International Journal of Systematic and Evolutionary Microbiology*, *59*(12), 3095–3099. https://doi.org/10.1099/ijs.0.012013-0

Arahal, D. R., Carmen Marquez, M., Volcani, B. E., Schleifer, K. H., & Ventosa, A. (1999). *Bacillus marismortui* sp. nov., a new moderately halophilic species from the Dead Sea. *International Journal of Systematic Bacteriology*, *49*(2), 521–530. https://doi.org/10.1099/00207713-49-2-521

Argüello, J. M., Eren, E., & González-Guerrero, M. (2007). The structure and function of heavy metal transport P1B-ATPases. *BioMetals*, *20*(3–4), 233–248. https://doi.org/10.1007/s10534-006-9055-6

Arora, P. K., Srivastava, A., & Singh, V. P. (2010). Application of monooxygenases in dehalogenation, desulphurization, denitrification and hydroxylation of aromatic compounds. *Journal of Bioremediation & Biodegradation*, *01*(03). https://doi.org/10.4172/2155-6199.1000112

Arulazhagan, P., & Vasudevan, N. (2011). Biodegradation of polycyclic aromatic hydrocarbons by a halotolerant bacterial strain *Ochrobactrum* sp. VA1. *Marine Pollution Bulletin*, *62*(2), 388–394. https://doi.org/10.1016/j.marpolbul.2010.09.020

Bhardwaj, N., Kumar, B., Agrawal, K., & Verma, P. (2020). Bioconversion of rice straw by synergistic effect of in-house produced ligno-hemicellulolytic enzymes for enhanced bioethanol production. *Bioresource Technology Reports*, *10*, 100352.

Bonfá, M. R. L., Grossman, M. J., Mellado, E., & Durrant, L. R. (2011). Biodegradation of aromatic hydrocarbons by Haloarchaea and their use for the reduction of the chemical oxygen demand of hypersaline petroleum produced water. *Chemosphere*, *84*(11), 1671–1676. https://doi.org/10.1016/j.chemosphere.2011.05.005

Boone, D. R., Whitman, W. B., & Koga, Y. (2001) Order III. Methanosarcinales ord. nov. In: Boone, D. R., Castenholz, R. W., & Garrity, G. M. (eds.) *Bergey's manual of systematic bacteriology, vol 1, 2nd edn, The Archaea and the deeply branching and phototrophic Bacteria* (pp. 287–289). Springer, New York.

Bosecker, K. (1997). Bioleaching: Metal solubilization by microorganisms. *FEMS Microbiology Reviews*, *20*(3–4), 591–604. https://doi.org/10.1111/j.1574-6976.1997.tb00340.x

Coombs, J. M., & Barkay, T. (2005). New findings on evolution of metal homeostasis genes: Evidence from comparative genome analysis of bacteria and archaea. *Applied and Environmental Microbiology*, *71*(11), 7083–7091. https://doi.org/10.1128/AEM.71.11.7083-7091.2005

Corral, P., Amoozegar, M. A., & Ventosa, A. (2020). Halophiles and their biomolecules: Recent advances and future applications in biomedicine. *Marine Drugs*, *18*(1). MDPI AG. https://doi.org/10.3390/md18010033

Cui, H. L., & Qiu, X. X. (2014). *Salinarubrum litoreum* gen. nov., sp. nov.: A new member of the family Halobacteriaceae isolated from Chinese marine solar salterns. *Antonie van Leeuwenhoek*, *105*, 621. https://doi.org/10.1007/s10482-014-0117-4

Dalvi, S., Azetsu, S., Patrauchan, M. A., Aktas, D. F., & Fathepure, B. Z. (2012). Proteogenomic elucidation of the initial steps in the benzene degradation pathway of a novel halophile, *Arhodomonas* sp. strain rozel, isolated from a hypersaline environment. *Applied and Environmental Microbiology*, *78*(20), 7309–7316. https://doi.org/10.1128/AEM.01327-12

Dastgheib, S. M. M., Amoozegar, M. A., Khajeh, K., & Ventosa, A. (2011). A halotolerant *Alcanivorax* sp. strain with potential application in saline soil remediation. *Applied Microbiology and Biotechnology*, *90*(1), 305–312. https://doi.org/10.1007/s00253-010-3049-6

de la Haba, C., Sánchez-Porro, M. C., & Marquez Antonio Ventosa, R. R. (2011). 3.1 Taxonomy of Halophiles. In: Horikoshi, K. (ed.) *Extremophiles handbook*. Springer, Tokyo. https://doi.org/10.1007/978-4-431-53898-1_13

de Lourdes Moreno, M., Sánchez-Porro, C., Piubeli, F., Frias, L., García, M. T., & Mellado, E. (2011). Cloning, characterization and analysis of cat and ben genes from the phenol degrading halophilic bacterium *Halomonas organivorans*. *PLoS ONE*, *6*(6). https://doi.org/10.1371/journal.pone.0021049

Dzionek, A., Wojcieszyńska, D., & Guzik, U. (2016). Natural carriers in bioremediation: A review. *Electronic Journal of Biotechnology*, *23*, 28–36. https://doi.org/10.1016/j.ejbt.2016.07.003

Edbeib, M. F., Wahab, R. A., & Huyop, F. (2016). Halophiles: Biology, adaptation, and their role in decontamination of hypersaline environments. *World Journal of Microbiology and Biotechnology*, *32*(8). Springer Netherlands. https://doi.org/10.1007/s11274-016-2081-9

Ellis, D. G., Bizzoco, R. W., & Kelley, S. T. (2008). Halophilic Archaea determined from geothermal steam vent aerosols. *Environmental Microbiology*, *10*(6), 1582–1590. https://doi.org/10.1111/j.1462-2920.2008.01574.x

Erakhrumen, A. A. (2015). Concentrations of heavy metals in untreated produced water from a crude oil production platform in Niger-delta, Nigeria. *Journal of Research in Forestry, Wildlife and Environment*, *7*(1), 89–101.

Erdoğmuş, S. F., Mutlu, B., Korcan, S. E., Güven, K., & Konuk, M. (2013). Aromatic hydrocarbon degradation by halophilic archaea isolated from ÇamaltI Saltern, Turkey. *Water, Air, and Soil Pollution*, *224*(3). https://doi.org/10.1007/s11270-013-1449-9

Eslami, M., Amoozegar, M. A., & Asad, S. (2016). Isolation, cloning and characterization of an azoreductase from the halophilic bacterium *Halomonas elongata*. *International Journal of Biological Macromolecules*, *85*, 111–116. https://doi.org/10.1016/j.ijbiomac.2015.12.065

Fathepure, B. Z. (2014). Recent studies in microbial degradation of petroleum hydrocarbons in hypersaline environments. *Frontiers in Microbiology*, *5*(4). https://doi.org/10.3389/fmicb.2014.00173

Feng, T. C., Cui, C. Z., Dong, F., Feng, Y. Y., Liu, Y. D., & Yang, X. M. (2012). Phenanthrene biodegradation by halophilic *Martelella* sp. AD-3. *Journal of Applied Microbiology*, *113*(4), 779–789. https://doi.org/10.1111/j.1365-2672.2012.05386.x

Gaonkar, S. K., & Furtado, I. J. (2018). Isolation and culturing of protease- and lipase-producing *Halococcus agarilyticus* GUGFAWS-3 from marine *Haliclona* sp. inhabiting the rocky intertidal region of Anjuna in Goa, India. *Annals of Microbiology*, *68*(12), 851–861. https://doi.org/10.1007/s13213-018-1391-6

Gibbons, N. E. (1974). Family V. Halobacteriaceae fam. nov. In: Buchanan, R. E. & Gibbons, N. E. (eds.) *Bergey's manual of determinative bacteriology* (8th edn., pp 269–273). Williams and Wilkins, Baltimore.

Gómez-Villegas, P., Vigara, J., Romero, L., Gotor, C., Raposo, S., Gonçalves, B., & Léon, R. (2021). Biochemical characterization of the amylase activity from the new haloarchaeal strain *Haloarcula* sp. Hs isolated in the odiel marshlands. *Biology*, *10*(4). https://doi.org/10.3390/biology10040337

Goswami, R. K., Agrawal, K., Shah, M. P., & Verma, P. (2021). Bioremediation of heavy metals from wastewater: A current perspective on microalgae-based future. *Letters in Applied Microbiology*, *2021*, 13564. https://doi.org/10.1111/lam.13564

Goswami, R. K., Agrawal, K., & Verma, P. (2022). An exploration of natural synergy using microalgae for the remediation of pharmaceuticals and xenobiotics in wastewater. *Algal Research*, *64*, 102703.

Guha Mazumder, D. N. (2015). Health Effects Chronic Arsenic Toxicity. In *Handbook of arsenic toxicology* (pp. 137–177). Elsevier Inc. https://doi.org/10.1016/B978-0-12-418688-0.00006-X

Haferburg, G., & Kothe, E. (2007). Microbes and metals: Interactions in the environment. *Journal of Basic Microbiology*, *47*(6), 453–467). https://doi.org/10.1002/jobm.200700275

Hassan, H. A., & Aly, A. A. (2018). Isolation and characterization of three novel catechol 2,3-dioxygenase from three novel haloalkaliphilic BTEX-degrading Pseudomonas strains. *International Journal of Biological Macromolecules*, *106*, 1107–1114. https://doi.org/10.1016/j.ijbiomac.2017.08.113

Hrycay, E. G., & Bandiera, S. M. (2015). Monooxygenase, peroxidase and peroxygenase properties and reaction mechanisms of cytochrome P450 enzymes. *Advances in Experimental Medicine and Biology*, *851*, 1–61. https://doi.org/10.1007/978-3-319-16009-2_1

Huijbers, M. M. E., Montersino, S., Westphal, A. H., Tischler, D., & van Berkel, W. J. H. (2014). Flavin-dependent monooxygenases. *Archives of Biochemistry and Biophysics*, *544*, 2–17). Academic Press Inc. https://doi.org/10.1016/j.abb.2013.12.005

Hung, K. S., Liu, S. M., Tzou, W. S., Lin, F. P., Pan, C. L., Fang, T. Y., Sun, K. H., & Tang, S. J. (2011). Characterization of a novel GH10 thermostable, halophilic xylanase from the marine bacterium *Thermoanaerobacterium saccharolyticum* NTOU1. *Process Biochemistry*, *46*(6), 1257–1263. https://doi.org/10.1016/j.procbio.2011.02.009

Hutcheon, G. W., Vasisht, N., & Bolhuis, A. (2005). Characterisation of a highly stable α-amylase from the halophilic archaeon *Haloarcula hispanica*. *Extremophiles*, *9*(6), 487–495. https://doi.org/10.1007/s00792-005-0471-2

Jafari, N., Rezaei, S., Rezaie, R., Dilmaghani, H., Khoshayand, M. R., & Faramarzi, M. A. (2017). Improved production and characterization of a highly stable laccase from the halophilic bacterium *Chromohalobacter salexigens* for the efficient delignification of almond shell bio-waste. *International Journal of Biological Macromolecules*, *105*, 489–498. https://doi.org/10.1016/j.ijbiomac.2017.07.055

Karigar, C. S., & Rao, S. S. (2011). Role of microbial enzymes in the bioremediation of pollutants: A review. *Enzyme Research*, *2011*(1). https://doi.org/10.4061/2011/805187

Khandeparker, R., Verma, P., & Deobagkar, D. (2011). Author version. *New Biotechnology*, *28*(6), 814–821.

Kumar, B., Bhardwaj, N., Agrawal, K., & Verma, P. (2020). Bioethanol Production: Generation-Based Comparative Status Measurements. In: *Biofuel production technologies: Critical analysis for sustainability* (pp. 155–201). Springer, Singapore.

Kumar, B., & Verma, P.(2020). Enzyme mediated multi-product process: A concept of bio-based refinery. *Industrial Crops and Products*, *154*, 112607.

Kunte, H. J., Lentzen, G., & Galinski, E. A. (2014). Industrial production of the cell protectant ectoine: Protection mechanisms, processes, and products. *Current Biotechnology*, *3*(1), 10–25.

Lee, H. S. (2013). Diversity of halophilic archaea in fermented foods and human intestines and their application. *Journal of Microbiology and Biotechnology*, *23*(12), 1645–1653. https://doi.org/10.4014/jmb.1308.08015

Lefebvre, O., & Moletta, R. (2006). Treatment of organic pollution in industrial saline wastewater: A literature review. *Water Research*, *40*(20), 3671–3682. Elsevier Ltd. https://doi.org/10.1016/j.watres.2006.08.027

Li, X., & Yu, H. Y. (2013). Halostable cellulase with organic solvent tolerance from *Haloarcula* sp. LLSG7 and its application in bioethanol fermentation using agricultural wastes. *Journal of Industrial Microbiology and Biotechnology*, *40*(12), 1357–1365. https://doi.org/10.1007/s10295-013-1340-0

Li, X., & Yu, H. (2014). Characterization of an organic solvent-tolerant lipase from *Haloarcula* sp. G41 and its application for biodiesel production. *Folia Microbiologica*, *59*, 455–463. https://doi.org/10.1007/s12223-014-0320-8. Epub 2014 May 2.

Malik, A. D. & Furtado I. J. (2019). *Haloferax sulfurifontis* GUMFAZ2 producing xylanase-free cellulase retrieved from Haliclona sp. inhabiting rocky shore of Anjuna, Goa-India. *Journal of Basic Microbiology*, *59*, 692–700.

Manikandan, M., Pašić, L., & Kannan, V. (2009). Purification and biological characterization of a halophilic thermostable protease from *Haloferax lucentensis* VKMM 007. *World Journal of Microbiology and Biotechnology*, *25*(12), 2247–2256. https://doi.org/10.1007/s11274-009-0132-1

Mesbah, N. M., & Wiegel, J. (2017). A halophilic, alkalithermostable, ionic liquid-tolerant cellulase and its application in in situ saccharification of rice straw. *Bioenergy Research*, *10*(2), 583–591. https://doi.org/10.1007/s12155-017-9825-8

Misal, S. A., & Gawai, K. R. (2018). Azoreductase: A key player of xenobiotic metabolism. *Bioresources and Bioprocessing*, *5*(1). Springer. https://doi.org/10.1186/s40643-018-0206-8

Mnif, S., Chamkha, M., & Sayadi, S. (2009). Isolation and characterization of *Halomonas* sp. strain C2SS100, a hydrocarbon-degrading bacterium under hypersaline conditions. *Journal of Applied Microbiology*, *107*(3), 785–794. https://doi.org/10.1111/j.1365-2672.2009.04251.x

Mohamedin, A. H., Mowafy, A. M., Ghanim, A. A., & History, A. (2018). Potential applications of some moderate halophilic bacteria. *Egyptian Journal of Aquatic Biology and Fisheries*, *22*(5). www.ejabf.journals.ekb.eg

Munson, M. A., Nedwell, D. B., & Embley, T. M. (1997). Phylogenetic diversity of Archaea in sediment samples from a coastal salt marsh. *Applied and Environmental Microbiology*, 63(12), 729–4733. https://doi.org/10.1128/aem.63.12.4729-4733.1997

Nair, L. G., Agrawal, K., & Verma, P. (2022a). An overview of sustainable approaches for bioenergy production from agro-industrial wastes. *Energy Nexus*, 100086.

Nair, L. G., Agrawal, K., & Verma, P. (2022b). An insight into the principles of lignocellulosic biomass-based zero-waste biorefineries: A green leap towards imperishable energy-based future. *Biotechnology and Genetic Engineering Reviews*, 1–51.

Ng, W. V., Kennedy, S. P., Mahairas, G. G., Berquist, B., Pan, M., Dutt Shukla, H., Lasky, S. R., Baliga, N. S., Thorsson, V., Sbrogna, J., Swartzell, S., Weir, D., Hall, J., Dahl, T. A., Welti, R., Goo, Y. A., Leithauser, B., Keller, K., Cruz, R., … Dassarma, S. (2000). *Genome sequence of Halobacterium species NRC-1*. www.pnas.orgcgidoi10.1073pnas.190337797

Nie, Y., Chi, C. Q., Fang, H., Liang, J. L., Lu, S. L., Lai, G. L., Tang, Y. Q., & Wu, X. L. (2014). Diverse alkane hydroxylase genes in microorganisms and environments. *Scientific Reports*, *4*. https://doi.org/10.1038/srep04968

Nie, Y., Fang, H., Li, Y., Chi, C. Q., Tang, Y. Q., & Wu, X. L. (2013). The genome of the moderate halophile *Amycolicicoccus subflavus* DQS3-9A1T reveals four alkane hydroxylation systems and provides some clues on the genetic basis for its adaptation to a petroleum environment. *PLoS ONE*, *8*(8). https://doi.org/10.1371/journal.pone.0070986

Nzila, A., Musa, M. M., Sankara, S., Al-Momani, M., Xiang, L., & Li, Q. X. (2021). Degradation of benzo[a]pyrene by halophilic bacterial strain *Staphylococcus haemoliticus* strain 10SBZ1A. *PLoS ONE*, *16*(2). https://doi.org/10.1371/journal.pone.0247723

Oie, C. S. I., Albaugh, C. E., & Peyton, B. M. (2007). Benzoate and salicylate degradation by *Halomonas campisalis*, an alkaliphilic and moderately halophilic microorganism. *Water Research*, *41*(6), 1235–1242. https://doi.org/10.1016/j.watres.2006.12.029

Oren, A. (2010). Industrial and environmental applications of halophilic microorganisms. *Environmental Technology*, *31*(8–9), 825–834. Taylor and Francis Ltd. https://doi.org/10.1080/09593330903370026

Oren, A. (2002). Diversity of halophilic microorganisms: Environments, phylogeny, physiology, and applications. *Journal of Industrial Microbiology and Biotechnology*, *28*(1), 56–63. https://doi.org/10.1038/sj/jim/7000176

Park, C., & Park, W. (2018). Survival and energy producing strategies of Alkane degraders under extreme conditions and their biotechnological potential. *Frontiers in Microbiology*, *9*(5). Frontiers Media S.A. https://doi.org/10.3389/fmicb.2018.01081

Patowary, K., Patowary, R., Kalita, M. C., & Deka, S. (2016). Development of an efficient bacterial consortium for the potential remediation of hydrocarbons from contaminated sites. *Frontiers in Microbiology*, *7*(7). https://doi.org/10.3389/fmicb.2016.01092

Paul, V. G., & Mormile, M. R. (2017). A case for the protection of saline and hypersaline environments: A microbiological perspective. *FEMS Microbiology Ecology*, *93*(8). Oxford University Press. https://doi.org/10.1093/femsec/fix091

Pérez-Pomares, F., Bautista, V., Ferrer, J., Pire, C., Marhuenda-Egea, F. C., & Bonete, M. J. (2003). α-Amylase activity from the halophilic archaeon *Haloferax mediterranei*. *Extremophiles*, *7*(4), 299–306. https://doi.org/10.1007/s00792-003-0327-6

Raghavan, T. M., & Furtado, I. (2000). Tolerance of an estuarine halophilic archaebacterium to crude oil and constituent hydrocarbons. *Bulletin of Environmental Contamination and Toxicology*, (*2000*)(65), 725–731.

Rezaie, R., Rezaei, S., Jafari, N., Forootanfar, H., Khoshayand, M. R., & Faramarzi, M. A. (2017). Delignification and detoxification of peanut shell bio-waste using an extremely halophilic laccase from an *Aquisalibacillus elongatus* isolate. *Extremophiles*, *21*(6), 993–1004. https://doi.org/10.1007/s00792-017-0958-7

Roy, A., Datta, S., Barman, M., Bhattacharyya, S., Chandra, B., & Viswavidyalaya, K. (2020). *Partitioning of arsenic in low and high arsenic accumulating rice cultivars*. https://www.researchgate.net/publication/342834505

SadrAzodi, S. M., Shavandi, M., Amoozegar, M. A., & Mehrnia, M. R. (2019). Biodegradation of long chain alkanes in halophilic conditions by *Alcanivorax* sp. strain Est-02 isolated from saline soil. *3 Biotech*, *9*(4). https://doi.org/10.1007/s13205-019-1670-3

Santorelli, M., Maurelli, L., Pocsfalvi, G., Fiume, I., Squillaci, G., la Cara, F., del Monaco, G., & Morana, A. (2016). Isolation and characterisation of a novel alpha-amylase from the extreme haloarchaeon *Haloterrigena turkmenica*. *International Journal of Biological Macromolecules*, *92*, 174–184. https://doi.org/10.1016/j.ijbiomac.2016.07.001

Schreck, S. D., & Grunden, A. M. (2014). Biotechnological applications of halophilic lipases and thioesterases. *Applied Microbiology and Biotechnology*, *98*(3), 1011–1021. Springer Verlag. https://doi.org/10.1007/s00253-013-5417-5

Sellés Vidal, L., Kelly, C. L., Mordaka, P. M., & Heap, J. T. (2018). Review of NAD(P) H-dependent oxidoreductases: Properties, engineering and application. *Biochimica et Biophysica Acta - Proteins and Proteomics*, *1866*(2), 327–347. Elsevier B.V. https://doi.org/10.1016/j.bbapap.2017.11.005

Semrau, J. D. (2011). Bioremediation via methanotrophy: Overview of recent findings and suggestions for future research. *Frontiers in Microbiology*, *2*(10). Frontiers Research Foundation. https://doi.org/10.3389/fmicb.2011.00209

Singh, D., & Gupta, N. (2020). Microbial Laccase: A robust enzyme and its industrial applications. *Biologia*, *75*(8), 1183–1193. Springer. https://doi.org/10.2478/s11756-019-00414-9

Singh, R., Hu, J., Regner, M. R., Round, J. W., Ralph, J., Saddler, J. N., & Eltis, L. D. (2017). Enhanced delignification of steam-pretreated poplar by a bacterial laccase. *Scientific Reports*, *7*. https://doi.org/10.1038/srep42121

Siroosi, M., Amoozegar, M. A., Khajeh, K., & Dabirmanesh, B. (2018). Decolorization of dyes by a novel sodium azide-resistant Spore laccase from a halotolerant bacterium, *Bacillus safensis* sp. strain S31. *Water Science and Technology*, *77*(12), 2867–2875. https://doi.org/10.2166/wst.2018.281

Smith, A. T., Smith, K. P., & Rosenzweig, A. C. (2014). Diversity of the metal-transporting P1B-type ATPases. *Journal of Biological Inorganic Chemistry*, *19*(6), 947–960. https://doi.org/10.1007/s00775-014-1129-2

Srivastava, P., & Kowshik, M. (2013). Mechanisms of metal resistance and homeostasis in Haloarchaea. *Archaea*, *2013*. https://doi.org/10.1155/2013/732864

Tapilatu, Y. H., Grossi, V., Acquaviva, M., Militon, C., Bertrand, J. C., & Cuny, P. (2010). Isolation of hydrocarbon-degrading extremely halophilic archaea from an uncontaminated hypersaline pond (Camargue, France). *Extremophiles*, *14*(2), 225–231. https://doi.org/10.1007/s00792-010-0301-z

Tapingkae, W., Tanasupawat, S., Parkin, K. L., Benjakul, S., & Visessanguan, W. (2010). Degradation of histamine by extremely halophilic archaea isolated from high salt-fermented fishery products. *Enzyme and Microbial Technology*, *46*(2), 92–99. https://doi.org/10.1016/j.enzmictec.2009.10.011

Taran, M. (2011). Poly (3-Hydroxybutyrate) production from crude oil by *haloarcula* sp. IRU1: Optimization of culture conditions by taguchi method. *Petroleum Science and Technology*, *29*(12), 1264–1269. https://doi.org/10.1080/10916466.2010.499405

Tchounwou, P. B., Yedjou, C. G., Patlolla, A. K., & Sutton, D. J. (2012). Heavy metal toxicity and the environment. *EXS*, *101*, 133–164. https://doi.org/10.1007/978-3-7643-8340-4_6

Uthandi, S., Saad, B., Humbard, M. A., & Maupin-Furlow, J. A. (2010). LccA, an archaeal laccase secreted as a highly stable glycoprotein into the extracellular medium by *Haloferax volcanii*. *Applied and Environmental Microbiology*, *76*(3), 733–743. https://doi.org/10.1128/AEM.01757-09

van Beilen, J. B., Funhoff, E. G., van Loon, A., Just, A., Kaysser, L., Bouza, M., Holtackers, R., Röthlisberger, M., Li, Z., & Witholt, B. (2006). Cytochrome P450 alkane hydroxylases of the CYP153 family are common in alkane-degrading eubacteria lacking integral membrane alkane hydroxylases. *Applied and Environmental Microbiology*, *72*(1), 59–65. https://doi.org/10.1128/AEM.72.1.59-65.2006

van Beilen, J. B., Neuenschwander, M., Smits, T. H. M., Roth, C., Balada, S. B., & Witholt, B. (2002). Rubredoxins involved in alkane oxidation. *Journal of Bacteriology*, *184*(6), 1722–1732. https://doi.org/10.1128/JB.184.6.1722-1732.2002

Wagner, N. L., Greco, J. A., Ranaghan, M. J., & Birge, R. R. (2013). Directed evolution of bacteriorhodopsin for applications in bioelectronics. *Journal of the Royal Society Interface*, *10*(84). https://doi.org/10.1098/rsif.2013.0197

Wainø, M., & Ingvorsen, K. (2003). Production of β-xylanase and β-xylosidase by the extremely halophilic archaeon *Halorhabdus utahensis*. *Extremophiles*, *7*(2), 87–93. https://doi.org/10.1007/s00792-002-0299-y

Wang, G., Kennedy, S. P., Fasiludeen, S., Rensing, C., & DasSarma, S. (2004). Arsenic resistance in *Halobacterium* sp. strain NRC-1 examined by using an improved gene knockout system. *Journal of Bacteriology, 186*(10), 3187–3194. https://doi.org/10.1128/JB.186.10.3187-3194.2004

White, C., Sayer, J. A., & Gadd, G. M. (1997). Microbial solubilization and immobilization of toxic metals: Key biogeochemical processes for treatment of contamination. *FEMS Microbiology Reviews, 20*(3–4), 503–516. https://doi.org/10.1111/j.1574-6976.1997.tb00333.x

Zeng, X. C., Yang, Y., Shi, W., Peng, Z., Chen, X., Zhu, X., & Wang, Y. (2018). Microbially mediated methylation of arsenic in the arsenic-rich soils and sediments of Jianghan Plain. *Frontiers in Microbiology, 9*(7). https://doi.org/10.3389/fmicb.2018.01389

Zgórska, A., Trząski, L., & Wiesner, M. (2016). Environmental risk caused by high salinity mine water discharges from active and closed mines located in the Upper Silesian Coal Basin (Poland). In *Proceedings IMWA 2016*, Freiberg/Germany, Drebenstedt, Carsten, & Paul, Michael (eds.) (pp. 85–92).

Zhang, T., Datta, S., Eichler, J., Ivanova, N., Axen, S. D., Kerfeld, C. A., Chen, F., Kyrpides, N., Hugenholtz, P., Cheng, J. F., Sale, K. L., Simmons, B., & Rubin, E. (2011). Identification of a haloalkaliphilic and thermostable cellulase with improved ionic liquid tolerance. *Green Chemistry, 13*(8), 2083–2090. https://doi.org/10.1039/c1gc15193b

Zhang, W. Y., Meng, Y., Zhu, X. F., & Wu, M. (2013). *Halopiger salifodinae* sp. nov., an extremely halophilic archaeon isolated from a salt mine. *International Journal of Systematic and Evolutionary Microbiology, 63*(PART10), 3563–3567. https://doi.org/10.1099/ijs.0.050971-0

# Index

## E



## F



## G



## H